Practical Computer
Data Communications

Applications of Communications Theory
Series Editor: R. W. Lucky, *AT & T Bell Laboratories*

A Continuation Order Plan is available for this series. A continuation order will bring delivery of
each new volume immediately upon publication. Volumes are billed only upon actual shipment.
For further information please contact the publisher.

Practical Computer Data Communications

William J. Barksdale
South TEC Associates
Huntsville, Alabama

PLENUM PRESS • NEW YORK AND LONDON

Library of Congress Cataloging in Publication Data

Barksdale, William J.
  Practical computer data communications.

  (Applications of communications theory)
  Includes bibliographical references and index.
  1. Data transmission systems. I. Title. II. Series.
TK5105.B356   1986   004.6   86-16981
ISBN-13: 978-1-4684-5166-5   e-ISBN-13: 978-1-4684-5164-1
DOI: 10.1007/978-1-4684-5164-1

© 1986 Plenum Press, New York
Softcover reprint of the hardcover 1st edition 1986
A Division of Plenum Publishing Corporation
233 Spring Street, New York, N.Y. 10013

This book is dedicated

to

Sandy, Claire,

and

all Data Communications Professionals

Preface

Several years ago when I began consulting full time, I quickly discovered that despite three advanced academic degrees my practical industrial experience had some significant gaps. It thus was necessary initially to spend considerable (nonbillable) time collecting and organizing a great deal of essential information on the various aspects of modern data communications. The task was made more difficult by the highly interdisciplinary nature of the field, with the required information scattered throughout the vast international literature of telecommunications, computers, electrical engineering, military systems, mathematics, operations research, optimization, speech processing, and the murky world of legal and regulatory policy. Although there were a number of fine books and periodicals in each of these specialized disciplines, I was unable to find a single comprehensive text that covered the entire field at even a modestly attractive technical and mathematical level.

After going to the trouble of organizing all this diverse material for my clients and students, it seemed rather natural to put it into book form and thus share it with those professionals working with computer data communications who need a comprehensive coverage of the subject at a level immediately applicable to their work and yet easily accessible for self-study. The project was facilitated by an agreeable publisher and an incredibly understanding and cooperative family, and *Practical Computer Data Communications* is the result.

The text has been written to facilitate self-study by practitioners, with the main focus on the fundamentals of data communications and their effective execution in practice. This is in sharp contrast to many existing texts which are either so theoretical that the applications will not be known for years hence, or so product-oriented as to suggest a vendor's catalog. Much of the material originated in my public seminar, so the text is suitable for use with seminars and in-house training programs. The actual manuscript was developed in conjunction with an evening college course sequence in computer science that I teach annually, so the book can also be used as a primary undergraduate text in computer science or as a supplementary text in a more theoretical engineering communications course.

To paraphrase an old cliché, the Mona Lisa was not painted by a committee—or a computer—and the same is true of the writing of this text, although that is as far as the analogy should be taken! The author is solely responsible for any and all technical and organizational content, good and bad, and there is no convenient coauthor upon whom to blame all the errors. However, several people did make significant contributions, directly or indirectly, and deserve recognition. First and foremost is my wife, who not only typed and proofread the entire manuscript in her limited spare time, but did so despite my many idiosyncracies and consistently illegible handwriting. Thanks are also due to Harry Yedid of Motorola and Byron Driver of Complexx Systems for several luncheon discussions and helpful comments on the drafts, and also to Bobby Boykin of Timeplex, who first convinced me that such a book was needed. My aunt, Miss Eloise Andrews, proofread most of the manuscript and thus deserves credit for its being written in correct English prose rather than the Lower Slobbovian dialect that I normally use.

President Ray Watson of the Southeastern Institute of Technology provided consistent support, and much of the research was done at the Redstone Scientific Information Center through the cooperation of director James Clark and his staff. I also owe a great deal of thanks to each one of my long-suffering students at Southeastern, who put up with the developing manuscript for class notes and endured my generally unscintillating evening lectures on empty stomachs, particularly Lynn Vandeberg of Boeing, Dala Bass of Teledyne, and Von Burton of the Marshall Space Flight Center. Finally, I would like to give credit to my parents and to my old high school mathematics teacher, Mrs. Virginia Haggard, who first convinced me that mathematics could be exciting despite the hard and sometimes frustrating intellectual work involved.

Bill Barksdale
Huntsville, Alabama

Contents

PART II: POINT-TO-POINT DATA LINKS

Chapter 5.   Modems and Dialers   109

Chapter 6.   The DTE/DCE Physical Interface   137

Chapter 7.   Data Link Protocol   171

PART III: WIDE AREA NETWORKS

PART IV: SPECIALIZED DATA NETWORKS

Chapter 12.   Network Security   419

Chapter 13.   Future Network Trends   437

Answers to True/False Questions   449

Introduction

1.1. Benefits and Objectives

This book is about the technical aspects of computer communications over data links and networks. In it we address this broad and dynamic subject from a unified point of view that should provide a large cross section of readers with a relatively complete picture of the field. Although each chapter covers a different aspect of data communications, all are carefully integrated in order to make the material flow in a coherent yet easy-to-read manner. This continuity also allows the more theoretical material to be included only where actually needed, with most of the book easily accessible by those who have never studied, or have long since forgotten, the calculus and intermediate differential equations.

As noted in the Preface, computer data communications is a highly interdisciplinary field, employing techniques and equipment from many different areas of human endeavor. People entering or currently working in such a field often find it frustrating and time consuming to locate and understand all the diverse technical material they will eventually need to know to function effectively and advance professionally. A major objective of this book is to provide the "big picture" that such people need in a single compact source and at a readily accessible but meaningful technical level.

Many different technical professionals involved with computer data communications can benefit directly from reading this book. For the *new entrant* to the field, or anyone changing jobs within it, there is an overview of the entire field plus essential details of the specific area in which they expect to begin working. For the *practicing engineer* there is a review of the basic theory plus a great deal of practical detail on how data networks and equipment operate and are designed. For the *computer scientist* there is a wealth of detailed information on what happens once the data leave the internal computer bus. And for the *technician* there is not only specific information on the operational details of the equipment, but also a broad overview that will provide valuable insight into just how it all works together.

The book will also be valuable to several other categories of professionals. For *managers* of computer centers and networks, there is a great deal of results-oriented reference material that has been carefully organized and concisely written, all in a single source that can be read over a long week-end. Technical *salespersons* will find a complete, straightforward, easy-to-read (except perhaps Chapters 3 and 9) presentation of technical insight that can greatly enhance their ability to understand and serve their customers. Finally, for professional *educators* and their *students* this book offers a direct, practical insight into the way theory and practice come together in the workplace to provide services and earn profits. We are obviously attempting to reach a broad audience; but then computer data communications is a broad subject offering attractive career opportunities to many different kinds of people.

The selection of material and the way it is presented is based on many years of personal experience in both practice and teaching. Particular emphasis was placed on professional self-study, an area that is too often ignored once we leave school and settle down to a steady pace on the proverbial treadmill. However, to get the full benefit that this book offers, considerable commitment and effort will be required. There is simply no way around the old adage that "learning requires effort." To this end each following chapter includes a list of 25 True/False questions to help the reader review and identify areas that may need further work. Each chapter also includes a list of carefully selected and annotated references to guide the reader to more specialized literature that is consistent with the level of the text material. These sample quizzes and references will also, of course, be helpful to instructors in developing their own course material.

The techniques presented have been carefully selected and organized to provide a complete introduction to the field as it exists today. Only those topics with direct application in practice have been included, while many interesting but more academic subjects were omitted. Since the theoretical

backgrounds and "mathematical maturity" of data communications practitioners vary greatly, most of the material requires only a knowledge of
algebra. Although a full understanding of a few areas in Chapters 3 and 9
requires a little calculus, this material (or even the entire chapters) can be
skipped with little loss of continuity by those who do not need the technical
detail. For the more technical readers, however, this material provides not
only valuable theoretical insight, but also a vital bridge to the literature for
more advanced studies.

Computer data communications, as we shall see, is not only very
complex but is extremely dynamic, and any professional in the field who
is unwilling to make an ongoing personal investment of time, money, and
intellectual effort is destined to be left in a cloud of "technological dust."
Sometimes it seems to be a lot like shaving: no matter how well you do it,
you have to get up the next day and start all over again! However, the
personal and financial rewards are well worth the effort required, and there
is great opportunity in the field for those who are willing to keep abreast.
Used well, this book should be very helpful in making the transition from
school to the working world, and then in maintaining that crucial balance
between theory and practice that is essential for professional growth over
the span of one's career. The British philosopher Alfred North Whitehead
said it well:

> The tragedy of the world is that those who are imaginative have but slight
> experience, and those who are experienced have feeble imaginations. Fools act
> on imagination without knowledge; pedants act on knowledge without imagina-
> tion or experience.

A final major objective of *Practical Computer Data Communications* is to
contribute in some meaningful way to the rectification of this condition in
the field of computer data communications.

1.2. Organization

There are four major parts to this book, each containing three closely
related chapters. Beginning with essential background material in Part I,
each succeeding part builds on the preceding parts, and similarly for the
included chapters.

The material of Part I, Data Communications Fundamentals (Chapters
2–4), is designed to provide essential background regarding the environment
in which data networks operate, the mathematical theory upon which their
operation is based, and the telephone system which is by far the most
important communications network today. We focus predominantly on
those aspects which are germane to the later material and then only at an
appropriate tutorial level of detail. In Part II, Point-to-Point Data Links

(Chapters 5–7), we go to great lengths to establish the most fundamental data communications entity: the point-to-point data link. Starting with telephone line modems, we establish the physical link for simply passing bits. In particular, we look closely at the terminal/modem (DTE/DCE) interface in all its ramifications. Then we add the rules for transferring coherent data characters in the form of data link protocols. The material in Parts I and II is sufficient for a large proportion of the data communications applications that have been used in the past.

However, in the future an increasing proportion of data traffic will be carried by digital networks, and this is the focus of Part III, Wide Area Networks (Chapters 8–10). We first extend data links to form simple networks by using both conventional and statistical multiplexers. Then switching and concentrating capabilities are added in the general form of communications processors. The nature of operational networks is illustrated with several practical network examples (SITA, ARPANET, TELENET, etc.). Next, the actual design of networks is explored both to provide an understanding of why existing networks are the way they are and of how they can be designed, improved, and expanded. With the component and design issues established we add the remaining essential topic, network architecture. Using the ubiquitous OSI Reference Model as a guide, the various layers of protocol that comprise modern network architectures and allow effective data transfer are described with particular emphasis on current standards. Then some practical architectures are described (ISO/X.25, DNA, SNA).

Finally, in Part IV, Specialized Data Networks (Chapters 11–13), we first consider Local Area Networks and their unique characteristics and capabilities. Then we look at network security and encryption, which are fast becoming vital considerations in many commercial networks. We conclude the text with a look towards the future and some considerations involved in sending digital speech, data, and other kinds of digital traffic over an Integrated Services Digital Network.

As this brief survey indicates, *Practical Computer Data Communications*, in a single source, blends a diverse collection of essential subjects into a single mosaic that is computer data communications today. It is an exciting young field in which there is great opportunity for both intellectual and financial reward. If the futurists are anywhere close to correct, information will be the means by which most people will soon be earning their living. In that Information Age, the very infrastructure will be the digital networks of computers and terminals that will interlace the globe and provide the masses of people with a standard of living unimagined in the past. This book describes just the tip of that iceberg; but perhaps, in some small way, it will be of lasting benefit to those of you who will design and engineer our future communications systems.

I

Data Communications Fundamentals

Overview of Data Communications

2.1. Introduction and Historical Notes

The fundamental purpose of this chapter is to provide the reader with a suitable *framework* upon which to place the more detailed material of the following chapters. In order to develop a practical working knowledge of a highly complex and interdisciplinary subject like data communications it is imperative to initially understand some elementary terminology, some of the historical basis for the field, the major categories and components involved, and the current environment within which it must exist. This chapter treats all of these aspects so as to provide some preliminary perspective on practical computer data communications.

It is important to point out, however, that this overview is hardly sufficient to do much in practice. A handful of buzz words and some historical anecdotes may suffice for a 20-minute talk before the Chamber of Commerce, but not a computer club made up of savvy engineers. Nevertheless, the rather superficial "Big Picture" presented in this chapter is absolutely essential to the deeper understanding that follows, and it should not be omitted by anyone unfamiliar with the data communications field.

A *data communications network* is a system for conveying information in digital form between one location and one or more remote locations. Depending on the context, *remote* may mean across a building or across the world and beyond. The information is represented for transmission as *data*, and in deference to common usage and clarity we shall use the word in both the *plural* and *singular collective* sense.

Modern data communications is highly interdisciplinary with roots in such fields as telephony, computer science, electronics, and information retrieval. It is also international in nature, with most governments and major firms worldwide deeply involved with their own data networks. Consequently, the *terminology* used in practice is something of a patchwork from many disciplines, and can be a real barrier not only to those just entering the field but also to those who have been around awhile. Every attempt has been made to make the terminology consistent throughout this book, even at the expense of deviating occasionally from the specialized literature. A great many acronyms are also used in data communication. Although exasperating to learn, they are essential to reading the literature and, once familiar, do save time. We use acronyms, too, but always defined at the first occurence and often again subsequently for clarity. Although we do not include a glossary, there are several good ones in the literature [1-5].

Historically, data communications goes back to the invention of the telegraph by Samuel Morse in 1838. Although the telegraph is strictly manual, many of the principles and terms of early telegraphy have carried over to today. The first telephone patents were issued in 1876 to Alexander Graham Bell, but it was not until Lee DeForest invented the vacuum tube amplifier in 1906 that transcontinental telephony was possible. Much of the key work in this century was done by Bell Laboratories, including sampled communications theory in 1920, pulse code modulation in 1938, information theory by Claude Shannon in 1948, and time division multiplexing for telephony in 1950. Also of enormous importance, of course, was the invention of the transistor in 1948 by John Bardeen, William Shockley, and Walter Brattain, also at Bell Laboratories. This, more than any other event, made possible affordable computing power and the incredibly sophisticated hardware required to build the data networks of today.

The AT&T T-carrier system, introduced in 1962, was originally used to carry voice channels digitally internal to the telephone network. However, it soon began carrying increasing amounts of nonvoice data and became the basis for the Dataphone™ Digital Service (DDS) digital network in 1975. Beginning in the early 1970s several value-added networks began operation over leased telephone lines. Sophisticated digital switching and protocol services were added using specially designed minicomputers at internal network junctions. In the mid-1970s satellite networks also began offering digital services, primarily to large-volume users.

One of the true turning points in data communications history was the 1968 *FCC Carterfone decision*, which allowed users to connect non-AT&T equipment to their telephone lines. This opened up an entirely new *interconnect industry* to make various terminal equipment, much of it oriented to data transmission. More recently, on January 1, 1984, in settlement of a 1974 antitrust suit (really dating back to 1949) the regional telephone operating companies were split from the giant AT&T, resulting in a massive deregulation and decentralization of the U.S. domestic telephone system.

Today there is a continuing growth in the amount of data being transmitted throughout the world, and a commensurate transition from analog to digital networks. Much of this evolution is taking place within the framework of what is called the Integrated Services Digital Network (ISDN) concept, designed to provide a full range of services to users over a highly sophisticated, uniform, all-digital network.

This book is about the *technical* aspects of these data networks: their component parts, how they are put together, the way they are operated, and how they are used. Before we become too mired in the details of networks, it is only fair to point out that they are not an end in themselves, but rather only exist to provide services to the end users, whether human or machine. Networks can be used in an unlimited variety of ways. The following list indicates some of the more common applications:

- *Source data entry and collection*—point of sale systems, inventory control, reservations systems.
- *Real-time data acquisition and control*—numerical process control, remote pipeline control, alarm monitoring.
- *Information retrieval*—credit verification, police record checks, bibliographic retrieval.
- *Remote job entry*—mailing list services, introductory computer programming courses.
- *Interactive time sharing*—technical design, text editing, scientific computation.
- *Interprocessor data exchange*—file sharing, data base transfers.

Certainly there are many other important examples, but our concern here is predominantly with the network per se rather than its specific uses. Thus in the next section we begin to organize and classify practical computer data communications networks.

2.2. Network Classification

A good way to begin understanding any new subject is to classify it, and computer data networks are no exception. There are many ways in which data networks can be categorized, but one of the most fundamental is according to its conceptual layout—i.e., *topology*. In general, networks consist of a number of geographically dispersed *nodes* connected together by communication *links*. The network topology gives the interconnections between these nodes and links.

As Fig. 2.1 shows, the topology may be a *star* such as a group of terminals each connected directly to a central computer. A *ring* or *loop* topology has exactly two links per node. *Tree* topologies have all nodes connected together by links, but do not contain any loops. A *bus* network is much like a ring except that it is not closed, so the two end nodes are connected by only one link apiece. The *multipoint* or *multidrop* topology is somewhere between a bus and a tree, with end users connected to a main

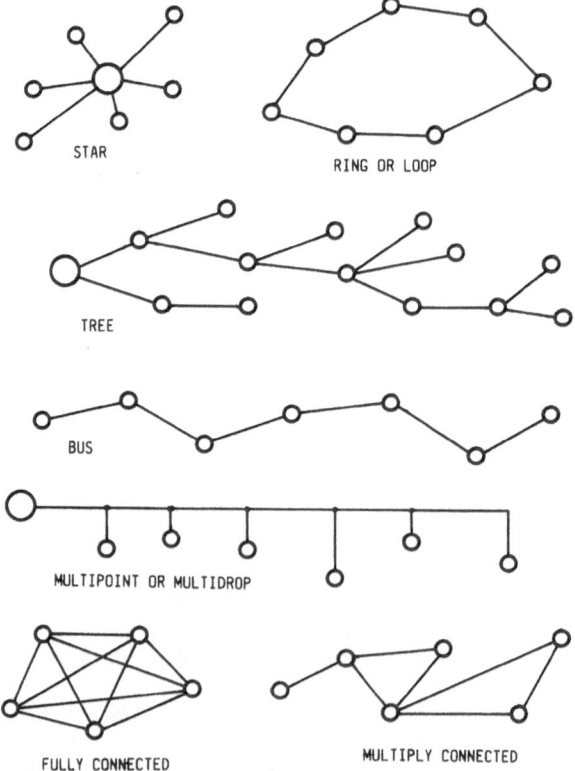

Figure 2.1. Network topologies.

line through individual links. Typically the main line is connected to, and controlled by, a central computer. For some requirements, such as extreme reliability, it is desirable to have a direct link between every pair of nodes, leading to the *fully connected,* or *complete,* topology. Finally, there are all kinds of other topological combinations which we shall lump together and call *multiply connected* networks.

Networks are also classified by their *geographic span. Local* networks are limited to one site, typically a building or building complex. They may cover a few miles at most, and typically much less. *Metropolitan* networks might cover a city, or perhaps a large government installation involving many square miles. *Regional* and *national* networks cover large segments of a country, or perhaps the entire country, while *international* networks span more than one country, often many.

There are many different *transmission media* that can be used to carry traffic on network links. *Wire pairs* are perhaps the most common, and have historically been the choice for local and smaller metropolitan connections. A *coaxial cable* can carry more traffic than a wire pair, and can cover long distances if the signals are periodically amplified along the way. It is most appropriate for metropolitan traffic and shorter regional routes. Special types of coaxial cable are also used for undersea cables. *Microwave radio* is common for long-haul regional and national network links due to its cost effectiveness. *Satellite links* are also used for long-haul transmission, particularly over rough terrain and for transcontinental communications. To remain stationary over the earth they must be placed in very high earth orbits, which results in a relatively long transmission delay that can be a problem in some applications. Finally, *optical fibers* of very pure glass are an increasingly important medium for data communications. They can carry huge amounts of traffic in digital form, and are immune to most external sources of interference. These lightguide links have applications at all geographical ranges, including undersea cables.

In order for a network to have many different possible end-to-end connections, some type of internal *switching* is usually required. Figure 2.2 shows the three major categories of switching for data networks. With *circuit switching* a dedicated physical path is set up when a call is placed, then maintained intact for the duration of the call regardless of whether or not traffic is continuously being sent. An example circuit is shown in part (a) of the figure, along with a typical distance and time relationship. Note that for short messages a relatively long internode signaling and connect time is required just to get the switching set up before any data can be sent, and additional time is required to disconnect the circuit. However, once data transfer begins the full channel is always available with no interruptions. Circuit switching is the familiar technique used for voice telephone calls, with the circuit being set up according to the number dialed.

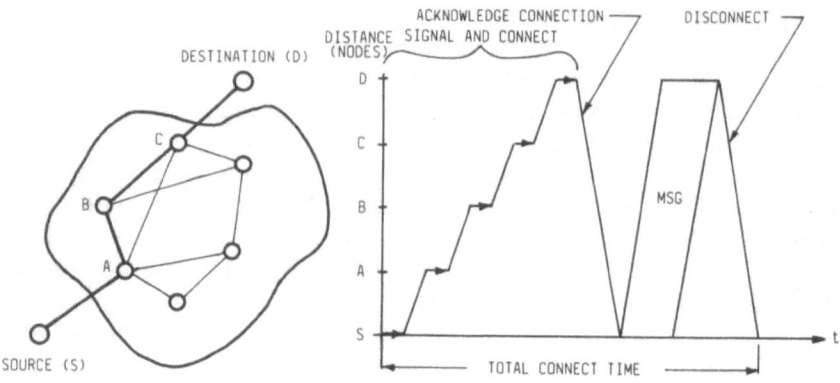

(a) Circuit Switching

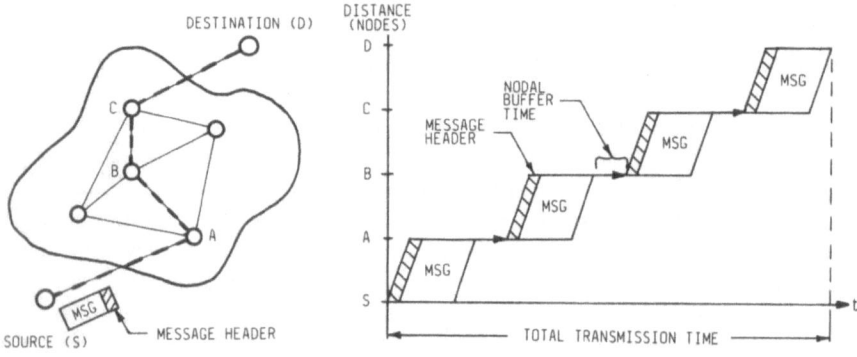

(b) Message Switching

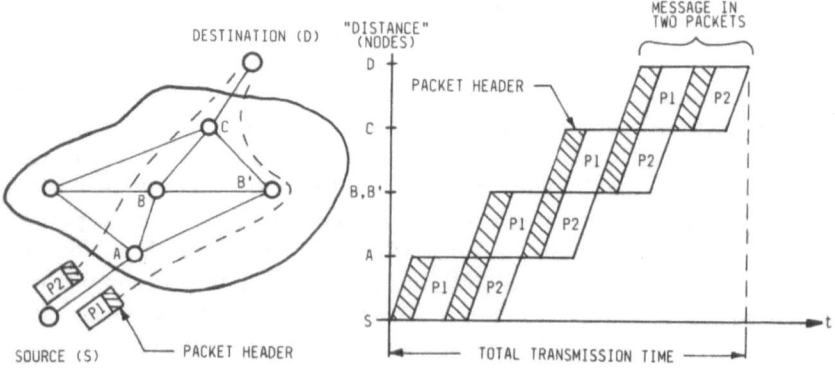

(c) Packet Switching

Figure 2.2. Network switching techniques.

Message switching requires a more "intelligent" network with nodes capable of reading the message addresses and temporarily storing entire messages, typically on magnetic disk memories. To the message the originator prefixes a *message header* containing the ultimate destination address. As Fig. 2.2b shows, each network node in turn stores, i.e., *buffers*, the entire message, reads the address and selects the next node, waits until the link to the next node becomes available, and then recalls and transmits the message. This is the way early telegraph system operators functioned. At each node the message can be logged and checked for errors, and the outgoing links can be nearly fully utilized since they are shared and each message must wait until the appropriate link is available. On the other hand, messages are stored at nodes for indeterminate times which can be exceedingly long in some circumstances such as when there is a rush of higher priority traffic. Large amounts of memory are required to buffer long messages, and the disk read/write time is often significant. Message switching networks also require relatively sophisticated "store-and-forward" computer switches at each node, but rapid advances in technology are making this requirement steadily less stringent.

The remaining switching technique, called *packet switching*, is illustrated in Fig. 2.2c with "distance" indicating either path through node B or B'. This is also a store-and-forward scheme, but now the nodes store small data *packets* in the *main memory* of the computer switches. This greatly reduces the nodal delay since mechanical disk operations are not required. However, such speed does not come free. Long messages must initially be divided into multiple packets, and each packet requires an individual header containing routing and packet sequencing information. The latter is required since, at the destination, the packet contents must be reassembled in the correct order into the original message before delivery to the user. Depending on the network, all packets of a message may follow the same route or they may be routed independently. In either case there is a "pipeline" effect, since the entire message is never buffered at once at internal nodes. Packet switching also requires considerable switch complexity as well as devices to *assemble* and *disassemble* the packets. There is also more potential overhead, since each packet must have a header. However, packet switching is very fast, and can also provide good security and survivability since messages can be split up and sent over alternate routes. Most large modern data networks are now packet switched.

While the switching method applies primarily to the network nodes, there are also several key classifications that primarily relate to links. First is the *direction of traffic flow*. There are three cases as shown in Fig. 2.3. A *simplex* link as in part (a) always carries traffic in the same direction; if two-way communication is desired, then a second link must be used for the return direction. Other links may operate in either direction, but not

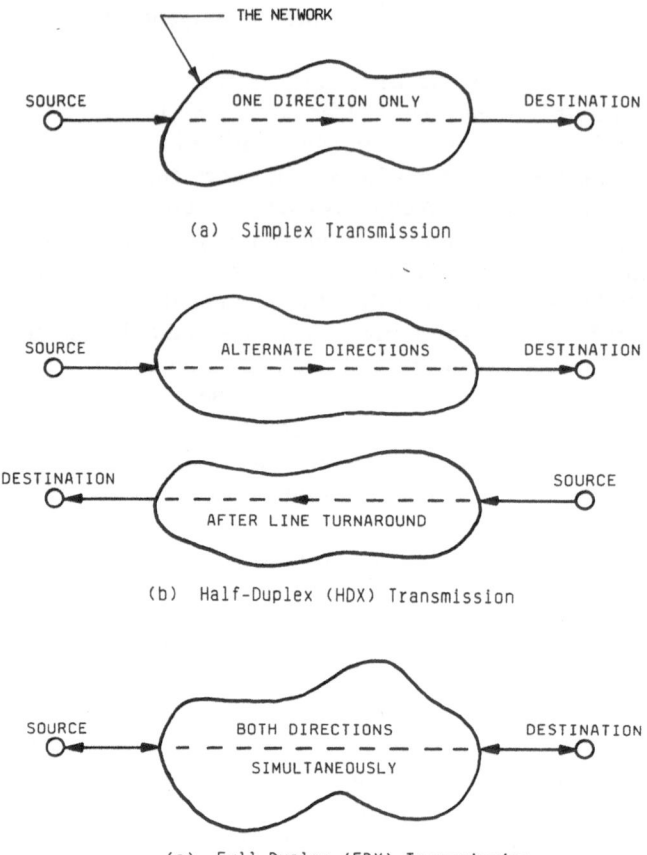

(a) Simplex Transmission

(b) Half-Duplex (HDX) Transmission

(c) Full-Duplex (FDX) Transmission

Figure 2.3. Direction of transmission.

simultaneously. To reverse the direction of traffic flow, such *half-duplex* (HDX) links must be formally *turned around* by some preagreed procedure, as part (b) indicates. This, of course, takes time and requires coordination. The final case, in part (c), is *full-duplex* (FDX). Here simultaneous transmission in both directions is possible.

A related link criterion is the *data rate* that can be carried, usually indicated in *bits per second* (bps). The term usually refers to the maximum rate that can be carried, but the user may choose to send at a lower average rate. Closely related to data rate is the analog *bandwidth* in hertz (Hz) of analog lines from which digital links are built. Bandwidth is analogous to the number of traffic lanes in a highway tunnel, with the data rate corresponding to the maximum number of cars per hour that can pass through for a fixed speed limit.

For telephone lines the common denominator is the familiar *voice-grade line* with a nominal bandwidth from 300 to 3000 Hz. Depending on the way data links are built from such lines, data rates up to 19,200 bps are technically feasible today. *Sub-voice-grade lines* are normally created by subdividing a voice-grade line, while *wideband* lines are available with different bandwidths extending well above 3000 Hz.

Voice-grade telephone lines also come in two basic types. *Dial-up* (switched, public) lines go through the telephone circuit switches and a connection must be established by dialing for each use. An alternative for heavy users requiring only point-to-point or multidrop connection is the *leased* (dedicated, private) line, which is permanently wired through the network and around the switches. This allows a more uniform quality of service which is always available and must be paid for whether used or not. Dial-up lines connect to the user's premises with one pair of wires (*two-wire* lines), while leased lines normally have two pairs (*four-wire* lines), allowing one pair for transmission in each direction.

Another important classification of data networks is with regard to the *protocols*, or rules of operation, that are used to ensure smooth and reliable communications. Even relatively simple networks require a hierarchy of such rules, so essentially all modern protocols are layered. There are protocols between users, between end terminal equipment, across the network per se, among network nodes, and at the most basic electrical and hardware levels.

A key aspect of network protocols is the issue of keeping the functions of the network "in step," i.e., *synchronized*. There are many kinds of synchronization at the various protocol layers, but the most common is that required for a link to operate properly. The simplest case is *asynchronous*,

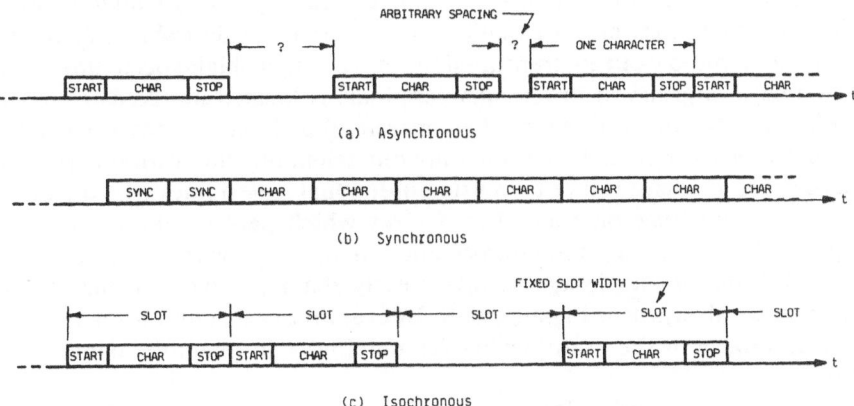

Figure 2.4. Synchronization methods.

where typically each character is appended with fixed start and stop bits, then transmitted whenever the link becomes available. Although the bits within each character are sent at a predetermined data rate, the characters can start at any time as Fig. 2.4a shows. In contrast, *synchronous* transmission provides a continuous stream of bits. Many characters are normally sent together as a block within a *frame* containing addressing, control, and error checking fields. Finally, there is an intermediate case known as *isochronous* transmission in which characters with start and stop bits are sent but must begin at distinct points in time, normally with spacing (slots) corresponding to one character. Figure 2.4 shows all three of the above cases.

Other network classifications are also possible, but ours are sufficient to establish a workable framework and establish the basic terminology. We extend these results in the next section by describing the major components from which data networks are constructed.

2.3. Network Components

Topologically, data networks are composed of only nodes and links. However, within these two broad areas there are many variations. We have already seen that lines can have different bandwidths, media, directional capabilities, and protocols. To convert analog lines such as voice-grade telephone lines into digital links capable of handling data, a conversion device called a *modem* (modulator–demodulator) or data set is used. Intuitively, modems take data to be transmitted and convert it into audio tones that are more efficiently passed through analog channels, and they also convert received tones back to the normal data form for delivery to the terminal. Thus modems effectively convert analog lines to digital links. There are many networking situations where it is highly desirable to combine a number of low-speed transmissions on a single high-speed link. For example, if a group of order-entry terminals is located hundreds of miles from the computer with which they operate, then huge cost savings can be realized over the years by leasing only one telephone line with one pair of high-speed modems rather than using individual leased lines with a pair of low-speed modems on each. The devices which perform this combining operation are called *concentrators* and *multiplexers*, with the distinction between the two no longer always clearly defined. Like modems, these devices are always used in pairs, with, for example, each multiplexer actually performing both the multiplexing (combining) and demultiplexing functions.

The more complex data network functions are typically performed by special purpose minicomputers or high-end microcomputers known collectively as *communications processors.* Store-and-forward switching is a major application of these processors at the nodes. They are also used for large *remote concentrators.* A third application is as *front-end processors* (FEP), which are collocated with mainframe host computers and perform the tedious and time-consuming communications tasks that the hosts would otherwise have to do. These functions include concentration, protocol operations, polling, error control, network management, and code conversions. An FEP should ideally do all the necessary processing so as to pass data to and from the host in exactly the form the host can use most efficiently. Most FEPs are programmable and can easily be configured for each particular application.

Many sizable networks have some type of *network control system* to monitor, manage, and handle errors and failures throughout the network. Network control is particularly important when networks span great distances, involve complex equipment from different vendors, and must have high availability. Typical functions include collecting and analyzing performance statistics, detecting and correcting errors, setting off alarms for significant errors, and testing the network periodically and whenever changes are made. Network control can be done either *in-band* by intermixing control signals with normal traffic, or *out-of-band* over separate low-speed secondary channels that parallel the normal channels. In either case there is a central control center from which the entire network can be directly accessed for control purposes.

The last major network component that we shall mention here is the *terminal.* Although sometimes considered as user equipment rather than part of the network itself, the whole point of the network is to connect remotely located terminals together. Furthermore many high-level network protocol functions are commonly performed by terminal equipment, so we shall include them as a bona fide network component. In the broad sense there are many kinds of terminal equipment, ranging from large mainframe computers, through specialized manufacturing and office machines, down to the familiar video display terminals and simple security sensors. They may be inherently digital devices, such as disk storage devices, or may convert analog signals such as voice to and from digital form for transmission. Traffic flow may also be highly *bursty,* such as with an interactive point-of-sale cashier terminal, or it may be very *steady* as with long file transfers. In any case, we shall use the word "terminal" to include practically any user-oriented device that sends or receives traffic through a network.

In later chapters, particularly Chapter 8, we shall elaborate on these and other network components. The basic terminology we have established here will be helpful for these more detailed studies. We now shift our focus from the technical aspects of data networks to the environment within which they must operate.

2.4. Standards and Regulation

Large communications networks can be incredibly complex. In fact, the pervasive domestic telephone network, which seems so simple to us as daily users, is often referred to as the world's most complex machine. For modern data communications networks the hardware detail required for data links, switches, concentrators, terminal interface equipment, and network control devices is hard for most people to imagine. If the software for operating these hardware devices and executing the network protocol is also considered, the possible variations among networks appear infinite. Furthermore, it is often essential for networks to intercommunicate. The developers and operators of data networks are inevitably able to achieve much higher performance at lower costs if there is a choice of vendors for the many different components that make up a network. Second sources can not only keep greedy vendors from exploiting the buyer, but can provide continuing support in the event that other suppliers discontinue production.

With this rationale, the need for standardization and regulation should be obvious. Standards fall into two basic categories. There are the obvious ones that are somehow developed and then formally promulgated by recognized standards organizations. Alternatively and equally important, are the *de facto* standards that ensue naturally from the practices of one dominant vendor. In reality these two categories are by no means distinct, since formal standards have been historically developed by simply adopting the predominant de facto standards with only minor modifications. In fact, the committees that develop formal standards are nearly always composed of representatives from those organizations having major stakes in the outcome.

Internationally, the two most important standards organizations for data communications are the CCITT† branch of the *International Telecommunications Union* (ITU) and the *International Standards Organization* (ISO). Founded in 1865, the ITU and CCITT are headquartered in Geneva, Switzerland. Over 150 nations are represented in CCITT, primarily by their telephone companies of Post, Telegraph, and Telephone (PTT) administrations. In the U.S. concerned parties are formally represented by the Department of State since there is no national PTT.

The CCITT standards are called *recommendations*, and are augmented and modified at plenary sessions held every four years. Plenary sessions results are actually international treaties, and are promulgated as a set of books with a different color for each period, e.g., red (1960), blue (1964), white (1968), green (1972), orange (1978), yellow (1980), and again red

---

† This stands for Comité Consultatif Internationale de Télégraphique et Téléphonique, in French.

(1984). Between plenary sessions standards are hammered out by various *Study Groups* made up of interested parties and experts, which may be subdivided into Working Parties and Rapporteur Groups. Candidate recommendations begin as working papers which are used to develop draft recommendations. The final draft is then submitted to the plenary for official approval. There are also several series of recommendations, with each identified by a different letter, e.g., V.3 defines the International Alphabet Number 5, which is nearly identical to the common ASCII character code. Of particular interest here are the V.-series dealing with data communications over the analog telephone network and X.-series dealing with data communications networks.

The ISO is also located in Geneva and deals with standards in many areas including information processing. Consequently it has gradually become increasingly involved with network protocols, and has worked closely with CCITT on these. ISO began in 1946 and has one representative organization from each member country. For the U.S. this is a trade organization known as ANSI. Unlike CCITT, ISO is not a governmental organization. Technical work is done in *Technical Committees* composed of Subcommittees and Working Groups. Preliminary working drafts of proposed standards go from a Draft Proposal to a Draft International Standard which should·be in essentially final form. Once approved by a vote of the members, it becomes an ISO International Standard.

There are also many national standardization organizations in the U.S. and elsewhere. In Europe the *European Computer Manufacturers Association* (ECMA) is a major source of standards for data networks. Domestically, the *American National Standards Institute* (ANSI) is run by the Computer and Business Equipment Manufacturers Association. Besides the OSI standards, ANSI promulgates many of its own, and is currently involved with the interconnection of telecommunications systems. Another important U.S. trade group is the *Electronic Industries Association* (EIA), which has established important physical interface standards for data communications equipment. The *Institute of Electrical and Electronic Engineers* (IEEE) is a major professional organization and publishes many electrical standards, including recently the key ones for local networks. Some computing standards relating to computer communications are produced by the *Association for Computing Machinery* (ACM).

Finally there is the federal government, of which the *National Bureau of Standards* (NBS) within the Department of Commerce is the major standardization entity. Most of the NBS standards relating to data networks are produced as *Federal Information Processing Standards* (FIPS) by the *Institute of Computer Sciences and Technology* (ICST) division of NBS. Many FIPS standards are just adoptions of other standards, but others are entirely developed by NBS.

In addition to these formal standards, there are many de facto vendor standards that are universally accepted and widely used by other vendors, often competitors. The two dominant sources of such data communications standards in the U.S. are IBM and AT&T. Both have had pervasive influence owing to their enormous customer bases and the technical excellence of their internal standards. In fact, it is not uncommon practice for standards organizations to adopt AT&T and IBM de facto standards with only minor changes.

Data communications is also affected by government regulation. At the federal level the *Federal Communications Commission* (FCC) grants licenses and approves tariffs for services proposed by common carriers of communications traffic. In effect these carriers are given a legal monopoly which is then regulated by the FCC. Carriers are also affected by court decisions primarily relating to antitrust and patent laws, and more broadly by congressional legislation. At the state level communications carriers are subject to state laws and *Public Service Commissions*, with policies varying considerably from state to state. Sometimes there are also local regulations concerning such matters as zoning and easements, but these are seldom a problem.

Perhaps the most significant regulatory case in recent years was the divestiture of AT&T in 1984, as a result of the court settlement of the Justice Department antitrust suit originally brought in 1949. In order to nullify a 1956 *Consent Agreement* restricting it from performing any data processing on telephone network traffic, AT&T sought relief from the FCC through its rulings, from the Congress through an attempted rewrite of the 1934 communications act, and from the courts through litigation.

In recent years there has been a definite trend in the U.S. and other countries toward increased deregulation, and this has somewhat lessened the need for regulation. However, so long as large networks exist and the public is dependent on them, regulation and standardization will remain important and valuable dimensions in the overall computer data communications picture.

We have now completed our introductory survey. With the resulting framework and terminology, we now begin our detailed chapter-by-chapter study of the many varied and fascinating aspects of *Practical Computer Data Communications.*

References

1. Abbreviations and acronyms, *Data Commun.* (Buyer's Guide Issue) 166–171 (1980).
2. Datacomm glossary penetrates the jargon, *EDN* **26**, 115–124 (1981).
3. M. Abrams and I. W. Cotton, *Computer Networks: A Tutorial* (4th ed.), IEEE Computer Society Press, Silver Springs, Maryland (1984), pp. 487–497.

4. Data communications glossary, *Data Commun.* **14**, 97–128 (1985).
5. *Glossary of Telecommunications Terms* (FED-STD 1037), Naval Publications and Forms Center, Philadelphia (1980).

Suggested Readings

James E. Brittain (ed.), *Turning Points in American Electrical History*, IEEE Press, New York (1977).

A unique collection of 64 key historical papers including many in communications and computing.

J. Brooks, *Telephone: The First Hundred Years*, Harper and Row, New York (1975).

The corporate history of AT&T to 1975, containing much insight into why the telephone system evolved as it did.

R. Cole, *Computer Communications*, Springer-Verlag, Berlin (1982).

A relatively modern introduction to a broad scope of data network topics, including network protocols, from a European point of view.

D. R. Doll, *Data Communications*, Wiley-Interscience, New York (1978).

Another very readable introduction to data communications links, equipment, and procedures.

J. Martin, *Introduction to Teleprocessing*, Prentice-Hall, Englewood Cliffs, New Jersey, (1972).

An easy-to-read introduction to simple data links with all the basics, although a bit dated now.

IEEE Communications Magazine (Special Issue on Divestiture: Two Years Later) **23** (12) (1985).

This issue contains an extensive collection of lively and often highly opinionated papers on the still controversial issue of the AT&T deregulation, written by many of the key participants in the legal and managerial proceedings.

IEEE Communications Magazine (Special Issue on 100 Years of Communications Progress) **22** (5) (1984).

A fascinating perspective on the history of communications technology over the past century as it all developed worldwide.

Check Your Understanding of Chapter 2—True or False?

1. The ECMA is the branch of the ITU that promulgates data communications standards.
2. AT&T and IBM are major sources of standards in the United States today.
3. The domestic telephone network is a fully connected network.
4. The bit rate is immaterial for asynchronous terminals.
5. Network control systems operate either in-band or out-of-band.

6. U.S. government data communications standards are promulgated by the NBS.
7. Transcontinental telephone calls were made possible by the vacuum tube amplifier.
8. Bus and multidrop networks have the same topology.
9. Loop and ring networks have the same topology.
10. Simplex lines must be "turned around" in order to have HDX communications.
11. In large packet networks, packets can be buffered for days on disk at nodes.
12. The landmark Supreme Court Carterfone decision opened up the interconnect industry.
13. The major benefit of multiplexing is a savings in line costs.
14. The CCITT is a major telecommunications regulatory agency worldwide.
15. Packet switching is always faster than circuit switching for data.
16. The AT&T DDS system was introduced in 1975 for carrying voice on interoffice trunks.
17. Bandwidth and data rate are technically the same thing, and so are used interchangeably.
18. Modems and data sets are the same thing in practice.
19. Traffic can flow in both directions simultaneously on FDX links.
20. The telephone was invented in 1776 by Alexander Graham Bell.
21. An FEP may also function as a remote concentrator for the same host.
22. Regional telephone companies are regulated by the FCC in the U.S.
23. Digital transmission is covered by the CCITT D.-series of recommendations.
24. The telegraph system was one of the earliest message-switched networks.
25. The terminology in data communications is always clear and unambiguous.

Communications Signals and Signal Processing

Objectives

o To provide an essential theoretical background for understanding data communications equipment and systems.
o To provide the necessary mathematical tools for this text.
o To develop some basic terminology for data communications.

Outline

3.1. Introduction

Computer data communications systems today rely almost entirely on electrical signals to represent the information being conveyed. Such representations may be analog voltages or currents, electromagnetic radio waves, stored electrical charges, digital voltage pulses, or the states of digital electronic gating circuits. Even with lightguide transmission, signals must be converted to and from electrical form at each end of the optical channel for processing and interfacing to the terminal equipment. Thus a solid understanding of electrical signals, filtering, and modulation is a fundamental prerequisite to any meaningful understanding of data communications.

		TIME	
		Continuous	Discrete
AMPLITUDE	Continuous	Analog	Sampled-Data
	Discrete	Quantized Staircase	True Digital

Figure 3.1. Classification of electrical signals.

In this text we shall only devote one chapter to this essential theory, which is hardly enough to do the subject justice. However, we have made every effort to cover all the topics germane to the following chapters. Although much of this latter material is accessible at an introductory level without the theory, a good theoretical grasp of it will greatly increase the depth of understanding and effectiveness of application in practice. In the long run, the fundamentals in this chapter are well worth the effort required to master them.

Communications signals can be classified in many ways, but for our purposes signals may be considered *continuous* or *discrete* in either *amplitude* or *time*. Continuous means the signal may range over a continuum of amplitudes or times, while discrete means the amplitudes or times are restricted to a set of distinct possible values. The resulting four-way classification is given in Fig. 3.1.

To illustrate, consider the "Tepid Tub Temperature Transducer" system shown in Fig. 3.2. The sensor produces an *analog signal* (continuous in amplitude and time) that varies with the water temperature. After being amplified and transmitted, the electrical signal looks like $s_1(t)$ shown in the figure. To monitor the temperature automatically and let us know when the temperature is just right (tepid), $s_1(t)$ is transmitted over to our home computer. However, this signal must be converted into digital form before it can be processed by the computer. Conversion is done in several steps. First, the signal is periodically sampled in time at *sampling interval T_s* with the resulting *sampled-data signal* (continuous in amplitude, discrete in time) shown as $s_2(nT_s)$ for $n = 0, 1, 2, \ldots$. Next $s_2(nT_s)$ is stretched out by the *hold* circuit for the full sampling interval to get another continuous analog signal $s_3(t)$. This *staircase signal* has a continuous range of possible step sizes. In this form the signal can be quantized, thus rounding off the amplitude steps to the nearest allowed (integer) voltages. We will call this a *quantized staircase signal* (discrete in amplitude, continuous in time), shown as $s_4(t)$. Finally, the signal is encoded as a three-bit binary number $s_5(n)$ which is clocked into the computer at *frequency $f_s = 1/T_s$*. The result, $s_5(n)$, is a true *digital signal*, which can be represented as a voltage or as a binary code (twos complement in the figure).

This formal taxonomy introduces some elementary signal terminology. However, it should be noted here that in practice the term "digital" is

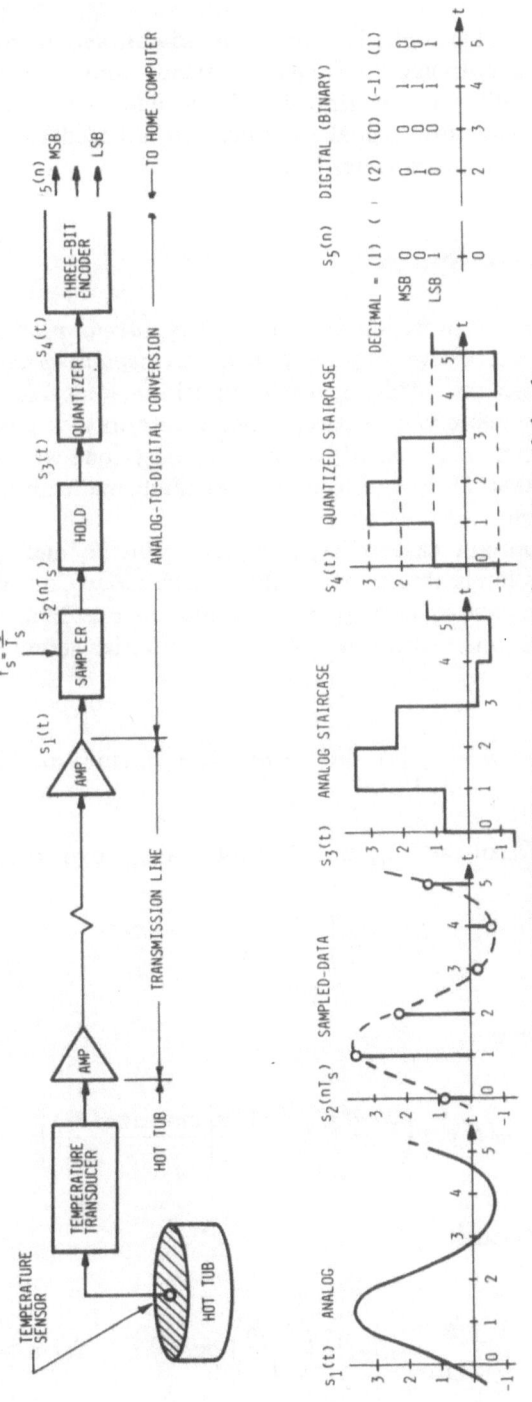

Figure 3.2. The Tepid Tub Temperature Transducer system and signals.

sometimes loosely also applied to both sampled-data and quantized staircase signals. We shall try to avoid any such confusion, and occasionally use the term "true digital" when clarification is needed. An alternative to the preceding time-waveform signal classification is based on the frequency composition, which we consider next.

3.2. Analog Signal Spectra

Signals can be represented by *time-domain waveforms* as in the preceding section, or alternatively by their *frequency-domain spectra*. The signal spectrum indicates the relative signal composition at all frequencies. There is a one-to-one relationship between the waveform of a communications signal and the signal spectrum; by knowing either one we can completely determine the other. The relationship is given mathematically by the classical *Fourier transform*.

In the important case of repeating, or *periodic*, analog signals the spectral information is obtained from the *Fourier series*. Consider the simple periodic square wave $s_0(t) = s_0(t + T_0)$ shown in Fig. 3.3a, with a period of T_0 sec and a pulse width of $\tau < T_0$. The Fourier series [1, 2] has the general form

$$s_0(t) = A_0 + \sum_{n=1}^{\infty} [A_n \cos(2\pi nt/T_0) + B_n \sin(2\pi nt/T_0)]$$

where the coefficients A_0, A_n, and B_n are the average, or dc, value

$$A_0 = \frac{1}{T_0} \int_{-T_0/2}^{T_0/2} s_0(t) \, dt = \frac{A\tau}{T_0}$$

the sine coefficients

$$A_n = \frac{2}{T_0} \int_{-T_0/2}^{T_0/2} s_0(t) \cos\left(\frac{2\pi nt}{T_0}\right) dt = \frac{2A\tau}{T_0}\left[\frac{\sin(\pi\tau n/T_0)}{(\pi\tau n/T_0)}\right] \qquad \text{for } n > 0$$

the cosine coefficients

$$B_n = \frac{2}{T_0} \int_{-T_0/2}^{T_0/2} s_0(t) \sin\left(\frac{2\pi nt}{T_0}\right) dt = 0 \qquad \text{for } n > 0$$

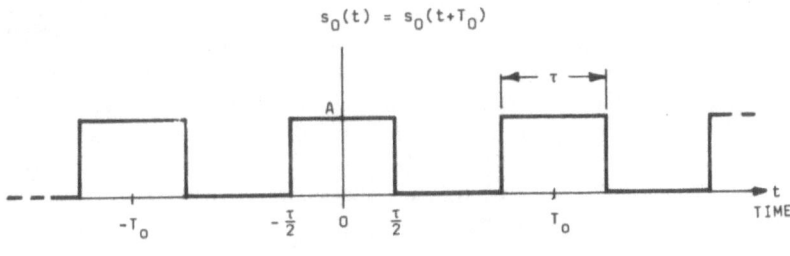

(a)   Time-Domain Square Wave

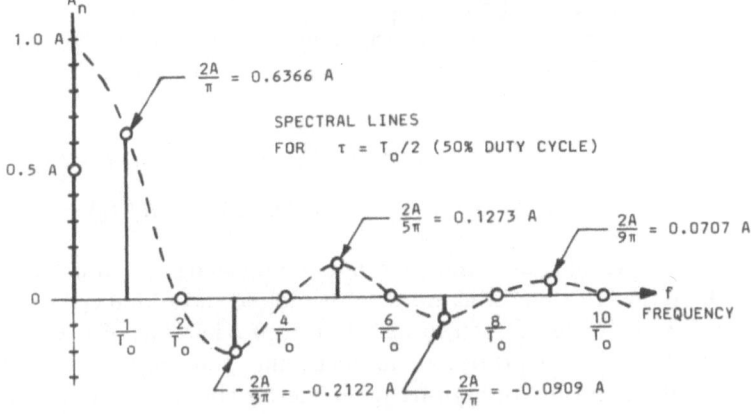

(b)   Line Spectrum of Square Wave

Figure 3.3. Periodic signal and line spectrum.

Therefore, the Fourier series representation for $s_0(t)$ is

$$s_0(t) = \frac{A\tau}{T_0} + \sum_{n=1}^{\infty} \left[\frac{2A\tau}{T_0} \frac{\sin(\pi\tau n/T_0)}{(\pi\tau n/T_0)} \right] \cos\left(\frac{2\pi n}{T_0} t \right)$$

We have represented the time-domain waveform of Fig. 3.3a as the sum of an infinite number of harmonically related sinusoids (cosines in this particular case) that are weighted by the corresponding *Fourier coefficients* A_n.

The two representations of $s_0(t)$ are completely equivalent. If we know the *fundamental period* T_0, or the *fundamental frequency* $f_0 = 1/T_0$, then the Fourier coefficients provide a complete description of the signal $s_0(t)$. In theory we could graphically construct the time waveform by adding up the infinite number of sinusoids in the series point by point. Rather than draw the waveform of $s_0(t)$ on the time axis, we could equivalently indicate the Fourier coefficients graphically on the frequency axis. This representation, called the *line spectrum* of $s_0(t)$, is shown in Fig. 3.3b for the particular case where $\tau = T_0/2$. Such a square wave that is "on" 50% of the time is known as a *50% duty cycle square wave*; in general, the duty cycle is the percentage of the period that the signal is nonzero. Notice that the spectral lines occur at multiples of the fundamental frequency $f_0 = 1/T_0$, and have amplitudes given by the Fourier coefficients. In the general case there may be an infinite number of lines with both nonzero sine and cosine terms in the series. The spectral lines are then complex valued, indicating a magnitude and a phase. The line at zero frequency is the A_0, or dc, term and represents the average signal level.

It is instructive to graphically construct an approximation to the periodic waveform $s_0(t)$ from the first few Fourier series components. We shall consider only the first four nonzero terms, since this series converges rapidly to a square wave. For the previous 50% duty cycle square wave case the first few terms are written out as

$$s_0(t) = 0.5000A + 0.6366A \cos{(2\pi t/T_0)} - 0.2122A \cos{(6\pi t/T_0)}$$
$$+ 0.1273A \cos{(10\pi t/T_0)} - 0.0909A \cos{(14\pi t/T_0)} + \cdots$$

The coefficients are, of course, the same values indicated by the line spectrum of Fig. 3.3b. If the first four terms are actually plotted and added up pointwise as in Fig. 3.4, the result is seen to be a fair approximation of the original waveform $s_0(t)$. The distortion is caused by the truncation of the series; specifically, the exclusion of high-frequency terms above the fifth harmonic

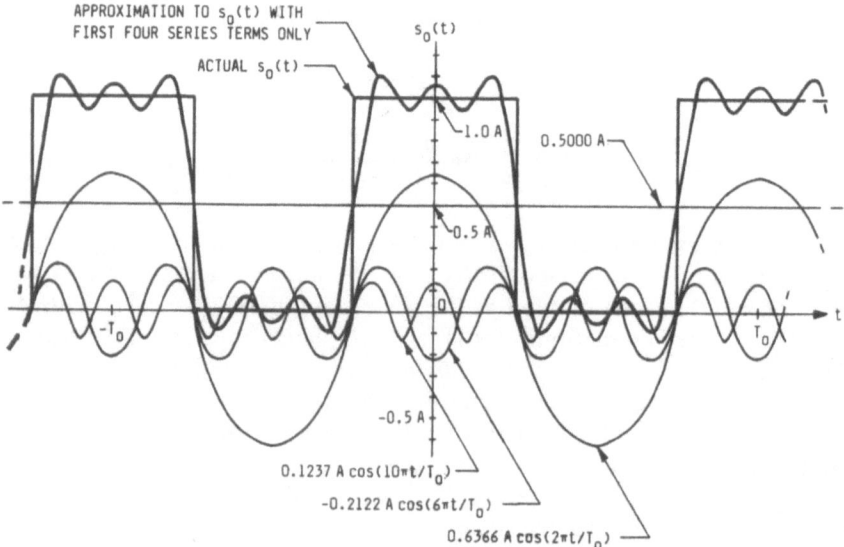

Figure 3.4. Square wave approximation by Fourier series terms.

at $f = 5f_0 = 5/T_0$. This type of truncation effect often occurs in data communication channels that must *bandlimit* the signals for efficient transmission. Communications systems can also cause distortion by attenuating low-frequency components, e.g., by using transformer coupling.

As already noted, in more general Fourier series there will be both sine and cosine terms, in which case a spectral line will have both a *magnitude* and a *phase* characteristic. Thus two plots will be required to completely represent the spectral information. It is both convenient and common practice to represent such values mathematically with complex numbers; here we shall use the symbol $j = \sqrt{-1}$ (pure mathematics usually uses i) for the *complex operator*. Thus, for any complex number x having real part x_p and imaginary part x_q, we can write the equivalent expressions

$$x = x_p + jx_q = (x_p^2 + x_q^2)^{1/2} e^{j \tan^{-1}(x_q/x_p)} = |x| e^{j\angle x}$$

where $|x|$ indicates the *magnitude* (or absolute value) of x, and $\angle x$ is the *phase angle* of x, ranging from $-180°$ to $+180°$ (phase will almost always be taken modulo $360°$). We can, of course, also plot these complex values in either polar or rectangular form in the complex plane.

It is often convenient to express the Fourier series in terms of magnitude and phase rather than in the previous rectangular form, in which case we can write the equivalent expression for an arbitrary periodic waveform $s_0(t) = s_0(t + T_0)$ as

$$s_0(t) = A_0 + \sum_{n=1}^{\infty} [A_n \cos (2\pi nt/T_0) + B_n \sin (2\pi nt/T_0)]$$

$$= C_0 + \sum_{n=1}^{\infty} C_n \cos (2\pi nt/T_0 + \theta_n)$$

The relationship among A_n, B_n and C_n, θ_n is easily found from the respective rectangular and polar representations,

$$C_0 = A_0 \quad \text{and} \quad C_n = (A_n^2 + B_n^2)^{1/2}$$

with

$$\theta_n = -\tan^{-1}(B_n/A_n)$$

Note the minus sign with the inverse tangent expression for θ_n.

To study this polar representation of the signal spectra consider the previous square wave of Fig. 3.3a, but now shifted in time by $\tau/4$ sec as shown in Fig. 3.5a. Denoting this signal by $s_1(t)$, we see that it is related

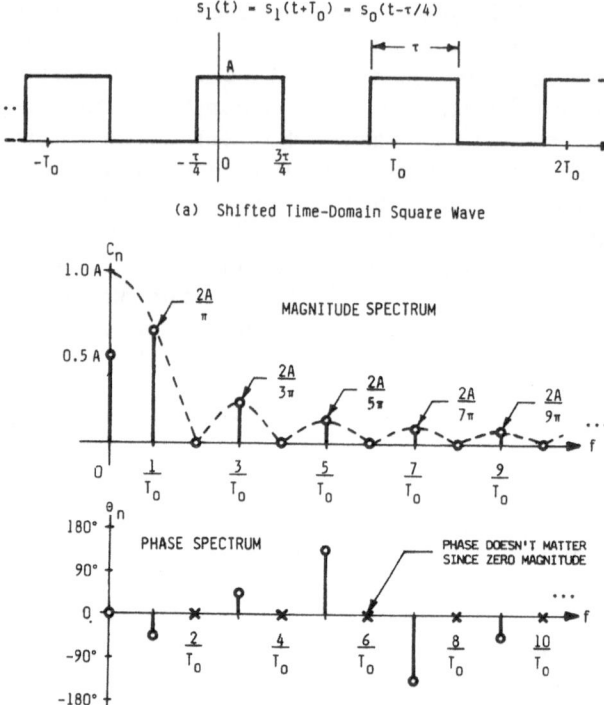

(a)   Shifted Time-Domain Square Wave

(b)   Spectrum of Shifted Square Wave

Figure 3.5. Time-shifted periodic signal and spectrum.

to $s_0(t)$ by the formula $s_1(t) = s_0(t - \tau/4)$, which is just a 0.25τ-sec time delay. The Fourier series can now be computed exactly as before,

$$A_0 = \frac{1}{T_0} \int_{-\tau/4}^{3\tau/4} A \, dt = A\tau/T_0, \qquad \text{average or dc value}$$

$$A_n = \frac{2}{T_0} \int_{-\tau/4}^{3\tau/4} A \cos\left(2\pi n t / T_0\right) dt = \left[2 \cos\left(\frac{\pi\tau n}{2T_0}\right) \frac{A\tau}{T_0} \frac{\sin(\pi\tau n / T_0)}{(\pi\tau n / T_0)} \right],$$

$$n > 0$$

$$B_n = \frac{2}{T_0} \int_{-\tau/4}^{3\tau/4} A \sin\left(2\pi n t / T_0\right) dt = \left[2 \sin\left(\frac{\pi\tau n}{2T_0}\right) \frac{A\tau}{T_0} \frac{\sin(\pi\tau n / T_0)}{(\pi\tau n / T_0)} \right],$$

$$n > 0$$

Hence, the complete Fourier series representation is given by

$$
\begin{aligned}
s_1(t) &= \frac{A\tau}{T_0} + \sum_{n=1}^{\infty} \left\{ \left[2 \cos\left(\frac{\pi\tau n}{2T_0}\right) \frac{A\tau}{T_0} \frac{\sin\left(\pi\tau n/T_0\right)}{(\pi\tau n/T_0)} \right] \cos\left(\frac{2\pi n}{T_0}t\right) \right. \\
&\quad \left. + \left[2 \sin\left(\frac{\pi\tau n}{2T_0}\right) \frac{A\tau}{T_0} \frac{\sin\left(\pi\tau n/T_0\right)}{(\pi\tau n/T_0)} \right] \sin\left(\frac{2\pi n}{T_0}t\right) \right\} \\
&= \frac{A\tau}{T_0} + \sum_{n=1}^{\infty} \left[\frac{2A\tau}{T_0} \frac{\sin\left(\pi\tau n/T_0\right)}{(\pi\tau n/T_0)} \right] \cos\left(\frac{2\pi n}{T_0}t - \frac{\pi\tau n}{2T_0}\right)
\end{aligned}
$$

which is easily shown to be exactly $s_0(t - \tau/4)$.

If the first few terms of this series are written out explicitly for the case of $\tau = T_0/2$, the magnitude and phase terms become apparent for each harmonic component of the series:

$$
\begin{aligned}
s_1(t) &= 0.5000A + 0.6366A \cos\left(2\pi t/T_0 - \pi/4\right) \\
&\quad - 0.2122A \cos\left(6\pi t/T_0 - 3\pi/4\right) \\
&\quad + 0.1273A \cos\left(10\pi t/T_0 - 5\pi/4\right) + \cdots \\
&= 0.5000A + 0.6366A \cos\left(2\pi t/T_0 - \pi/4\right) \\
&\quad + 0.2122A \cos\left(6\pi t/T_0 + \pi/4\right) \\
&\quad + 0.1273A \cos\left(10\pi t/T_0 + 3\pi/4\right) + \cdots
\end{aligned}
$$

Each term of this final series has nonnegative magnitude and corresponding phase angle for each harmonic. This is just the form that can be represented by a complex number in polar form, $C_n e^{j\theta_n} = C_n \cos\theta_n + jC_n \sin\theta_n$, for any spectral line at $f = nf_0 = n/T_0$, $n = 0, 1, 2, 3, \ldots$. Figure 3.5b shows this line spectrum represented as a magnitude and a phase plot.

An arbitrary periodic signal $s_0(t)$ can also be represented by a *complex*, or *two-sided*, *Fourier series* of the form

$$
s_0(t) = \sum_{n=-\infty}^{\infty} S_n e^{j2\pi nf_0 t}, \qquad \text{where } f_0 = 1/T_0
$$

By Euler's identity we have $e^{j2\pi nf_0 t} = \cos\left(2\pi nf_0 t\right) + j\sin\left(2\pi nf_0 t\right)$, and the *complex* Fourier coefficients S_n are given by

$$
S_n = \frac{1}{T_0} \int_{-T_0/2}^{T_0/2} s_0(t)\, e^{-j2\pi nf_0 t}\, dt, \qquad -\infty < n < \infty
$$

This is the form commonly used in theoretical communications work. The relationship between the trigonometric and complex Fourier series coefficients is easily shown to be

$$S_0 = A_0$$

$$S_n = \tfrac{1}{2}(A_n^2 + B_n^2)^{1/2}\, e^{-j\tan^{-1}(B_n/A_n)}, \qquad n > 0$$

$$S_n = S_{-n}^*, \qquad n < 0$$

where the asterisk denotes complex conjugate. For an example of the complex Fourier series consider again the square wave $s_0(t)$ of Fig. 3.4a. Substituting directly we have

$$S_n = \frac{1}{T_0}\int_{-\tau/2}^{\tau/2} A\, e^{-j2\pi n f_0 t}\, dt = \frac{A\tau}{T_0}\frac{\sin\,(2\pi n f_0 \tau/2)}{(2\pi n f_0 \tau/2)} = \frac{A\tau}{T_0}\frac{\sin\,(\pi\tau n/T_0)}{(\pi\tau n/T_0)}$$

Therefore, the Fourier series is

$$s_0(t) = \sum_{n=-\infty}^{\infty}\left[\frac{A\tau}{T_0}\frac{\sin\,(\pi\tau n/T_0)}{(\pi\tau n/T_0)}\right] e^{j2\pi n f_0 t}$$

which, with $\tau = T_0/2$, simplifies to the following concise form:

$$s_0(t) = \sum_{n=-\infty}^{\infty}\left[\frac{A}{2}\frac{\sin\,(n\pi/2)}{(n\pi/2)}\right] e^{j2\pi n f_0 t}$$

The corresponding line spectrum has lines at frequencies $f = n/T_0 = nf_0$, for $-\infty < n < \infty$. Figure 3.6 shows this two-sided line spectrum, along with the corresponding one-sided spectrum. The envelope is obtained by

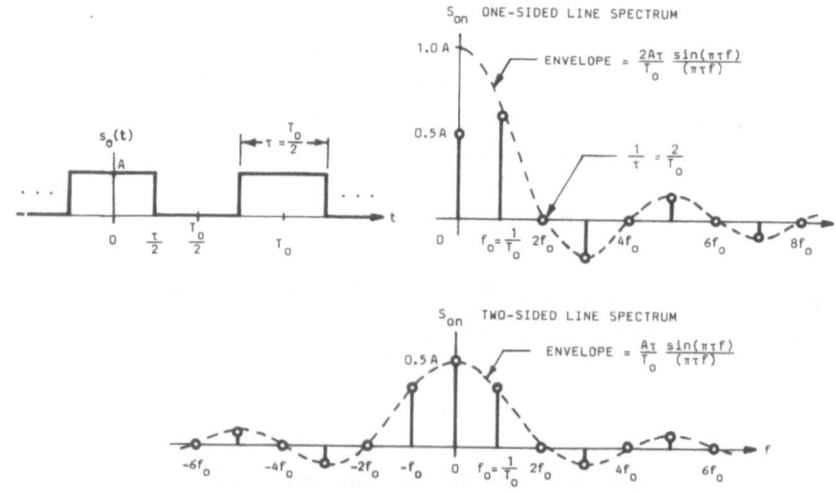

Figure 3.6. One-sided and two-sided line spectra.

simply replacing nf_0 by f in the Fourier coefficient expression. The envelope zero crossings are at $f = k/\tau = 2kf_0$ for $k = \pm1, \pm2, \pm3, \ldots$. This is appropriately called the *two-sided line spectrum* in contrast to the *one-sided line spectra* considered previously. The obvious fact that there are lines at *negative frequencies* is only a consequence of the mathematical model we have used for the signal. The physical interpretation is that we have, purely for mathematical convenience, divided the actual signal magnitude C_n at each nonzero frequency into two parts, putting half at each positive frequency $f = nf_0$ and the other half at the mathematical negative frequency $f = -nf_0$. Since $S_{-n} = S_n^*$, the phase at $-nf_0$ is the negative of that at $+nf_0$. In communications systems and data transmission it is usually easier to work with two-sided spectra than one-sided, particularly where modulation (spectral translation) is involved.

Now we wish to consider the effect of varying both the *pulse width* (decreasing τ) and the *pulse rate* (decreasing T_0). First let the pulse width, τ, be only one-fourth of the period, T_0; i.e., the duty cycle is 25%. Then the two-sided Fourier series is easily obtained by substituting into the general pulse train formula with $\tau = \frac{1}{4}T_0$. Denoting the waveform by $s_2(t)$, we get

$$s_2(t) = \sum_{n=-\infty}^{\infty} \left[\frac{A}{4} \frac{\sin(n\pi/4)}{(n\pi/4)} \right] e^{j2\pi nf_0 t}, \qquad \tau = \tfrac{1}{4}T_0$$

The spectral lines still occur at frequencies $f = n/T_0$, but now the "envelope" zero crossings are found at $f = k/\tau = 4kf_0$, $k \neq 0$. This is shown in Fig. 3.7b for comparison with the original 50% duty cycle line spectrum in Fig. 3.7a.

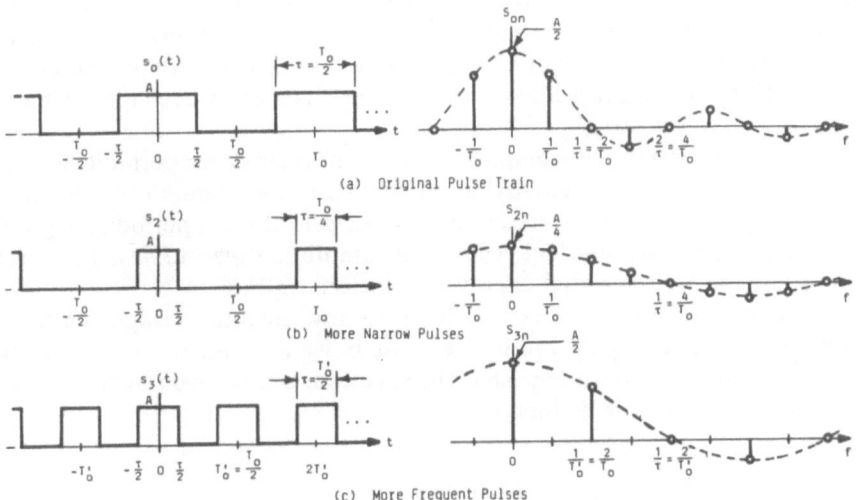

Figure 3.7. Time- and frequency-domain signal relationships.

Next consider the effect on the line spectrum of "speeding up" the preceding narrow pulses so that they occur twice as fast by letting the period now be $T_0' = \frac{1}{2}T_0$ with $\tau = \frac{1}{4}T_0 = \frac{1}{2}T_0'$. The overall effect is to double the pulse rate while keeping the duty cycle 50%. Denoting this last waveform by $s_3(t)$, we find the Fourier series representation to be

$$s_3(t) = \sum_{n=-\infty}^{\infty} \left[\frac{A}{2} \frac{\sin (n\pi/2)}{(n\pi/2)} \right] e^{j4\pi n f_0 t}, \qquad f_0 = \frac{1}{T_0} = \frac{2}{T_0'}$$

for which the signal and line spectrum are shown in Fig. 3.7c for comparison with the previous cases in (a) and (b).

Notice that the line spacing depends only on the fundamental period (T_0 or T_0'), i.e., on the *pulse repetition rate*, while the envelope of the spectral lines depends only on the pulse width τ. This simple case illustrates a much more general and fundamental property of all communications signals and spectra; namely, that narrow, rapidly changing signals tend to have a wider bandwidth (or significant spectral width) than more slowly changing ones. Conversely, slowly changing signals have relatively narrow bandwidths since the corresponding significant spectral lines are spaced close together. Intuitively, slowly changing signals have spectra concentrated at low frequencies, while fast changing signals have significant high-frequency spectral lines. "Compressing" the time-domain signal results in "expanding" the frequency-domain spectra, and vice versa! This argument illustrates the very important fact that to transmit at high data rates the signal must change correspondingly fast, and a wide bandwidth channel will thus be necessary to pass the wide signal spectrum without excessive distortion. Thus the data rate and required bandwidth are proportional. In fact, it is common practice to use the terms *data rate* and *bandwidth* interchangeably even though these two terms are not strictly the same thing.

Although it provides complete spectral information for periodic signals, the Fourier series is actually a time-domain representation. The true frequency domain representation of signals, periodic or aperiodic, is given by the *Fourier transform*. The result is the amplitude *spectral density*, or just the spectrum, of the signal expressed on a *per Hz* basis.

To extend our prior results let $s(t)$ now denote a *single* pulse of amplitude A and having width τ sec. The pulse is assumed to be centered at the origin as shown in Fig. 3.8a. The spectrum is now given by the *Fourier transform* $\mathscr{F}$, which is defined by

$$S(f) = \mathscr{F}[s(t)] = \int_{-\infty}^{\infty} s(t)\, e^{-j\omega t}\, dt, \qquad \omega = 2\pi f$$

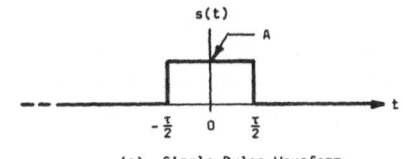

(a) Single Pulse Waveform

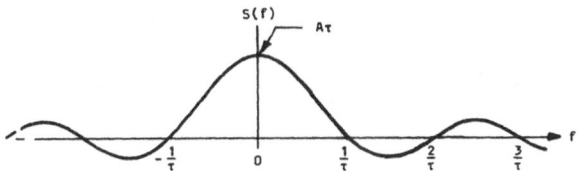

(b) Continuous Amplitude Spectral Density of Single Pulse

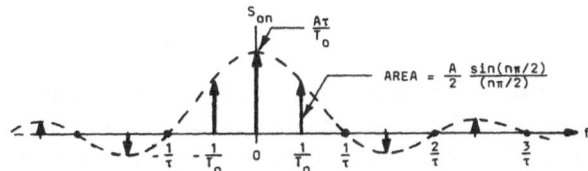

(c) Discrete Amplitude Spectral Density of Periodic Pulse Train

Figure 3.8. Spectral density.

where $\omega = 2\pi f$ is the radian frequency in rad/sec. For our pulse the spectrum is obtained directly from the formula,

$$S(f) = \int_{-\tau/2}^{\tau/2} A\, e^{-j\omega t}\, dt = A\tau \frac{\sin(\pi\tau f)}{(\pi\tau f)}, \qquad -\infty < f < \infty$$

which is plotted in Fig. 3.8b. Note that the zero crossings of the actual spectrum are now at $f = k/\tau$, $k = \pm1, \pm2, \pm3, \ldots$, and that there is a *continuum* of frequencies rather than the discrete harmonics of the periodic case. For aperiodic signals the spectrum is continuous in frequency, hence there is always an infinite number of frequencies and they are clearly not harmonically related.

$S(f)$ is a *spectral density*, e.g., V/Hz or A/Hz, while the line spectrum is just in volts or amperes. Furthermore, $S(f)$ is defined strictly in the frequency domain, while the line spectrum is actually part of the time-domain signal representation $s(t)$. For signals with distinct frequency components the amplitude spectral *density* is *infinite*, and we shall represent such cases mathematically by using the *unit impulse function*. Figure 3.8c shows two-sided amplitude spectral *density* for the 50% duty cycle pulse train $s_0(t)$ of Fig. 3.6. The *area* of the impulse is proportional to the amplitude of the corresponding harmonic, while the actual height is infinite, although it is common practice to draw the heights proportional to the areas for convenience.

The Fourier transform converts a time domain signal $s(t)$ to a unique† frequency-domain spectrum $S(f)$. There is also a unique inverse transform relation from $S(f)$ back to $s(t)$ given by the inverse Fourier transform

$$s(t) = \mathscr{F}^{-1}[S(f)] = \int_{-\infty}^{\infty} S(f)\, e^{+j\omega t}\, df, \qquad \omega = 2\pi f$$

where, again,

$$S(f) = \mathscr{F}[s(t)] = \int_{-\infty}^{\infty} s(t)\, e^{-j\omega t}\, dt$$

There are numerous tables of Fourier Transform pairs [3], but it is instructive to calculate some of the most common pairs here. For example, consider a time-delayed signal $s(t - t_0)$, where $S(f)$ is the known spectrum of the original (undelayed) signal $s(t)$. Then the spectrum can be calculated directly as follows:

$$\mathscr{F}[s(t - t_0)] = \int_{-\infty}^{\infty} s(t - t_0)\, e^{-j\omega(t - t_0 + t_0)}\, dt$$

$$= e^{-j\omega t_0} \int_{-\infty}^{\infty} s(\zeta)\, e^{-j\omega\zeta}\, d\zeta = e^{-j\omega t_0} S(f)$$

We now return to the impulse, or delta, function $\delta(x)$, which we *define* as a pulse with unit area and having zero width! By definition $\delta(x)$ has the property of unit area:

$$\int_{-\infty}^{\infty} \delta(x)\, dx = \int_{-\varepsilon}^{\varepsilon} \delta(x)\, dx = 1, \qquad \varepsilon > 0$$

In the time domain impulse functions are an idealization of narrow noise spikes, often referred to as impulse noise. In the frequency domain they may represent the spectral density of discrete frequency components, such

---

† To be precise the uniqueness is only within a set of zero measure; loosely this means waveforms may differ at distinct points and still be considered the same signal.

as in the Fourier series. Now consider the Fourier transform of a unit impulse signal $s(t) = \delta(t)$ at the origin,

$$\mathscr{F}[\delta(t)] = \int_{-\infty}^{\infty} \delta(t)\, e^{-j2\pi ft}\, dt = \lim_{\substack{\varepsilon \to 0 \\ \varepsilon > 0}} e^{-j2\pi f2\varepsilon} \int_{-\varepsilon}^{\varepsilon} \delta(t)\, dt = 1$$

This implies that the spectrum of the unit impulse is flat with equal amplitude spectral density at *all frequencies* from $-\infty < f < \infty$! Since the Fourier transform pair is unique, we also have

$$\mathscr{F}^{-1}[1] = \int_{-\infty}^{\infty} 1 e^{+j2\pi ft}\, df = \delta(t)$$

Mathematically, it follows that the integral of any complex exponential of the form $e^{\pm j2\pi xy}$ taken over the entire real line is an impulse, i.e.,

$$\int_{-\infty}^{\infty} e^{\pm j2\pi xy}\, dx = \delta(y)$$

As an example, consider a dc battery voltage $v(t) = 12.0$ V. Then the spectrum $V(f)$ of $v(t)$ is

$$V(f) = \int_{-\infty}^{\infty} 12 e^{-j2\pi ft}\, dt = 12\delta(f)$$

This is just an impulse at zero frequency, or dc, as expected. The total "amplitude" is 12.0 V, but the *amplitude spectral density* in V/Hz at $f = 0$ Hz is *infinite*.

To determine the spectral density of a periodic signal $s_0(t)$ it is only necessary to express it as a Fourier series and take the Fourier transform directly. Thus

$$\mathscr{F}[s_0(t)] = \mathscr{F}\left[\sum_{n=-\infty}^{\infty} S_n e^{j2\pi nf_0 t} \right]$$

$$= \sum_{n=-\infty}^{\infty} S_n \int_{-\infty}^{\infty} e^{-j2\pi(f-nf_0)t}\, dt = \sum_{n=-\infty}^{\infty} S_n \delta(f - nf_0)$$

This is analogous to the two-sided spectrum, with the lines replaced by impulses weighted by the (complex) Fourier coefficients S_n. To illustrate, consider again the specific case of the 50% duty cycle square wave of the last section. Calculation of the spectrum gives

$$S_0(f) = \mathscr{F}[s_0(t)] = \sum_{n=-\infty}^{\infty} \left[\frac{A}{2} \frac{\sin{(n\pi/2)}}{(n\pi/2)} \right] \delta(f - nf_0)$$

The spectral density $S_0(f)$ was shown previously in Fig. 3.8c. In the general case the spectrum will be complex, having a magnitude and a phase, so two plots will be required, with impulses indicating the magnitude spectral density.

We have now developed the basic spectral theory to understand data communications signals. To add some intuitive understanding to this theory, we next consider some general signal categories and their respective spectral properties. Figure 3.9 shows three such signals. Comparison of the two *baseband* signals in (a) and (b) indicates that the more slowly changing $s_a(t)$ has a relatively narrow spectrum concentrated around dc, while the more rapidly changing one $s_b(t)$ has a wider bandwidth with higher frequency components but is still around dc. Both spectra are continuous.

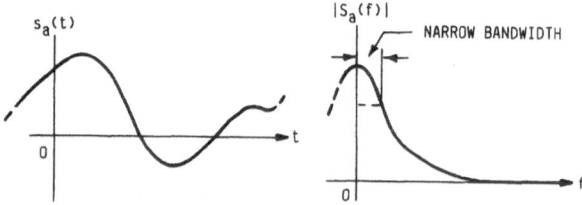

(a) Slowly-Changing Narrow Bandwidth Baseband Signal

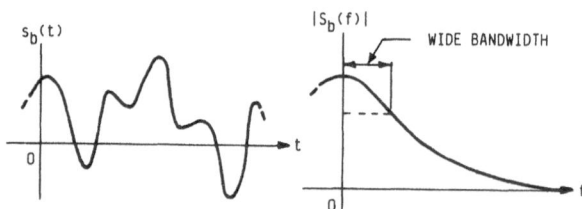

(b) Rapidly-Changing Wide Bandwidth Baseband Signal

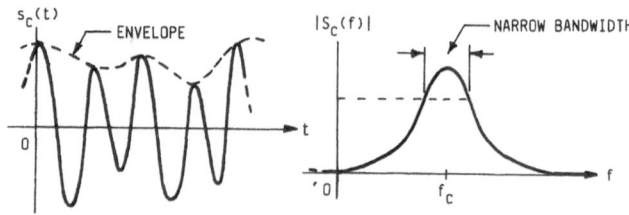

(c) Passband Signal Centered at Frequency f_c

Figure 3.9. General time-frequency relationships.

The third case, $s_c(t)$ in part (c), is more complicated and interesting. The corresponding spectrum is located about a nonzero frequency $f_c > 0$; hence this is not a baseband signal. Rather, it is called a *passband* signal with f_c being the *center frequency*. Such signals vary at a changing rate that averages out to the center frequency f_c; that is, they tend to "ring" about the center frequency. The rate at which the signal envelope can change depends on the spectral bandwidth, and not on the location of the center frequency. Thus we could increase f_c to make the signal "ring" faster without changing the envelope variation at all. Such linear shifting of the center frequency, and hence the spectrum, is a form of *modulation*. On the other hand, increasing the bandwidth without changing f_c allows the envelope to vary faster, but with the same "ringing" frequency. Intuitively, it should be possible to send more information during a given time using a signal that can vary quickly. This is indeed the case. Thus the bandwidth can be directly related to the capacity to transmit information in digital communications systems, and again implies why the terms "bandwidth" and "data rate" are used interchangeably.

Now that we have developed the fundamental concepts of analog signals and spectra, we turn to the closely related issues of signal filtering and the transmission of signals through channels in which there is only a limited amount of bandwidth. Since in data communications this bandwidth is usually expensive, it is important that signals be selected and shaped to ensure efficient use of the available bandwidth with acceptable error rates.

3.3. Analog Filters

In the most general communications sense we can consider a *filter* to be any frequency-selective system. Thus the output signal spectrum of a filter is obtained from the input signal spectrum by selectively modifying each frequency in a manner uniquely defined by the particular filter. Reshaping the spectrum obviously reshapes the signal waveform as well. This is necessary in data communications for many reasons, including noise removal, distortion correction, separation of different signals, and efficiently fitting signals into band-limited transmission channels.

In practice filters can be made from practically anything that will pass a signal and may have all kinds of frequency-selective characteristics. For example, they may be classified as electrical or mechanical, analog or digital, linear or nonlinear, active or passive, time-varying or time-invariant, or according to what frequencies are passed; e.g., low-pass, bandpass, high-pass, band-stop, all-pass, or multiband. Another common classification is by the classical mathematical approximating functions used to design the filters, e.g., Butterworth, Chebyshev, elliptic, or Bessel. The term "filter" clearly encompasses a huge range of possibilities, but for this introduction to analog communications it will be sufficient to restrict our consideration to the most important class of *linear time-invariant* (LTI) *electrical* filters. Later we shall extend our results to include selected digital (numerical) and time-varying filters as well.

Linear filters are nearly always easier to understand in the frequency domain, since the input–output relationships are much simpler than in the time domain. Any LTI filter can be characterized by its *frequency response* $H(f)$, which is just the response of the filter to sinusoidal inputs of unit amplitude taken over all frequencies of interest. Since an LTI filter can change only the magnitude and phase of a sinusoidal signal, we can display the entire frequency response $H(f) = |H(f)| e^{j \angle H(f)}$ with a pair of curves on the frequency axis. As with signal spectra, the phase is normally taken modulo 360°. Since, as we have seen, delay is directly related to phase shift,† *group delay* is sometimes used instead of phase. It is also common practice to express the magnitude logarithmically in *decibels* (dB), where by definition

$$|H(f)|_{dB} = 20 \log|H(f)|$$

with the frequency axis also being logarithmic.

Mathematically, LTI filtering is described in the frequency domain by point-by-point multiplication of the (complex-valued) input signal spectrum $S_i(f)$ by the (complex-valued) filter frequency response $H(f)$; hence the filter response signal spectrum $S_r(f)$ is given by

$$S_r(f) = H(f)S_i(f) \qquad \text{for all } f$$

Back in the time domain the LTI filtering corresponds to a *convolution* (denoted by $*$)

$$s_r(t) = h(t) * s_i(t) = \int_{-\infty}^{\infty} h(\xi)s_i(t-\xi)\, d\xi$$

where $h(t) = \mathscr{F}^{-1}[H(f)]$ is appropriately called the filter *impulse response* since it is the theoretical response of the filter to a unit impulse function input.‡ These relationships, shown in Fig. 3.10, only describe the *steady-state* response after all transients have died away.

---

† The relation is given mathematically by

$$\text{delay} = -\frac{d}{d\omega}\,(\text{phase shift}) = -\frac{1}{2\pi}\frac{d}{df}[\angle H(f)], \qquad \omega = 2\pi f$$

‡ It may be intuitively helpful to think of an impulse input as performing an "instantaneous frequency response," since the spectrum of $\delta(t)$ contains all frequencies with equal density.

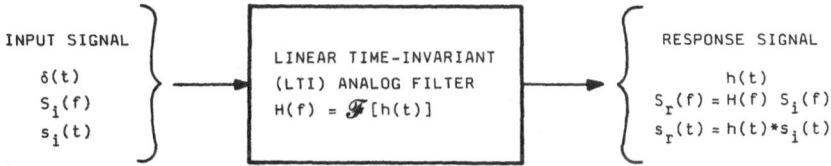

Figure 3.10. LTI filter steady-state relationships in the time- and frequency-domains.

Most data communications channels can be modeled satisfactorily as linear filters, so the above relationships indicate the importance of good spectral shaping of the transmitted signal. If the transmitted spectrum $S_i(f)$ does not match the channel frequency response $H(f)$, then the received signal $S_r(f)$ will be weak and distorted, with a poor *signal-to-noise ratio* (SNR). To the user the result is a high error rate or perhaps even a complete loss of signal in the noise.

There are many ways to specify, design, and construct electronic filters, but for digital telecommunications applications the standard building blocks are resistors, capacitors, and operational amplifiers. Figure 3.11 shows several representative filter realizations and the associated frequency responses. Part (a) is a simple first-order *passive low-pass filter* with a single resistor (R) and capacitor (C). Note that at $f = f_c$ the magnitude is reduced to $1/\sqrt{2}$ of its maximum value, while the phase shift is $-45°$. In general, any frequency where the magnitude of the frequency response has decreased to $1/\sqrt{2} = 0.7071$, or by $20 \log (1/\sqrt{2}) = -3.0103$ dB, is called a *cutoff* (break, corner, half-power, 3 dB) *frequency*, and is taken by convention to be the dividing line between a passband and a stop-band. Thus the bandwidth of this particular filter is f_c Hz, with f_c completely defined by R and C. As the frequency is increased the frequency response *rolls off* towards zero with the phase shift approaching $-90°$, as shown.

Figure 3.11b shows a second-order *active bandpass* filter based on an *operational amplifier*. The "op amp" is the universal building block for analog electronics, and nearly all high-performance communications filters employ one or more. In our example the filter passband is $1/\pi R_2 C$ Hz wide and peaks at the *center* frequency $f_0 = 1/2\pi(R_1 R_2 C^2)^{1/2}$ Hz with gain of R_2/R_1. Very low and very high frequencies are heavily attenuated and, in particular, no dc component is passed. The phase drops from $-90°$ to $-270°$, with exactly a $-180°$ shift at f_0, as might be anticipated from the minus sign in $H(f)$ and inverting input to the op amp. The third example is an *active all-pass filter*, which produces only a phase shift with no attenuation. Such "phase shifters" are important in sophisticated modulation processes and also for correcting undesirable group delay. By properly selecting the filter components R_1 and C, it is possible to provide any desired phase shift, hence delay, between $0°$ and $-180°$. By cascading two such active

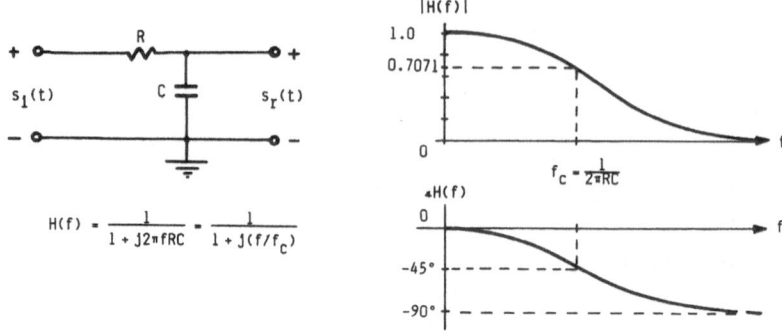

$$H(f) = \frac{1}{1 + J2\pi fRC} = \frac{1}{1 + J(f/f_c)}$$

(a)   Passive Analog Low-Pass Filter

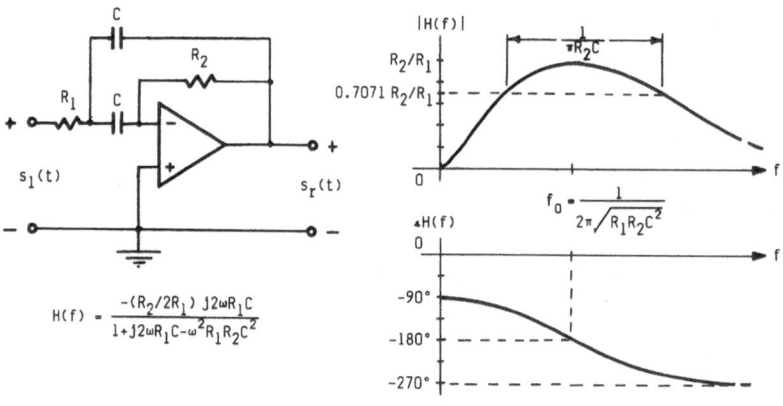

$$H(f) = \frac{-(R_2/2R_1)\ J2\omega R_1 C}{1 + J2\omega R_1 C - \omega^2 R_1 R_2 C^2}$$

(b)   Active Second-Order Analog Bandpass Filter

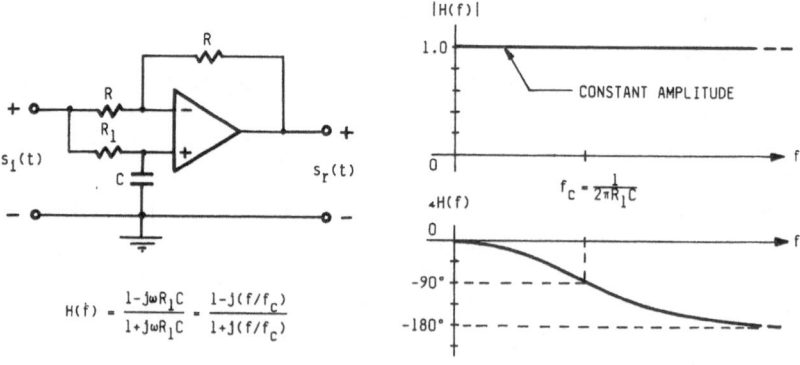

$$H(f) = \frac{1 - J\omega R_1 C}{1 + J\omega R_1 C} = \frac{1 - J(f/f_c)}{1 + J(f/f_c)}$$

(c)   Active Analog All-Pass Phase-Shift Filter

Figure 3.11. Typical analog filters.

stages the range of phase shift can be doubled. In fact, an important general property of operational amplifier filter design is the ability to build complex, high-quality filters by cascading several simple stages with negligible loading among stages.

Filter design is concerned with many criteria, but paramount among them are the *flatness of the passband* frequency response and the *roll-off of the stop-band*. Both of these parameters are directly related to the complexity, or *order*, of the filter. To illustrate this tradeoff, consider the classical low-pass Butterworth filter characteristics shown in Fig. 3.12a, which can be realized with a variety of passive and active circuits. The magnitude response is defined in terms of the cutoff frequency f_c and the filter order N. As shown, increasing N makes the passband flatness and the stop-band roll-off better, but at the cost of more filter stages. A logarithmic plot of the frequency response is shown in part (b) for comparison. Note that the attenuation is always -3 dB at f_c and the stop-band roll-off is -20 dB per octave times the filter order.

In practice, analog filter design lies somewhere between a catalog look-up technique and pure witchcraft, with computer-aided design techniques indicated for all but the simplest cases. In the final analysis, however, good filter design is as much an art as a science.

Having now introduced some fundamental filter concepts, we return to the input–output signal relationship with a simple and intuitive example. Thus, let the periodic square wave of Fig. 3.3 with a 50% duty cycle and period T_0 be applied as $s_i(t)$ to a low-pass filter as in Fig. 3.11a with cutoff frequency set to exactly $f_c = 1/T_0$. The input signal and one-sided line spectrum are shown in Fig. 3.13a, with the filter frequency response $H(f)$ in (b). We are, of course, interested in the filtered square-wave response $s_r(t)$. Since $s_i(t)$ consists of harmonically related sinusoids, the frequency response $H(nf_0)$ evaluated at each harmonic will give us the Fourier series, and hence the waveform, of the response $s_r(t)$.

Now the input signal was shown in the last section to be

$$s_i(t) = 0.5000 + 0.6366 \cos{(\omega_0 t)} - 0.2122 \cos{(3\omega_0 t)} + 0.1273 \cos{(5\omega_0 t)}$$

$$- 0.0909 \cos{(7\omega_0 t)} + \cdots$$

where $\omega_0 = 2\pi f_0$, and we have set $A = 1$ for convenience. The filter frequency response is

$$H(f) = \frac{1}{1 + j(f/f_c)} = \frac{1}{\sqrt{1 + (f/f_c)^2}}\, e^{-j\tan^{-1}(f/f_c)}$$

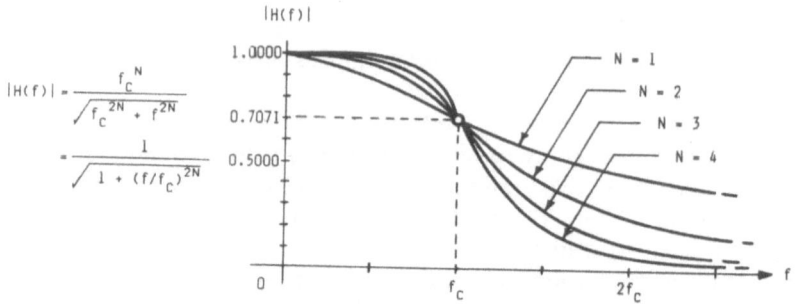

$$|H(f)| = \frac{f_c^N}{\sqrt{f_c^{2N} + f^{2N}}}$$

$$= \frac{1}{\sqrt{1 + (f/f_c)^{2N}}}$$

(a) Butterworth Filter Characteristics of Orders 1 through 4

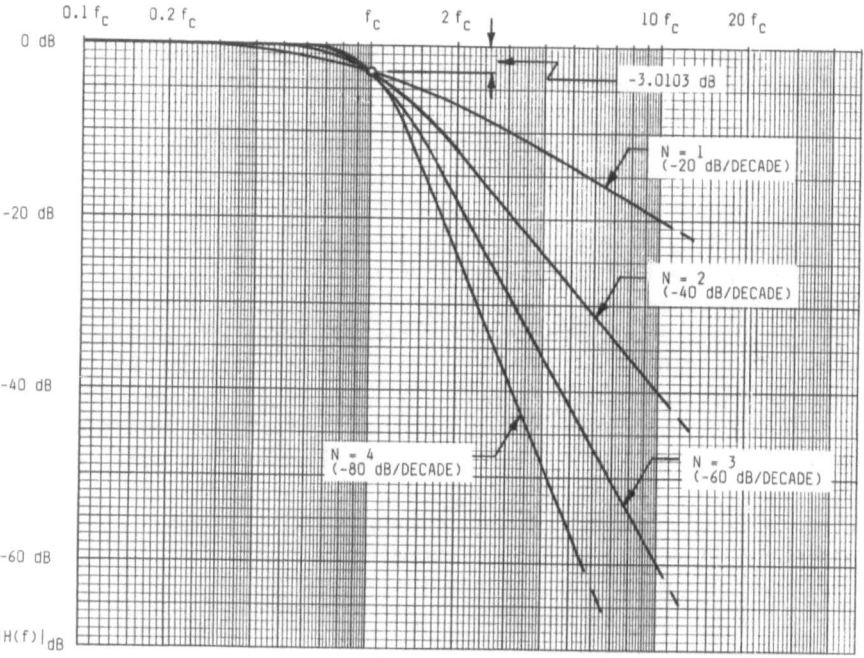

(b) Logarithmic Frequency Response Plot for the Butterworth Filter

which, at the frequencies $f = nf_0$, $n = 0, 1, 3, 5, 7, \ldots$, has the values

$$H(0) = 1.0000 e^{j0°}$$
$$H(f_0) = 0.7071 e^{-j45.0°}$$
$$H(3f_0) = 0.3162 e^{-j71.6°}$$
$$H(5f_0) = 0.1961 e^{-j78.7°}$$

$$H(7f_0) = 0.1414e^{-j81.9°}$$

By taking the Fourier transform of each input series term, multiplying by the corresponding frequency response, and upon inverse transforming the result, we obtain the output signal,

$$s_r(t) = 0.50000 + 0.4501 \cos(\omega_0 t - 45.0°) - 0.0671 \cos(3\omega_0 t - 71.6°)$$
$$+ 0.0250 \cos(5\omega_0 t - 78.7°) - 0.0129 \cos(7\omega_0 t - 81.9°) + \cdots$$
$$= 0.50000 + 0.4501 \cos[\omega_0(t - 0.125T_0)]$$
$$+ 0.0671 \cos[3\omega_0(t + 0.100T_0)]$$
$$+ 0.0250 \cos[5\omega_0(t - 0.044T_0)]$$
$$+ 0.0129 \cos[7\omega_0(t + 0.039T_0)] + \cdots$$

Several instructive points can be made from this result, which is shown in Fig. 3.13c. First, the average value of the input remains at 0.5, since the

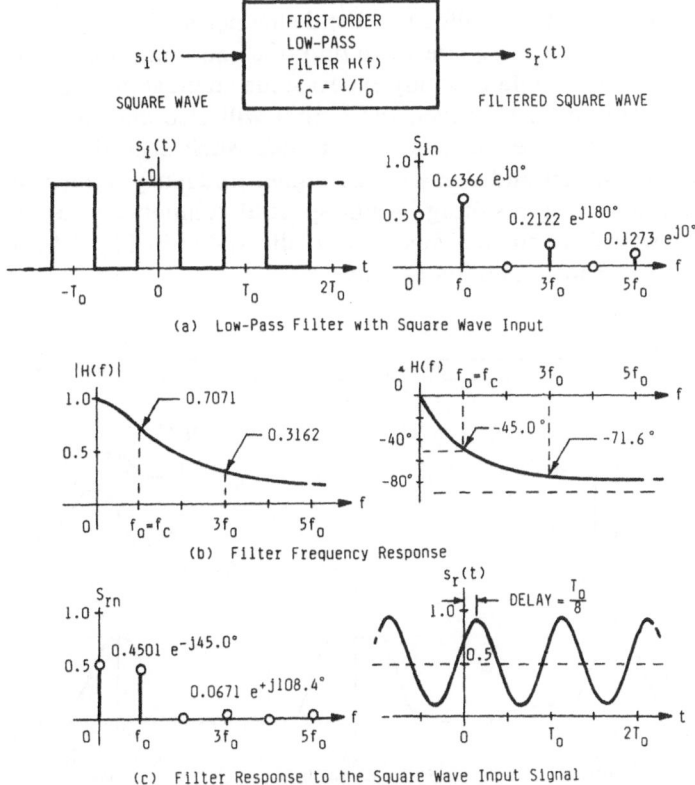

(a) Low-Pass Filter with Square Wave Input

(b) Filter Frequency Response

(c) Filter Response to the Square Wave Input Signal

Figure 3.13. Low-pass filtering.

filter dc response is unity. At the cutoff frequency the input component is attenuated to 0.7071 of its original value, and delayed 1/8 of the period. All higher harmonics are in the stop-band and are greatly attenuated relative to the dc value. As the waveform shows, the output looks essentially like a slightly distorted, delayed first harmonic plus the original input dc value. Note finally that no new frequencies result, which is an inherent characteristic of all linear (but not of nonlinear) filters.

The same general intuitive ideas depicted in Fig. 3.9 can be extended to all LTI filtering. For example, the output of narrowband low-pass filters will always change relatively slowly regardless of the input, since the high frequencies necessary for rapid variation will be filtered out. For bandpass filters the output will always be a bandpass signal that will "ring" at an average rate equal to the filter center frequency, with the possible envelope variation rate determind by the bandwidth.

We conclude this section by considering the fundamental relationship between pulse shape, channel bandwidth, and the maximum feasible pulse rate that can be transmitted. Suppose we want to send data pulses as fast as possible over a bandlimited baseband channel, such as a long wire cable. If we put a nice clean square wave into one end and observe what comes out the other, we would certainly expect some attenuation and perhaps also some random noise. However, the output will also appear *"smeared" in time* and may also be *"skewed" to one side.* Such distortion is due to the uneven attenuation and delay of the higher-frequency input components, thereby causing the resulting output spectral components to add up to a different waveform shape. This effect is illustrated in Fig. 3.14, where the spectral plots show magnitudes only.

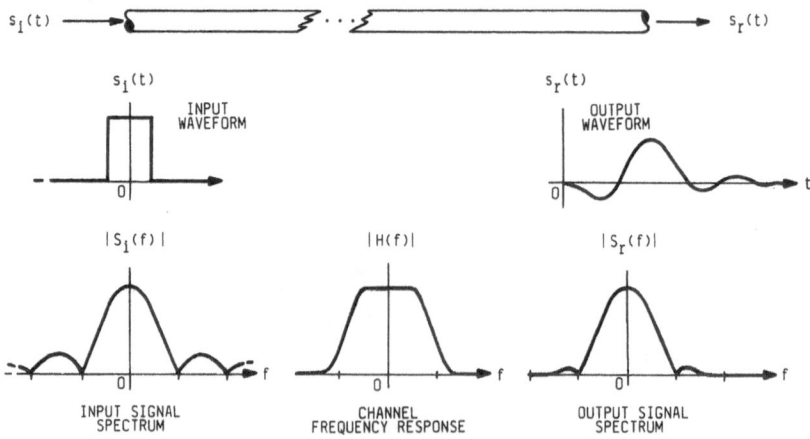

Figure 3.14. Effects of channel distortion.

Now the next pulse we send will be similarly distorted. To avoid distortion we could wait for the first to die away before sending the second; but this delay can severely limit the feasible data rate of the channel. The above distortion is appropriately called *intersymbol interference* (ISI). Fortunately, the effects of ISI can be greatly mitigated by properly shaping the input signal spectrum in such a way as to compensate for the channel distortion. To understand how this can be done depends on the following result. Let the baseband spectrum of the response signal have any odd symmetry about the half-amplitude bandwidth B, as shown in Fig. 3.15a for the cases of 0%, 50%, and 100% excess bandwidth beyond B Hz. Then the output waveforms corresponding to the input pulses will have zero crossings exactly every $t = k/2B$ sec,† $k = \pm 1, \pm 2, \pm 3, \ldots$. If we now let the transmitted pulse *heights* vary with the information being sent (this is called pulse amplitude modulation or PAM), then transmitting a new pulse at exactly every $T = 1/2B$ sec will result in *only one nonzero value* at one point in each T-sec interval, as Fig. 3.15b shows. Consequently, sampling at exactly this point will produce a sequence of samples that have amplitudes proportional to the heights of the transmitted PAM pulses. Therefore the original information can theoretically be recovered without distortion. Of

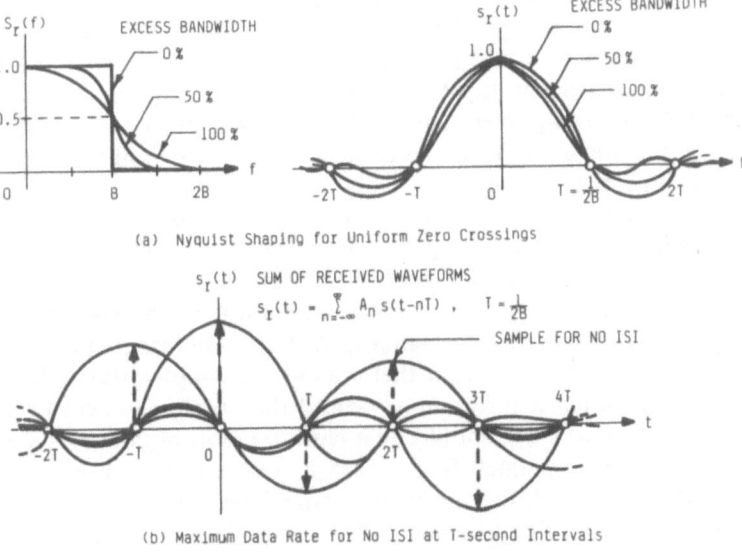

(a) Nyquist Shaping for Uniform Zero Crossings

(b) Maximum Data Rate for No ISI at T-second Intervals

Figure 3.15. ISI and Nyquist shaping.

† More generally, the spectrum $S_r(f)$ will have non-constant phase, in which case we require complex-conjugate symmetry about B Hz for equally spaced zero crossings of $s_r(t)$. This condition is known as Nyquist's First Criterion [4].

course, there will be lots of interference between the sample points, but, if the sampling is done perfectly, this will have no effect on the final result.

In practice when bandwidth utilization is important, the transmitted and received signals are carefully shaped with filters called *equalizers* that compensate for the nominal channel frequency response characteristics. This can be done at passband as well as baseband. Equalization can result in predistortion of the transmitted signal and/or correction of distortion in the received signal. Correct equalization is so crucial to the acceptable performance of high-speed modems that adaptive equalizers are used to adjust precisely for each channel used.

3.4. Digital Signals and Digital Signal Processing

With the ever improving cost-to-performance ratio of digital computing it has become increasingly attractive to process both analog and digital signals digitally. This can be done with either programmable or hardwired digital devices that allow nonlinear and time-varying filters to be realized in the same manner as LTI ones. Digital signal processing also provides a known level of accuracy depending on the word length, and can provide signal delays and storage times of any desired length.

The basic approach to digital filtering is to convert the physical signal to be filtered from a voltage or current into a *series of numbers,* which are then manipulated according to an *algorithm* that defines the particular filtering procedure employed. The resulting filtered numerical sequence may then be either converted back to a physical signal or perhaps be used in its existing form for storage or further digital processing.

Since signal conversion from analog to digital form is often the initial step in digital signal processing, we shall consider it first. The basic procedure, illustrated by the data acquisition system of Fig. 3.2, can now be described mathematically. In particular, we want to know how fast we have to sample, and what the digital signal spectra and filter frequency responses look like. Intuitively, we expect that we cannot sample arbitrarily slowly, since large amounts of information would then be lost between samples. We also expect that there should be a close relationship between a digital signal and the analog signal from which it is obtained, and that we could obtain either one from the other. In theory it will always be possible to exactly recover the original analog signal from the samples provided only that we sample "fast enough," which will be defined precisely by the Uniform Sampling Theorem.

To approach the sampling process mathematically we can multiply an analog signal $m(t)$ by a train of narrow sampling pulses so that the resulting output is zero everywhere except when the pulses occur, in which case the value is the same as $m(t)$. If the pulses are very narrow, then $m(t)$ can be considered as being essentially constant during the sample time and the result is a sampled-data signal. In order to simplify the mathematical analysis, it is convenient to let the narrow pulse train approach a train of weighted impulses in the limit. Although this *impulse sampling* is not exactly attainable in practice, it is a close approximation when the samples are narrow. Furthermore, if the signal is later quantized and encoded, the resulting numerical value will correspond to a single point in time. Thus let $s_\delta(t)$ be a uniformly spaced train of impulse functions, i.e.,

$$s_\delta(t) = \sum_{n=-\infty}^{\infty} \Delta\delta(t - nT_s)$$

where Δ is the area of each impulse, and corresponds to the area of the narrow sampling pulses that are approximated by the impulses. The repetition or *sampling rate* is $1/T_s = f_s$. This sampling scheme can be represented by conventional analog multiplication, as shown in Fig. 3.16.

In the *time domain* the sampled output $m_s(t)$ is given by

$$m_s(t) = m(t)s_\delta(t) = \sum_{n=-\infty}^{\infty} \Delta m(nT_s)\delta(t - nT_s)$$

Although this result is straightforward, the *frequency-domain analysis* is much more useful for data communications. Thus, let $M(f)$ be the spectrum of $m(t)$. To obtain the spectrum of $s_\delta(t)$ we first express it as a Fourier series since it is clearly periodic. The properties of $\delta(t)$ make the computation of the Fourier coefficients S_n trivial,

$$S_n = \frac{1}{T_s} \int_{-T_s/2}^{T_s/2} \Delta\delta(t) e^{-j\omega_s nt} \, dt = \frac{\Delta}{T_s}, \quad \text{where } \omega_s = 2\pi f_s$$

Therefore, the series can be written out as follows:

$$s_\delta(t) = \sum_{n=-\infty}^{\infty} S_n e^{j\omega_s nt} = \sum_{n=-\infty}^{\infty} \frac{\Delta}{T_s} e^{j\omega_s nt}$$

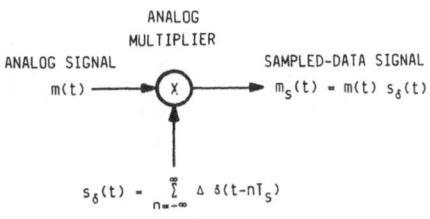

Figure 3.16. Impulse sampling model.

From this result the spectrum $M_s(f)$ of $m_s(t)$ is easily obtained,

$$M_s(f) = \mathscr{F}[m_s(t)] = \mathscr{F}\left[\sum_{n=-\infty}^{\infty} \frac{\Delta}{T_s} m(t) e^{j\omega_s nt}\right]$$

$$= \sum_{n=-\infty}^{\infty} \frac{\Delta}{T_s} \mathscr{F}[m(t) e^{j\omega_s nt}] = \sum_{n=-\infty}^{\infty} \frac{\Delta}{T_s} M(f - nf_s), \qquad f_s = 1/T_s$$

This result is important, for it says that the sampled signal spectrum is just the original analog signal spectrum weighted by Δ/T_s and repeated at integer multiples nf_s of the sampling frequency. This is illustrated in Fig. 3.17. Theoretically the index n extends to infinity; so the spectrum does indeed contain the very high-frequency components that we intuitively expected. In practice, of course, the spectrum eventually dies out with increasing frequency, with the attenuation related to the narrowness of the samples.

From Fig. 3.17b it is apparent that the original analog bandwidth is not changed in the spectral repetitions $M(f - nf_s)$, and this observation is the key to determining just what sampling "fast enough" really means. Thus suppose that in the previous example f_s is decreased, or equivalently T_s increased, so that samples are taken further apart in time. As Fig. 3.18 indicates, the spectral repetitions will eventually begin to overlap. This overlap first occurs when the sampling rate decreases to *just twice the highest frequency of the analog spectrum* $M(f)$, as in part (a). The spectral overlap causes a form of distortion known as *aliasing*, which, once it occurs, cannot in general be undone. The time-domain manifestation of aliasing is an inability to distinguish between the correct frequency components and lower

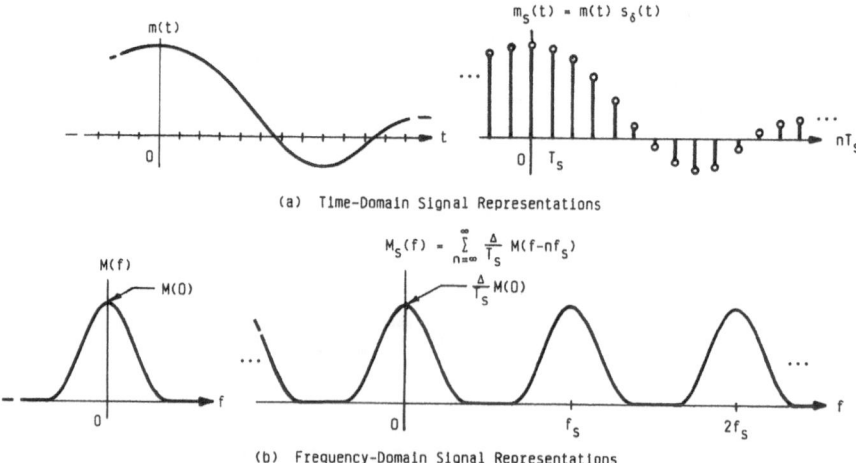

(a)   Time-Domain Signal Representations

(b)   Frequency-Domain Signal Representations

Figure 3.17. Sampled-data signal and spectrum.

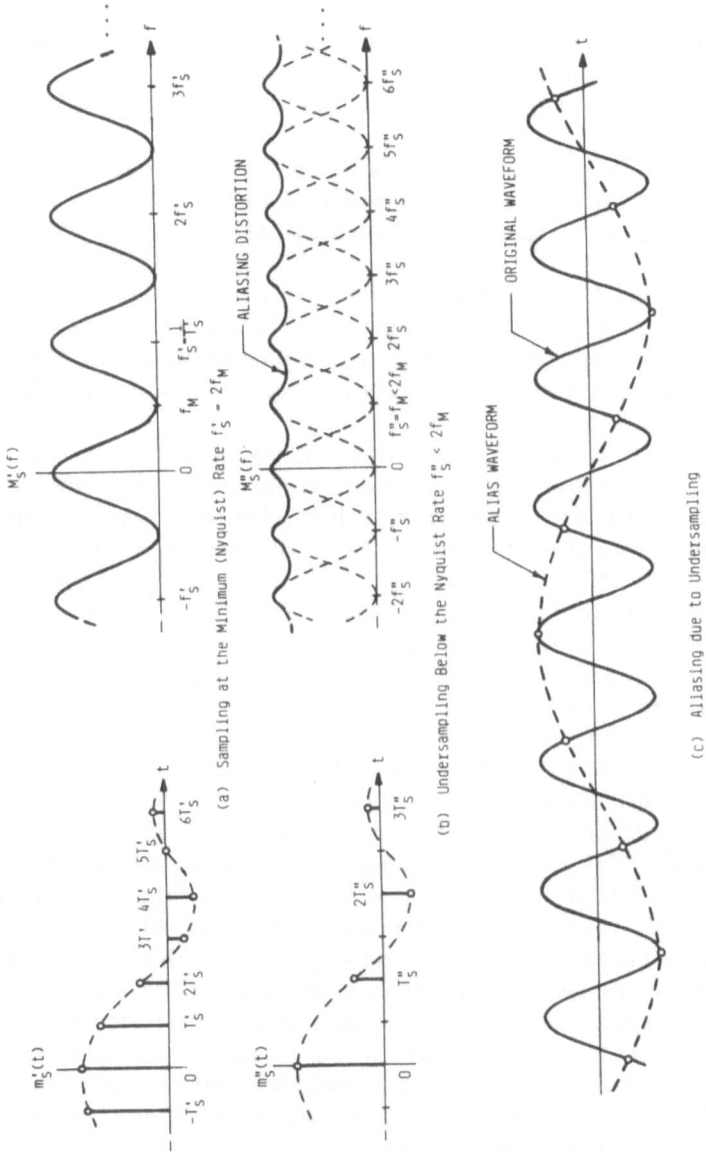

Figure 3.18. Undersampling and aliasing.

ones that "impersonate" them. This situation is shown in Fig. 3.18c, where clearly there are not enough samples to define uniquely the correct signal waveform. The corresponding spectrum $M_s''(f)$ with obvious aliasing distortion is shown in Fig. 3.18b.

For baseband signals, the minimum sampling rate possible without incurring aliasing is called the *Nyquist rate* f_N. The preceding intuitive ideas can now be formalized in a fundamental theorem due to Nyquist:

Theorem 3.1 (*Baseband Uniform Sampling.*) *A baseband signal* $m(t)$ *bandlimited to* f_M *Hz and sampled at a uniform rate of* $f_s > 2f_M = f_N$ ($T_s < 1/2f_M$) *can be exactly reconstructed from the samples.*

The theorem resolves the question of what "fast enough" means; we must get no less than two samples per cycle of every frequency component of the analog signal $m(t)$. For bandpass signals it is often possible to sample at rates much lower than $2f_M$ if a lower rate can be found for which there is no spectral overlap, i.e., aliasing.

To recover the original baseband analog signal from the samples we need only pass it through an ideal low-pass filter with bandwidth f_M. In practice, of course, there must be enough spacing for a nonideal filter to roll off. Such a *reconstruction filter* operates in the time domain by "smoothing in," or *interpolating*, the proper curve between samples. In the frequency domain it simply picks out the particular spectral repetition located at baseband and filters out all others, with the result being the original analog signal spectrum to within a constant scale factor. Figure 3.19 shows the frequency-domain spectra of equivalent analog and digital filtering operations, including sampling and reconstruction.

There are several important practical considerations related to sampling. First, it is usually desirable to pass the analog signal through a low-pass *antialiasing* filter before sampling to ensure that it is bandlimited to an acceptable $f_M < f_s/2$. Since allowance must be made for the roll-off of both antialiasing and reconstruction filters, practical sampling rates are commonly selected somewhat higher than the Nyquist rate. Although we used impulse sampling for mathematical convenience, any other periodic sampling waveform will also work. However, holding the sample during the digital coding process is actually linear filtering and, if the effect is not later removed, it will introduce distortion in the recovered waveform. Finally, since practical channels are inevitably bandlimited, often severely, the direct transmission of sampled signals can be highly inefficient unless some variation of Nyquist shaping is employed.

Having now developed the basic properties of digital signals, we next consider some practical transform techniques for efficiently obtaining the spectra of digital signals. First, recall from Fourier series theory that if a

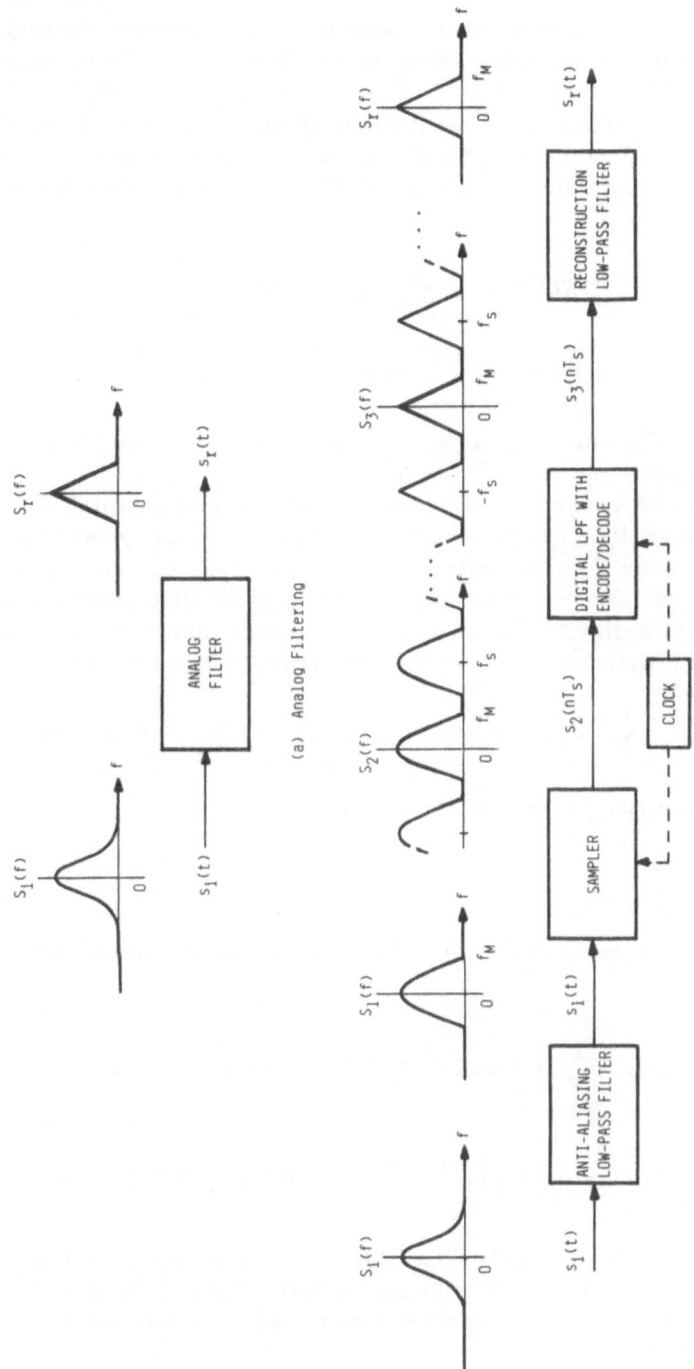

Figure 3.19. Signal spectra for analog and digital filtering.

time domain signal is periodic then its spectrum will be discrete. Conversely, we have seen for sampled-data signals that if the spectrum is periodic then the signal itself is discrete. This is easily shown mathematically by simply writing the Fourier series for the periodic spectrum in terms of the time samples $s(nT_s)$. Denoting the periodic spectrum in the conventional manner as $S(e^{j\omega T_s})$ with sampling frequency $f_s = 1/T_s$, we get the following transform pair:

$$S(e^{j\omega T_s}) = \sum_{n=-\infty}^{\infty} s(nT_s)\, e^{-j2\pi T_s f}$$

$$s(nT_s) = \frac{1}{f_s} \int_{-f_s/2}^{f_s/2} S(e^{j\omega T_s})\, e^{j2\pi n T_s f}\, df$$

Figure 3.20a–3.20c summarizes the signal-spectrum transforms for the three cases discussed so far.

It now follows logically that if a signal is both *periodic and discrete* in the time domain, then its spectrum will be both *discrete and periodic* in the frequency domain. For the particular case where $T_0 = T_s$, hence $f_s = Nf_0$ for some integer N (i.e., N samples per cycle in the time and frequency domains) the transform pair has a very simple form known as the *discrete Fourier series*. Mathematically we have for $s_0(nT_s) = s_0(nT_s + NT_s)$,

$$s_0(nT_s) = \sum_{m=-\infty}^{\infty} S_m\, e^{j(2\pi m/NT_s)nT_s} = \sum_{k=0}^{N-1} \left(\sum_{m=-\infty}^{\infty} S_{mN+k} \right) e^{j(2\pi/N)kn}$$

and, defining $s_0(nT_s) = s(n)$ and

$$\sum_{m=-\infty}^{\infty} S_{mN+k} = \frac{1}{N} S(k),$$

this reduces over one period to the simple *discrete Fourier transform* (DFT) pair,

$$s(n) = \frac{1}{N} \sum_{k=0}^{N-1} S(k)\, e^{j(2\pi/N)nk}, \qquad 0 \leq n \leq N-1$$

$$S(k) = \sum_{n=0}^{N-1} s(n)\, e^{-j(2\pi/N)kn}, \qquad 0 \leq k \leq N-1$$

This relationship is illustrated in Fig. 3.20d. Note that the time and frequency axes are indexed only with the integers n and k for simplicity, although the actual values are obviously related by $t = nT_s = n/Nf_0$ and $f = kf_0 = k/NT_s$, respectively.

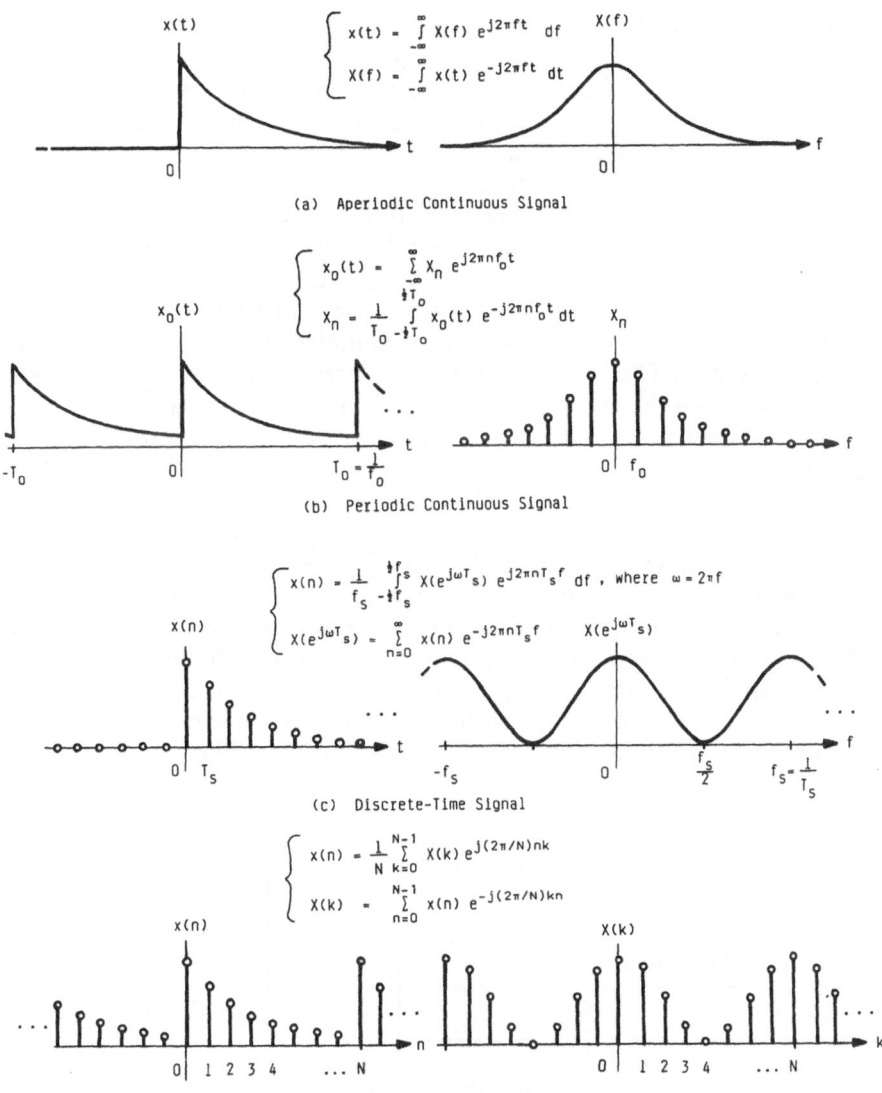

Figure 3.20. Signal and spectral relationships.

The preceding discrete Fourier transform is of enormous practical importance for two key reasons. First, the discrete nature makes it ideal for numerical evaluation by computer. In fact, it is possible to evaluate the transform summations in stages in such a way that the amount of actual multiplication required is very small, particularly when N is chosen to be a power of 2. This has led to a collection of simple but extremely fast numerical algorithms known collectively as the *fast Fourier transform* (FFT) [5]. The basic software and hardware approaches to spectral analysis with the FFT are shown in Fig. 3.21. The second reason is that the approximate spectra of both analog and nonperiodic waveforms can be calculated with the DFT. Analog signals must, of course, be sampled in accordance with the sampling theorem, and nonperiodic discrete signals must be truncated to obtain a finite record length of N samples. Then the DFT can be applied directly to get N *frequency samples* of the spectrum. The truncation can be a major source of error, and there are a number of *windowing* techniques to gradually taper the record ends so as to minimize this distortion according to various criteria [6]. Very long data sequences can also be handled with the DFT by carefully segmenting them into shorter records and processing each record individually, then combining the results to obtain the complete spectrum [7].

To process, or filter, digital signals we could conceivably use the sampled-data signal values directly, but this approach is seldom used. Rather, the samples are quantized and binary encoded to get a *numerical series*, which can then be operated on with some type of digital computing device. This operation, which of course determines the nature of the filtering, may be in the form of a software program, a hardwired circuit, or any other system capable of manipulating numerical values in the desired manner. Our concern here is not so much the actual filter hardware but rather the practical concepts that must underlie any implementation.

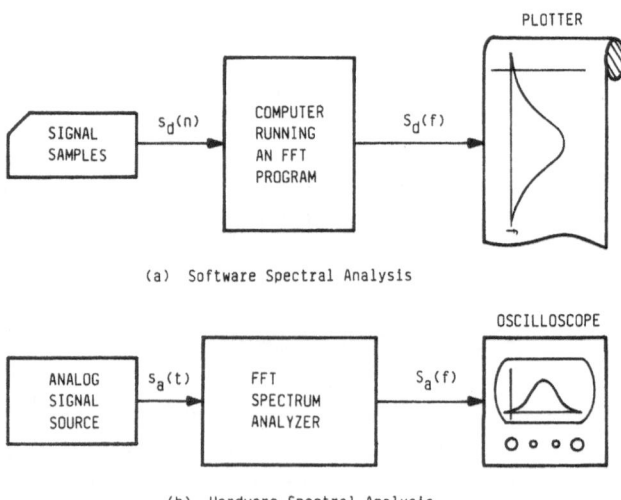

Figure 3.21. FFT spectrum analysis.

First note that if the input and output signals are going to be digital, then their spectra must be periodic, so we would expect the frequency response of a digital filter to also be periodic. This is in fact true, and in the usual case of the filter clock frequency equaling the signal sampling rate f_s, the spectrum repeats every $f = nf_s$ Hz, $n = 0, \pm 1, \pm 2, \ldots$. The digital filter response to a single *unit pulse sequence* $\delta(n)$, i.e., 1, 0, 0, $\ldots$, corresponds to the unit impulse response of an analog filter, and is consequently often referred to as the *digital filter impulse response* $h(n)$. If we denote the filter frequency response by $H(e^{j\omega T_s})$ then it is related to $h(n)$ by

$$H(e^{j\omega T_s}) = \sum_{n=-\infty}^{\infty} h(n)e^{-j2\pi n T_s f}$$

$$h(n) = \frac{1}{f_s} \int_{-f_s/2}^{f_s/2} H(e^{j\omega T_s})e^{j2\pi n T_s f} \, df$$

which is completely analogous to the previous discrete-time signal spectrum of Fig. 3.20c.

Digital filters can be *classified* in a number of ways. Perhaps the most common is according to whether the impulse response $h(n)$ settles to zero after a finite number of output samples, or never in theory remains at zero. The two categories are called *finite- and infinite-impulse response* (FIR and IIR) filters, respectively. A closely related classification is based on whether there is internal feedback (*recursive* filters) or not (*nonrecursive* filters).† As with analog filters, they may be classified as linear or nonlinear, as time-varying or time-invariant, by filter order, and by passband location and bandwidth. A particular frequency response may also be implemented in a number of different filter configurations.

There are several common approaches to digital filter design and realization. One straightforward approach for FIR filters is based on the DFT, or FFT. Since linear time-invariant digital filtering corresponds mathematically to *digital time-domain convolution,* which is equivalent to frequency-domain multiplication, we can filter by simply multiplying the input signal spectrum by the filter frequency response to get the output signal spectrum. Thus digital filtering can be done by segmenting the input $\bar{s}_i(n)$ and using the FFT to compute the segment spectra, then performing an inverse FFT to obtain the final output signal sequence, as shown in Fig. 3.22. As mentioned before, care must be taken in recombining the transformed segments to get the final output $\bar{s}_r(n)$.

IIR filters can be designed by using a mathematical mapping, such as the bilinear transformation, to convert an existing analog filter design to a

---

† Recursive filters are generally IIR and nonrecursive, FIR. The major exception is the *frequency-sampling filter,* which is recursive but also FIR owing to exact internal cancellations.

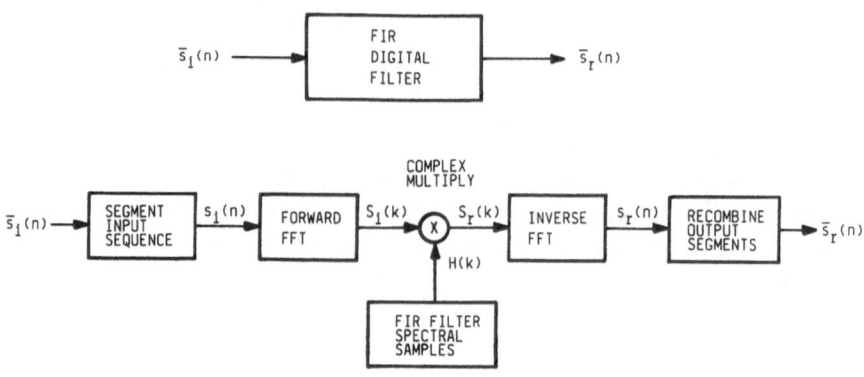

Figure 3.22. FFT digital filtering procedure.

digital one. Other mappings can also be defined to preserve the analog filter response to some particular input, usually the unit impulse or unit step [7].

The simplest but most important digital filter for data communications systems is the FIR *transversal filter.* Also known as the tapped delay line or moving average filter, each output sample is a weighted average of the most recent M inputs. Letting h_m be the weight of the mth tap, the response sample $s_r(n)$ is just

$$s_r(n) = \sum_{m=0}^{M-1} h_m s_i(n - m)$$

as shown in Fig. 3.23a. This expression is actually a *discrete convolution* of the input $s_i(n)$ and the filter impulse response h_m, and so the frequency response must be

$$H(e^{j\omega T_s}) = \sum_{m=0}^{M-1} h_m e^{-j\omega m T_s}$$

which is clearly periodic at the sampling frequency f_s.

Transversal filters have many nice properties. They are inherently stable, are very simple to implement in software or hardware, can produce practically any frequency response, can be built with linear phase and hence no delay distortion, and can be made time varying by simply adjusting the tap values at each iteration of the filter. Except for very simple cases, computer-aided design methods are used and many software design programs are available [8].

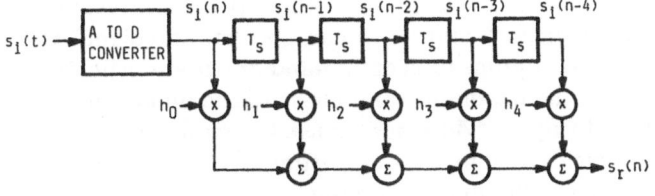

(a)   Transversal Nonrecursive FIR Digital Filter

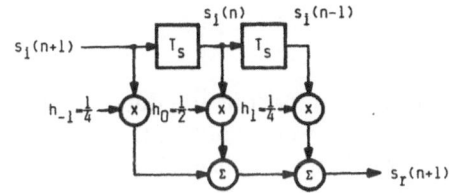

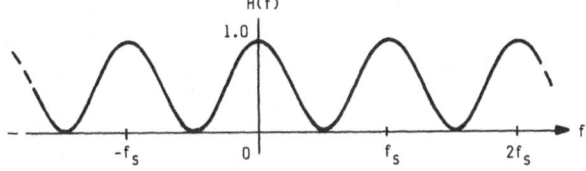

(b)   Simple Raised-Cosine Transversal Filter and Spectrum

Figure 3.23. The transversal digital filter.

As an example of a transversal filter design, consider the widely used *raised cosine* spectral shape defined by

$$H(e^{j\omega T_s}) = \tfrac{1}{2} + \tfrac{1}{2}\cos\left(2\pi T_s f\right) = \tfrac{1}{4}\,e^{-j2\pi T_s f} + \tfrac{1}{2} + \tfrac{1}{4}\,e^{+j2\pi T_s f}$$

This is exactly the Fourier series form we require, so the tap coefficients are readily identified to be $h_{-1} = \tfrac{1}{4}$, $h_0 = \tfrac{1}{2}$, and $h_1 = \tfrac{1}{4}$, with all other $h_m = 0$. The resulting transversal digital filter and periodic low-pass frequency response are shown in Fig. 3.23b. Note that here the input sample is denoted by $s_i(n + 1)$ vice $s_i(n)$ for notational convenience.

The close similarity between analog and digital signal processing should now be apparent. In data communications the particular choice often is up to the designer, and both approaches are commonly used in the same equipment. For highly accurate and sophisticated filtering requirements such as adaptive equalization, digital filtering is usually the only viable approach. On the other hand there are many simple and isolated filtering requirements such as noise removal where passive or active analog filters are entirely satisfactory and much cheaper than the digital alternative.

The important point to keep in mind about any type of linear filter is that it modifies the magnitude and phase of the input signal spectrum in a predetermined way, but does not change the input frequencies. In the next section we consider the process of shifting the signal frequencies, which is, in a general sense, the dual of filtering.

3.5. Modulation

In the majority of applications the data signals produced by a user's terminal equipment are not well suited for transmission over data communications channels. For example, the signal spectrum frequencies may not be within the available channel bandwidth or they may be located near the channel edges where there is severe distortion. Many communication channels will not pass signals at or near dc because of transformer coupling, and thus will greatly attenuate narrow baseband signals. Finally, it is often desirable to subdivide a large channel bandwidth to simultaneously accommodate multiple signals, either in the same or opposite directions. In all of these cases it is desirable to *translate*, and usually also *reshape*, the original data signal before transmission. The process by which this is done is called analog or digital *modulation*.

There are many types of modulation. The most basic approach corresponds mathematically to multiplication. The spectrum of an arbitrary baseband signal $m(t)$ can be translated to any nonzero frequency f_c by simply multiplying it by a sinusoidal *carrier* $s_c(t) = \cos(\omega_c t)$ to get the *modulated signal* $s_m(t) = m(t) \cos(\omega_c t)$, where $\omega_c = 2\pi f_c$. This is shown schematically in Fig. 3.24.

To understand how this *linear* modulation process works, assume first that $m(t)$ is a single tone of frequency $f_m \ll f_c$; i.e., $m(t) = A \cos(\omega_m t)$. Then $s_m(t)$ can be written as

$$s_m(t) = A \cos(\omega_m t) \cos(\omega_c t)$$

$$= \tfrac{1}{2}A \cos[2\pi(f_c - f_m)t] + \tfrac{1}{2}A \sin[2\pi(f_c + f_m)t]$$

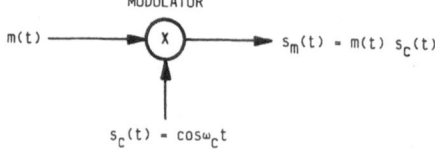

Figure 3.24. Basic modulation.

Although this result is centered at f_c, it contains neither f_c nor f_m but only their sum and difference. We have, in effect, translated the two-sided baseband spectrum of $m(t)$ up to f_c without distorting it (except for attenuation).

We can now extend this intuitive idea to any $m(t)$ with baseband spectrum $M(f)$. Using Euler's identity and taking the Fourier transform of $s_m(t) = m(t)\cos(\omega_c t)$ gives

$$S_m(f) = \mathscr{F}[m(t)\tfrac{1}{2}(e^{+j\omega_c t} + e^{-j\omega_c t})]$$

$$= \tfrac{1}{2}M(f - f_c) + \tfrac{1}{2}M(f + f_c)$$

The spectral component of difference frequencies $M(f - f_c)$ is located below f_c, and is consequently called the *lower sideband*, while the *upper sideband* $M(f + f_c)$ contains only sum frequencies. This process is descriptively known as *double-sideband suppressed-carrier* (DSB-SC) modulation.† Figure 3.25b shows the modulated signal and spectrum corresponding to the baseband signal in part (a). Note that the baseband information is carried in the envelope of $s_m(t)$ with the actual signal varying at f_c Hz.

To recover the original signal $m(t)$ we must *demodulate* $s_m(t)$. The basic procedure is based on our previous observation that whenever two

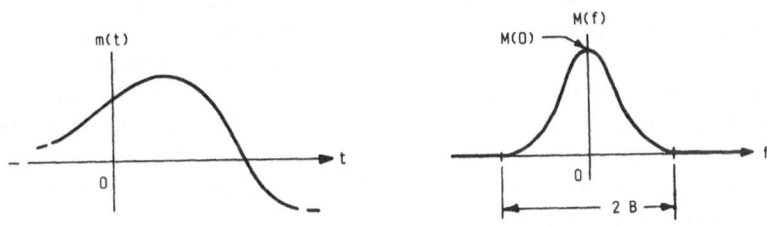

(a) Baseband Signal and Spectrum

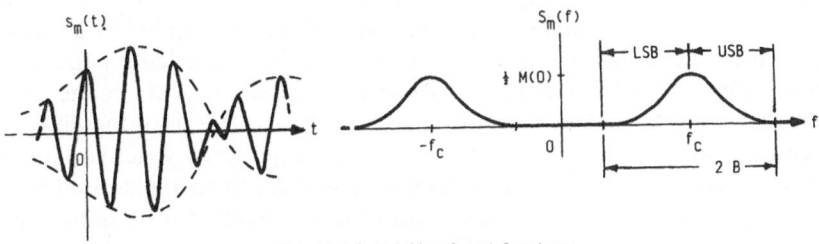

(b) Modulated Signal and Spectrum

Figure 3.25. DSB-SC modulation.

† Despite the name there may be a component of $s_m(t)$ at the carrier frequency if $m(t)$ has a distinct dc component, since $f_c \pm 0 = f_c$.

signals are multiplied we always get the sum and difference frequencies. Thus if we multiply $s_m(t)$ by $s_d(t) = \cos(\omega_c t + \varphi)$ *at the receiver* we obtain

$$s_m(t)s_d(t) = m(t) \cos(\omega_c t) \cos(\omega_c t + \varphi)$$

$$= \tfrac{1}{2}m(t)\cos\varphi + \tfrac{1}{2}m(t)\cos(2\omega_c t + \varphi)$$

Now the term $\tfrac{1}{2}m(t)\cos(2\omega_c t + \varphi)$ is centered at frequency $2f_c \gg f_m$ so it can be easily removed by a low-pass filter. The remaining term is proportional to $m(t)$, provided φ is not $\pm 90°$, a condition which the receiver can maintain.

A major disadvantage of DSB-SC modulation is that exactly the same information is transmitted in the upper and lower sidebands. Thus for long distance point-to-point applications, such as ship-to-shore and amateur radio, only one of the sidebands is commonly transmitted. This *single-sideband* (SSB) modulation has higher effective signal power at the cost of greater receiver complexity. SSB is also widely used in the analog telephone system for combining voice channels onto wideband trunks, since it allows twice as many conversations as does DSB modulation for the same bandwidth.

In the other direction, for many broadcast applications, such as commercial radio there is only one transmitter but many receivers. Here the paramount consideration is simple, cheap receivers rather than power. In such cases receiver demodulation is much easier (and cheaper) when both sidebands are transmitted, and a distinct carrier component is added. This *double-sideband with carrier* (DSB-C) modulation allows the original baseband signal to be recovered with a simple envelope detector circuit rather than some type of multiplier circuitry. It is the modulation technique used in common AM broadcast radios.

Figure 3.26 shows a conventional AM superheterodyne broadcast receiver and associated signal spectra. The demodulation is done in two steps. The received signal is first amplified and then multiplied by a sinusoid from the *local oscillator*, whose frequency f_{LO} is controlled by the tuning dial. It always tracks at a fixed *intermediate frequency*† f_{IF} above the desired station carrier frequency f_c as the spectrum shows. The resulting difference frequency around f_{IF} is passed through a fixed, high-quality, active bandpass *IF filter* centered at f_{IF}. The IF filter selects and amplifies the desired station signal (sensitivity) while rejecting all others (selectivity) as indicated by $S_{IF}(f)$ in the figure. Finally, the filter output is detected by a passive envelope detector and the resulting audio signal is amplified to drive the speaker. Although not directly used for data transmission, this broadcast receiver

---

† In most AM radios f_{IF} is 455 kHz.

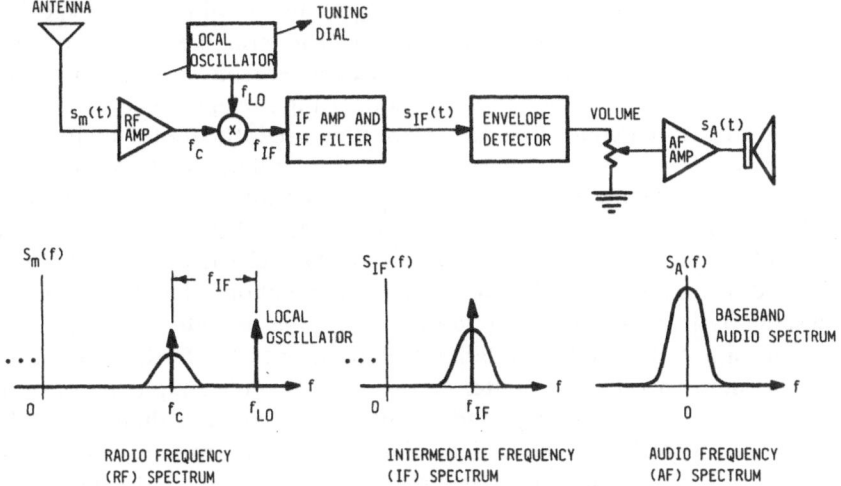

Figure 3.26. AM superheterodyne radio and signal spectra.

provides an excellent practical example of filtering and modulation. A good understanding of its operation will prove well worth the effort involved.

In addition to SSB and DSB it is also possible to send one complete sideband and only a part, or vestige, of the other. Such *vestigal-sideband* (VSB) modulation provides better bandwidth and power utilization than DSB and is easier to receive than SSB. It has been used in the past for data communications, but it is now used primarily for television video modulation. With *frequency* (FM) modulation the carrier frequency rather than its amplitude varies with $m(t)$. This provides good immunity to noise and allows the designer to trade bandwidth for signal-to-noise ratio. FM is widely used in all types of radio systems where noise is a problem and there is plenty of bandwidth. It is also quite common for low-speed data communications.

Armed with our introductory background in signals, filtering, and modulation, we now turn to one of the fundamental aspects of computer data communications, namely *digital modulation*. Digital modulation is distinguished from analog in that only a few different symbols are transmitted as opposed to a continuum of signal levels in the analog case. However, the key considerations are still matching the transmitted spectrum to the transmission channel and providing a viable demodulation procedure. A number of digital modulation techniques have been proposed and implemented in the past, but *frequency-shift keying* (FSK), *phase-shift keying* (PSK), and *quadrature amplitude-shift keying* (QASK) are the three dominant approaches in widespread use today.

Before we delve into the details of digital modulation a few historical notes on terminology are worth mentioning. The term "keying," like much of telecommunications terminology, is a carryover from earlier times when telegraph operators sent data in Morse code using a telegraph key. Although more sophisticated codes like ASCII are now used, the term "keying" has been retained and now connotes all types of digital modulation. The basic device for performing keying is the *modem* (modulator–demodulator) or, in the telephone terminology, data set. There are all types of modems for different frequency bands, data rates, and channel categories. Our focus here and in Chapter 4 will be on modems for telephone channels, although the same basic principles apply to the other types as well.

The simplest digital modulation technique is FSK. Hardware required to implement an FSK transmitter and receiver is quite simple, and can be incorporated in a single integrated circuit chip costing less than $25. FSK is also robust in that it has a high tolerance to noise and distortion. The major drawback is that it requires a relatively wide bandwidth, and thus is limited to relatively low speeds. For binary data, FSK shifts between two tones, as shown in Fig. 3.27a. If we let the higher frequency correspond to a binary 1 and the lower to a binary 0, we get the waveform shown. In the old telegraph terminology a 1 was called a *mark* and a 0 a *space*; hence, we still use the terms mark and space frequencies, respectively. The basic equation for an FSK signal is a constant-amplitude sinusoid with variable frequency,

$$s_{FSK}(t) = A \cos\left[2\pi f(t)t\right]$$

where the frequency $f(t)$ is either f_M for a mark or f_S for a space. Figure 3.27b gives the block diagrams for a simple FSK transmitter and receiver pair. The transmitter just switches an oscillator between the mark and space frequencies, and then filters the result to remove out-of-band harmonics. The receiver contains an input filter for noise removal and perhaps line equalization, then it just filters the signal with narrow bandpass filters centered at f_M and f_S. The filter outputs are then measured, with the largest indicating the received frequency. The result is cleaned up by an electronic slicer (hard limiter), and then sent to the terminal as the received binary data. The system is asynchronous, i.e., there is no need for the receiver to recover either a bit clock or the carrier reference. This is the main reason for its simplicity.

Phase-shift keying (PSK) is more complex than FSK, but it is also more efficient. For a given bandwidth and power level it is possible to transmit with PSK at a higher data rate. To understand why this is true, it is necessary to consider how PSK operates. The carrier frequency is held constant while the carrier phase is varied among several distinct values (typically four or

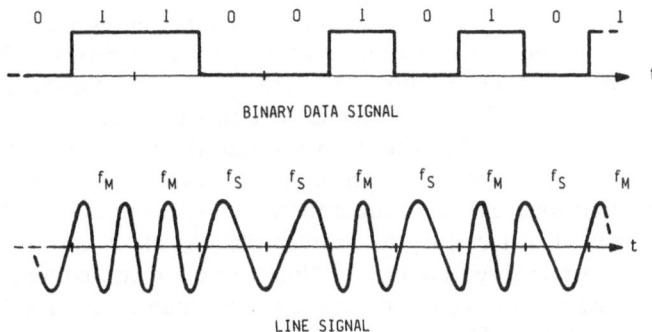

LINE SIGNAL

(a)   FSK Modulated Waveform

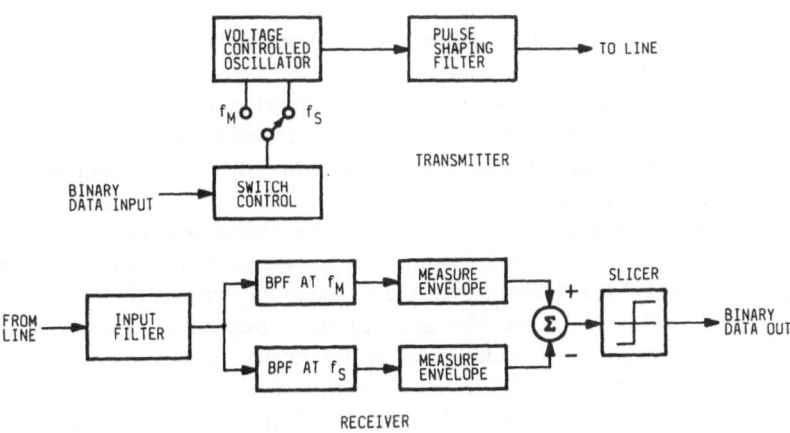

(b)   FSK Transmitter and Receiver

Figure 3.27. FSK modulation.

eight) to convey the desired binary information. The basic equation for a
PSK waveform is

$$s_{PSK}(t) = A \cos [2\pi f_c t + \theta(t)]$$

where the carrier phase $\theta(t)$ can have only the allowed phase values, e.g.,
for four-phase PSK these values might be 0°, 90°, 180°, 270°. Since there
are four phases, it takes two bits of data to select each; thus there is only
one *line symbol* of constant phase for each two binary data bits. These line
symbols are called *bauds*, and the line symbol rate is the *baud rate* (Bd).
As an example, sending data at 2400 bps using four-phase PSK results in
a baud rate of 1200 Bd on the line. This, then, is the main reason higher
data rates are possible with PSK.

Since PSK signals are transmitted at a constant amplitude, they tend to have good immunity to common amplitude noise. However, they are sensitive to phase noise such as jitter. This good performance does not come free, of course, but requires the receiver to know the transmitted carrier phase during each symbol in order to determine the proper signal phase. Thus the receiver must recover the *transmitted carrier*, both in frequency and in phase, and then use this synchronous reference to demodulate the PSK signal. It is also necessary to recover the *bit rate clock* so that the demodulated and recovered binary data bits can be identified and sent to the destination terminal or computer at the proper times. Thus *synchronous data* is transmitted using PSK.

There are a variety of ways to recover carrier and clock synchronization information, but the most common technique involves a closed-loop feedback circuit called the *phase-locked loop* (PLL). Figure 3.28 shows both a PSK signal and the block diagram of a representative PSK transmitter and receiver. A shift register is used to collect the binary data bits for phase encoding each baud and similarly for recovering the binary bits at the PSK receiver. Filtering is more critical than in FSK, since it is important to equalize the channel phase characteristic to insure accurate phase recovery at the receiver.

In practice it is much simpler to use the phase difference rather than absolute phase for PSK encoding, in which case we have *differential phase-shift keying* (DPSK). The receiver carrier recovery circuit then only has to recover the allowed phase *changes* but does not have to identify them absolutely. The phase decoder need only compare the current phase with the prior one to recover the transmitted bits. In fact, it is possible simply to delay the previous symbol by one baud time and use it directly to demodulate the current baud, thereby eliminating the need for any carrier recovery circuit in the receiver.

The third common type of digital modulation is *quadrature amplitude-shift keying* (QASK), which uses both amplitude and phase changes in the carrier to convey binary information.† By varying the carrier *amplitude* as well as the phase it is possible to pack even more bits into each baud than with PSK. Since only the baud rate is limited directly by channel bandwidth, this approach works so long as the baud symbols can be distinguished by the receiver; again, the price for this improved performance is increased complexity of the transmitter and receiver. In particular, it is necessary to adjust the received signal very carefully so that there is an accurate amplitude level for the received signal, regardless of the attenuation introduced by the transmission channel. This adjustment is done with a feedback circuit known in practice as the *automatic gain control* (AGC). Of course, there is

† The telephone company uses a particular variation of QASK called *quadrature amplitude modulation* (QAM), and this terminology is sometimes used for the more general QASK.

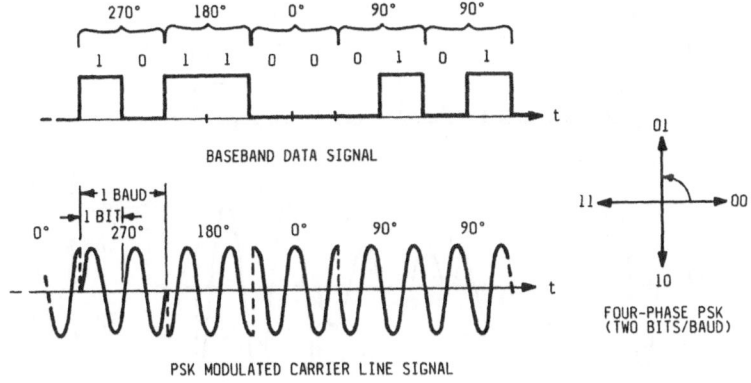

(a) Four-Phase PSK Waveform

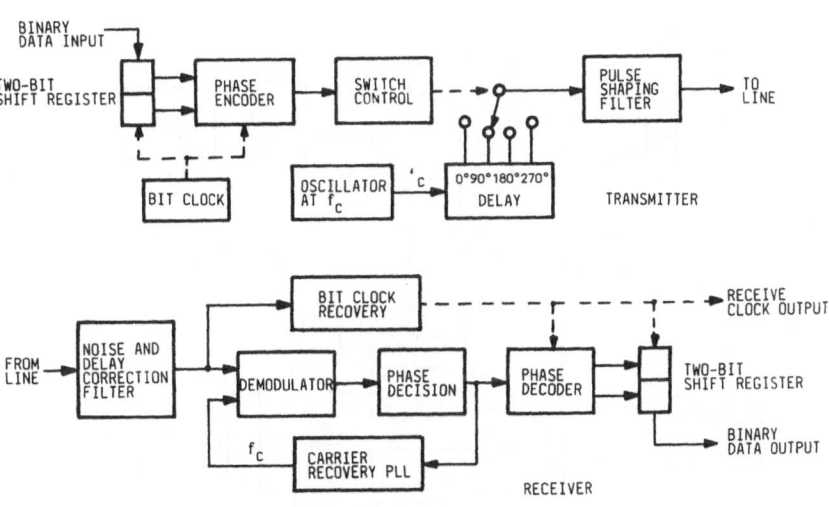

(b) Four-Phase PSK Transmitter and Receiver

Figure 3.28. PSK modulation.

still the requirement for accurate carrier phase and clock synchronization, although the carrier recovery is again considerably simplified by differential phase encoding.

For QASK it is not uncommon for four *bits* to be encoded into each *baud*; for example, by letting three bits select one of eight phases and the remaining bit select one of two amplitude levels. Figure 3.29a shows a typical QASK signal for the case of four phases and two amplitudes and thus three bits per baud. Part (b) shows the corresponding transmitter and

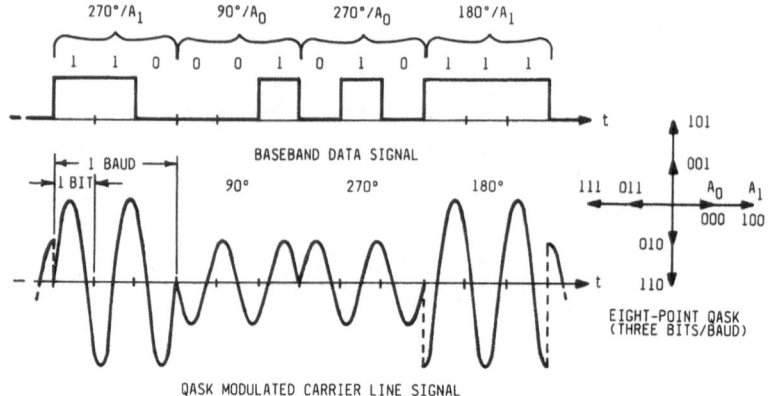

(a)   Eight-Point QASK Waveform

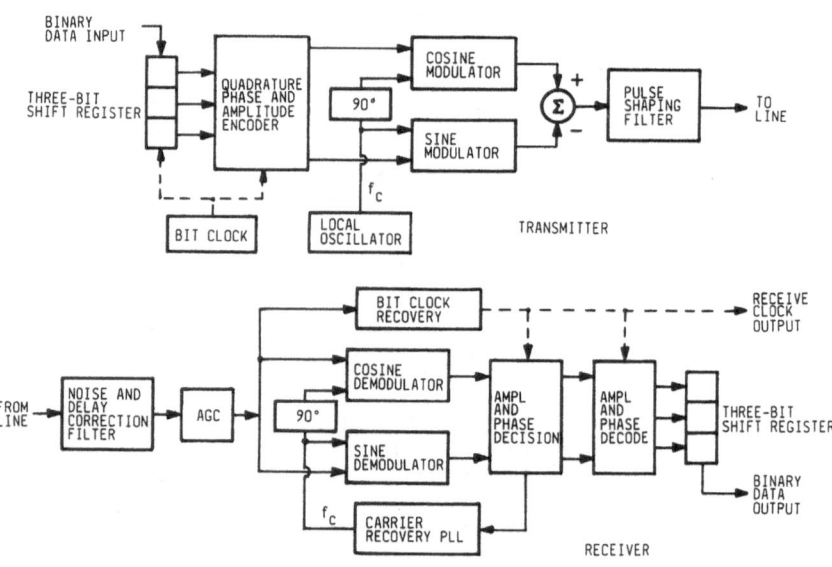

(b)   Eight-Point QASK Transmitter and Receiver

Figure 3.29. QASK modulation.

receiver block diagrams. Notice that in addition to the clock and carrier recovery circuits, there is also an AGC block at the receiver input.

The equation for the QASK signal is given by the following:

$$s_{QASK}(t) = A(t) \cos \left[2\pi f_c t + \theta(t)\right]$$

which indicates both amplitude and carrier phase variation. In practice it is usually easier to consider an *in-phase* (cosine) and a *quadrature* (sine) component of the carrier, in which case the above expression can be written as

$$s_{QASK}(t) = A_p(t) \cos (2\pi f_c t) - A_q(t) \sin (2\pi f_c t)$$

where the subscripts p and q denote in-phase and quadrature, respectively. This latter form is the basis for the transmitter and receiver block diagrams of Fig. 3.29b. As the bandwidth efficiency (bits/baud) is increased, the accuracy of the clock and carrier synchronization becomes increasingly critical. Therefore, QASK is usually reserved for high-speed transmission in situations where the additional cost of very accurate synchronization circuits can be justified.

There are several ways to improve the performance of PSK and QASK systems, particularly with regard to synchronization. *Data scramblers* are commonly used to ensure that the transmitted line signal spectrum is evenly spread out over the entire bandwidth and that there are no periodic data patterns having discrete spectral lines (impulses), since such patterns play havoc with the synchronization circuits. For example, the carrier tracking loop may try to track one of these periodic harmonics rather than the true carrier. Another technique for improving high-speed data transmission is the use of an *adaptive equalizer*, which is typically a transversal digital filter in which the tap gains are continually adjusted according to an optimal algorithm. Before data transmission begins, an initial *training pattern* is sent in order to allow the receiver to iteratively adjust its adaptive equalizer taps so as to optimally correct for the magnitude and phase characteristics of the particular line over which it happens to be operating. Once this equalizer training phase is completed, data transmission begins and the equalizer may continue to make small corrections as the line varies slowly over time. *Scrambling* is essential with automatic equalization to prevent the equalizer from tracking data patterns; i.e., attempting to change the transmitted signal spectrum along with the channel frequency response.

This material on digital modulation should serve to tie together the preceding material on signals, spectra, filtering, and modulation in general. It also provides an introduction to some of the more mundane but essential considerations such as timing, synchronization, line equalization, and scrambling. We shall return to the subject of modems in Chapter 5, but our approach there will be more from an operational standpoint.

In this second of three chapters on communications fundamentals we have covered a great deal of the theoretical material upon which all communications are ultimately based. Although it is certainly possible to purchase and install equipment with no regard for how it works, the specification, operation, and maintenance of data communications systems will be much more effective when the personnel concerned are familiar with the principles of operation. Even with turnkey systems, some theoretical knowledge can prevent one from being completely at the mercy of the vendor.

In the next chapter we turn to the domestic telephone system. Once almost entirely owned and operated by the monolithic American Telephone and Telegraph Company, it is now undergoing rapid change and decentralization resulting from dramatic changes in technology and regulatory policy. However, to the user it appears (more or less) as a single coherent network, and the basic internal characteristics and procedures are fairly standard. In particular, the nominal characteristics of the standard voice-grade telephone line are well modeled as a noisy linear bandpass filter, and are basically the same regardless of the supplier of telephone services.

References

1. E. A. Guillemin, *The Mathematics of Circuit Analysis*, Wiley, New York (1949), Chap. 7.
2. M. E. VanValkenburg, *Network Analysis* (3rd ed.), Prentice-Hall, Englewood Cliffs, New Jersey (1974), Chap. 15.
3. S. M. Selby, (ed.), *Standard Mathematical Tables* (16th ed.), The Chemical Rubber Company, Cleveland, Ohio (1968), pp. 468–483.
4. H. Nyquist, Certain topics in telegraph transmission theory, *Trans. AIEE* **47**, 617–644 (1928).
5. J. W. Cooley and J. W. Tukey, An algorithm for the machine calculation of complex Fourier series, *Math. Comput.* **19**, 297–301 (1965).
6. L. R. Rabiner and B. Gold, *Theory and Application of Digital Signal Processing*, Prentice-Hall, Englewood Cliffs, New Jersey (1975), Chap. 3.
7. A. V. Oppenheim and R. W. Schafer, *Digital Signal Processing*, Prentice-Hall, Englewood Cliffs, New Jersey (1975).
8. IEEE ASSP Digital Signal Processing Committee (ed.), *Programs for Digital Signal Processing*, IEEE Press, New York (1979).

Suggested Readings

G. R. Cooper and C. D. McGillem, *Methods of System and Signal Analysis*, Holt, Rinehart and Winston, New York (1976).

A less rigorous, application-oriented treatment of Fourier series and transforms as used in system and circuit analysis.

H. Taub and D. L. Schilling, *Principles of Communications Systems*, McGraw-Hill, New York (1971).

A fine old standard upper-division college text that includes data transmission and particularly PLLs at a moderate mathematical level.

R. E. Ziemer and W. H. Tranter, *Principles of Communications*, Houghton-Mifflin, Boston (1976).

A well written, modern introduction to analog and data communication systems theory.

R. W. Lucky, J. Salz, and E. J. Weldon, Jr., *Principles of Data Communications*, McGraw-Hill, New York (1968).

A classical collection of data communication theory from a telephone system viewpoint, particularly the first half on baseband transmission.

G. Daryanani, *Principles of Active Network Synthesis and Design*, Wiley, New York (1976).

One of the few comprehensive and yet completely practical books on analog filter design.

A. V. Oppenheim and R. W. Schafer, *Digital Signal Processing*, Prentice-Hall, Englewood Cliffs, New Jersey (1975).

The classic text on digital filters and the FFT, and still one of the best available.

Check Your Understanding of Chapter 3—True or False?

1. A typical noise spike should have a very narrow bandwidth.
2. After sampling, the resulting signal spectrum is always periodic and discrete.
3. FSK is the simplest digital modulation technique used by telephone line modems.
4. Delay and phase shift of a filter are the same thing.
5. The bandwidth of a passband signal determines what the carrier frequency must be.
6. If a signal has a continuous spectrum then it cannot be periodic.
7. Baud is the European (French) term for bit, and the two are the same thing.
8. Rounding a signal to the nearest allowed level is called qualification.
9. All-pass filters can be built, but are useless in practice.
10. Delaying a signal in time does not affect its magnitude spectrum.
11. Aliasing results when analog signals are grossly undersampled.
12. Two-sided phase spectra of real signals are always odd functions of frequency.
13. Most IIR filters are of the nonrecursive type.
14. The Uniform Sampling Theorem requires no more than two samples of any signal frequency.
15. PSK modems use differential phase encoding to simplify the receiver.
16. Superheterodyne AM radios convert every received station to the same frequency.
17. The FFT gives only the magnitude, and not the phase, spectrum of a signal.
18. Knowing the spectrum of a signal completely determines its waveform in theory.
19. Square pulses distorted by group delay will be skewed and smeared in time.
20. In theory, perfect Nyquist shaping can eliminate ISI at the receiver sample times.
21. Amplifying a signal by 0.7071 is equivalent to a gain of 3.0103 dB.
22. Sampling is closely related to DSB-SC modulation in concept.
23. Doubling the width of a square pulse will also double the practical bandwidth.
24. The average value of the output of a bandpass filter will have zero average value.
25. The output signal of a no-pass filter is nothing.

4

The Domestic Telephone System

4.1. Telephone System Organization

The present telephone system in the United States represents more than a century of evolution in voice telephone communications. Since the original telephone patents were issued to Alexander Graham Bell in 1876 there have been enormous technological strides in telecommunications, and today we enjoy a convenient and highly reliable system interconnecting practically every home and business in the country, as well as providing easy access to most foreign nations. The huge telephone network, which is often called the world's most complex machine, is an incredible system of electronic and mechanical switches, cable and radio links, network management gear, and subscriber premises equipment. It is certainly one of our most valuable national resources. Today, some of the telephone plant is very modern, but a great deal of older equipment is still in use. It is being

73

steadily replaced, but the huge capital costs involved makes replacement a slow process, particularly since most old equipment still works very well.

Until 1984 the domestic system was dominated by the giant AT&T, which supplied about 80% of the domestic telephones (around 150 million), which were used for some 750 million calls per day, including the bulk of long-distance service. The size is illustrated by the 1983 revenues of almost $70 billion and a depreciated capital plant of about $124 billion at historical cost. On January 1, 1984, this pervasive monopoly was split into seven independent regional holding companies comprised of the 22 old Bell Operating Companies, with the remaining "new" AT&T reduced to approximately one-fourth its old size. The actual court action was the culmination of decades of antitrust action, and in effect traded the old monolithic monopoly for much more competitive opportunities in information transmission and processing. The new arrangement subdivides each geographical operating company territory into a number of Local Access and Transport Areas (LATA). Within a LATA traffic is carried by the local telephone company, e.g., one of the regional Bell Operating Companies or one of some 1600 other smaller local firms. Any telephone traffic *between* LATAs must be carried by a long-haul carrier such as the new AT&T Communications division or other such carrier specified by the customer. Note that this inter-LATA traffic may be inter- or intrastate, depending on the particular LATAs involved.

The internal telephone network consists of a hierarchy of *switching* centers interconnected by high-capacity links called *trunks*. The switches that connect directly to the customer equipment are located in *local* (central, class 5, end) offices. The local office switches are part of the regional operating companies, and connect to customer telephones through a twisted wire pair called a *two-wire local loop*. The particular subscriber is uniquely identified by the last four digits of the telephone number, while the first three are for the local office switch. Local offices are connected through interoffice trunks, usually consisting of bundles of wire pairs within cables. In large cities *tandem offices* are sometimes used to facilitate the local communication switching between local offices. This local switching network is often called the *exchange plant*.

Long-distance calls are routed from the central office through the *toll plant*, which is normally accessed by first dialing the digit 1, followed by the area code if the destination is in a different calling area. The toll plant consists of a nationwide system of carrier modulated coaxial cables, microwave radio circuits, satellite links, and lightwave channels that connect a hierarchy of circuit switches, of which the *toll center* (or class 4 office) is the lowest. These toll offices connect with local offices via toll connect trunks, and toll offices are interconnected through primary, sectional, and regional offices. Figure 4.1 indicates some of the telephone hierarchy and

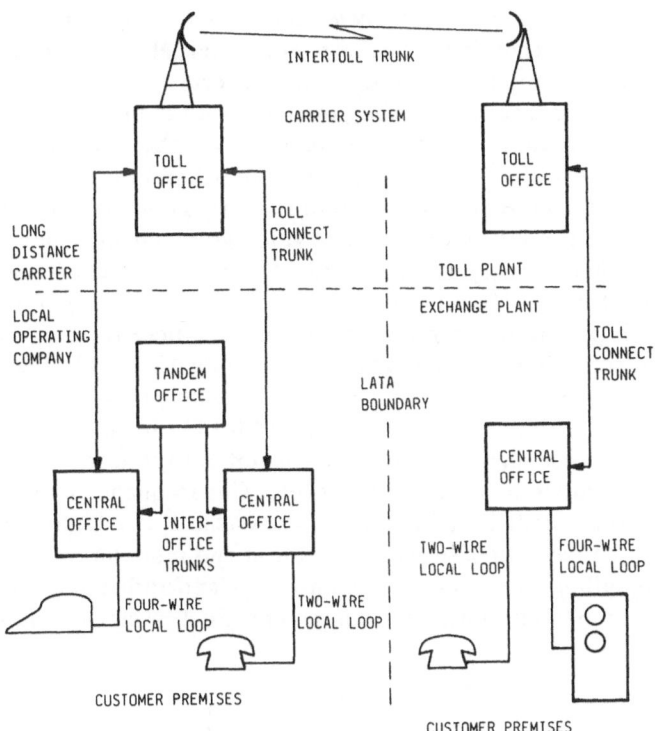

Figure 4.1. Telephone system structure.

terminology. It is interesting to note that although much of the present long-distance traffic today goes over microwave radio, there is a nationwide underground coaxial cable system, some of it hardened against nuclear attack (the *hardened route*), that is intended to provide highly reliable telephone communications throughout the country in the event of massive natural disaster or enemy attack.

4.2. The Analog FDM Carrier System

In the exchange plant where distances between caller and destination are usually on the order of a few miles, there is conventionally a physical wire path set up by the switching equipment. However, this approach becomes impractical for the toll plant due to the cabling cost over the large distances involved. The historically cost effective approach to toll trunks is to use a modulated carrier system that combines many voiceband circuits onto a wideband channel using *frequency-division multiplexing* (FDM). The wideband channels have been carried over coaxial cables, microwave radios, and satellite links in either analog or digital form.

Although some variation exists, there is a well-defined frequency multiplexing hierarchy for toll trunks. Figure 4.2 shows the AT&T FDM conventions and bandwidths. There are many standard and nonstandard ways to divide the *basic voice-grade channel* into *subvoice-grade channels*, depending on the particular applications. Twelve voice-grade channels are combined into a basic *channel group*, typically with SSB suppressed carrier modulation (although DSB has been used in some older equipment). Five channel groups multiplex into a *supergroup*, and ten of these form a *mastergroup* containing some 600 voice circuits or one analog TV circuit. Mastergroups may be frequency multiplexed further to form *jumbogroups*, which are finally combined to fill the transmission bandwidth of a large wideband carrier channel. Of course, the entire hierarchy is not always needed and there are wideband channels at various lower levels. This frequency multiplexed toll network is often called the *analog carrier system*. It is a high-quality FDM system that contains precise synchronization and modulation equipment as well as signalling procedures for controlling and indicating the status of the trunk circuits and customer equipment. Such systems require a huge

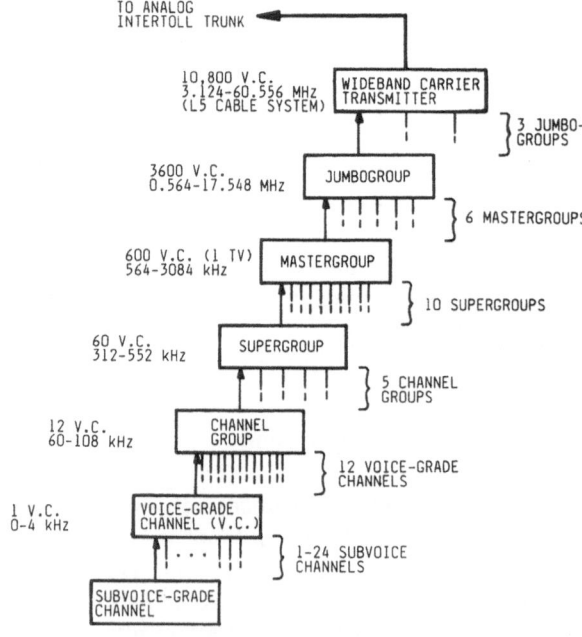

Figure 4.2. Analog FDM hierarchy.

amount of filtering to separate the many voice channels involved in each carrier grouping.

The analog carrier system uses separate channels to send and to receive transmissions, hence it is equivalent to a *four-wire* channel. Since the local loop is typically two-wire, it becomes necessary to convert from two to four wires before entering the carrier system. This conversion has historically been done at the central office by a clever transformer device known as a *hybrid*, which is located at the local switch. At the destination end, another hybrid converts the transmission back to two wires for the destination switch and local loop. Hybrids also allow the insertion of tones and voltages for signaling and for powering remote equipment. Transformer hybrids are now being replaced by electronic hybrids, which are cheaper and provide more accurate conversion.

To complete a call through the exchange and toll plants requires a number of signaling, controlling, and switching operations. For example, when a subscriber picks up the phone the central office must detect this off-hook condition, supply a dial tone, and then receive the dialed digits. Next a trunk route must be selected to the destination office and a ring tone produced at both ends. Once the called party answers, communications must be allowed to proceed uninterrupted and any necessary toll charges assessed until either end hangs up, at which time all trunks must be released. There are two basic approaches to modern signaling between offices. It may be done *in-channel*, i.e., over the same channel as the user traffic, by using battery voltage or tones. A newer technique is called *common channel†* *signaling* (CCS), where blocks of signaling information from a group of voice channels are combined by digital TDM and sent at 2400 bps over a separate voice-grade channel reserved for signaling. If this CCS channel is entirely separate from the corresponding voice channels, it is called nonassociated. CCS is much faster and more error resistant than older methods, and is fundamental to the future Integrated Services Digital Network (ISDN) concept that combines voice, data, and video on a common network.

Among the various local and remote switching offices, in-channel signaling is done in a variety of ways. E and M signaling is used for relatively short-haul trunks, with combinations of battery voltage, ground, and open circuit on two additional wires (the E and M leads) to indicate on- or off-hook, trunk idle or seized, and trunk clearing to both ends. Tone signaling is now more widely used, particularly on the carrier system where dc voltages are not feasible. Single-frequency (SF) tones may be in-band (typically 2600 Hz) or out-of-band (3700 Hz), and convey simple supervisory information such as hook status. Two-frequency (2VF) signaling sends supervisory and

† This is called Common Channel Interoffice Signaling (CCIS) by AT&T, and Signaling System No. 6 (SSN6) by CCITT. For evolving digital networks the CCITT SSN7 will be used in most cases.

address information with sequences of the two available tones, typically 2400 and 2600 Hz. The more common multifrequency (MF) signaling uses different pairs of six tones to send addresses in-band, much like the familiar dual tone multifrequency (DTMF) telephone dialing. Finally, audible tones are also widely used for status signals such as dial tone, ringing, recorded call, and busy signals on the trunk or at the called station.

Another integral part of the telephone network is circuit switching. Central office switches concentrate many subscriber lines onto relatively few outgoing trunk circuits, since it is assumed that only a few subscribers will use their telephones at any given time. If this is not the case, then the system becomes overloaded and some subscribers are blocked. If the switch itself is overloaded, there is just a long delay in receiving dial tone, while if all trunks are in use, a busy signal is returned.

The simplest type of switching is the classical *cord-board* system, where phone plugs are manually inserted to connect subscribers according to verbal requests to human operators ("number, please!"). The last such Bell system switch in the United States was only recently phased out in 1984, but the old names of *Tip* and *Ring* for the two phone plug contacts are still used to indicate the two wires of all local phone lines. In 1891 A. B. Strowger patented an electromechanical telephone switch that required no operator intervention.† These two-motion switches formed the basis for the popular *step-by-step* (S × S) offices, many of which are still in use today. However, they require considerable maintenance and are electrically noisy, and are consequently poorly suited for data transmission. S × S switches use *direct progressive control*, where each dialed digit operates the next stage of the switch until a complete path is established. An improved switching method that uses an array of switching points (crosspoints) to connect input and output lines is the *crossbar* (XBAR) switch. The crosspoints were originally wire contacts activated by metal bars, but were later replaced by much less noisy electronically activated reed relays. Today there are integrated transistor arrays that do the crosspoint switching electronically, which is much faster and quieter. A major feature of crossbar switches is *common control*, which is now done by a digital computer. Here the entire number is received and used to address a single crosspoint, which is a much faster procedure than the S × S method. It also allows subscriber numbers to be assigned to different physical lines, so the same phone number may often be retained after a subscriber moves to a new home or office.

The techniques discussed so far operate by creating a continuous physical path through the switch, and are consequently examples of *space-division* switches. An entirely different approach is called *time-division* switching, in which inputs are converted to digital bits and assigned specific

† Strowger was an undertaker and the local operator was the wife of his competitor, a classical conflict of interest which motivated the invention.

time slots in a TDM bit stream. The conversion device is typically an integrated circuit called a coder–decoder, or *codec*. In order to have FDX operation or to use more than one TDM stream, it is necessary to interchange the time slot contents in transit between the two communicating stations. This *time slot interchange* function is done by storing the transmitted slot contents in memory, then reading them back out at high speed in the interchanged order. This processing speed limits the number of lines a time-division switch can handle, since there must be one slot for each incoming and each outgoing line. However, large switches can be obtained by combining time-division (T) and space-division (S) stages, e.g., TSTST, so that the S stages basically move slots from one TDM bit stream to another. Digital switching is, of course, highly compatible with digital data transmission. Even for analog signals there is a steady trend toward moving the point at which the signal is digitized out towards the subscriber. The hybrid location for digital switches is on the subscriber line side, since the digitization and switching must be done in each direction separately. Note that this is not the case for analog space-division switching, where considerably fewer hybrids are located on the trunk side of the switch.

Computer control is used for all time-division switches and also for the more recent space-division ones. Known as *Electronic Switching Systems* (ESS) at AT&T, computer control provides fast execution of complex switching functions under program control. It also allows extensive features such as conference calls and call forwarding, and can be easily modified and reconfigured by simply loading a new program. ESS is now well established, being used on over half of the public lines in the United States.

4.3. The Digital TDM Carrier System

In 1962 AT&T introduced its T-carrier digital pulse code modulation (PCM) system for sending voice channels in digital form. This was the product of many years of research and development in PCM, and was originally intended mainly to transmit multiple conversations on the same interoffice trunk wire pairs that previously supported only one. Although much more bandwidth was required for PCM than for analog voice signals this fact was not always a serious limitation, e.g., on interoffice trunk circuits. And there were many compensating advantages, such as fewer and less stringent filtering requirements, the capability to send both data and voice, and high noise immunity. The latter resulted from the capability to completely regenerate the simple binary pulses before noise buildup caused bit errors, thereby removing any accumulated noise and distortion. *Regenerative* repeaters could be cascaded in long strings, but this was eventually limited by accumulated jitter. There was also a related requirement for very precise timing throughout the network to ensure that the proper pulses were always identified.

Even though initially used for sending digitized voice locally within the exchange plant, the T-carrier system was obviously well suited for the transmission of synchronous data. Schemes were soon developed to extend the range using microwave radio and coaxial cables. By the end of 1974 it was providing data transmission for the AT&T *Dataphone* ™ *Digital Service* (DDS), a true digital network offering leased connections from 2.4 to 56 kbps. Today, in addition to DDS, the T-carrier system forms the basis for a number of new public data transmission services as well as for many private networks.

The T-carrier system, like its analog counterpart, is hierarchial. At the lowest level is the single voice channel, which is sampled at 8000 Hz. Although it normally takes 12 to 14 bits to obtain acceptable quality PCM speech, the number can be reduced to only seven or eight through the use of companding. Here the voice signal is put through a nonlinear amplifier that has a gain greater than unity for small signals and begins to saturate for larger signals. This reduces the overall dynamic range of speech from around 50 dB to about 25 dB, and hence is called *compression*. At the receiving end the reconstructed analog signal is passed through another nonlinearity, or *expandor*, that restores the original amplitude relationships. Since each type of amplifier is required on a two-way circuit, the pair is called a *compandor*. Figure 4.3 shows this analog-to-digital conversion process, along with some key signals and corresponding spectra. Initially seven-bit coding was used, with an eighth bit providing signaling at 8 kbps for each channel. Later, to improve speech quality, the coding was shifted to the full eight bits on most samples and the signaling rate was greatly reduced. In either case, this single digital voice channel of 64 kbps constitutes one DS0 signal.

The basic layer of the T-carrier hierarchy is the *T1 level*, at which 24 DS0 signals are time-division multiplexed into a single 1.544 Mbps DS1 digital bit stream. The device that does the necessary filtering, sampling, companding, PCM encoding, and multiplexing is the *digital* (D) *channel bank*. It, of course, interfaces to the analog subscriber lines on one side and to the high-speed T1 line on the other. Figure 4.4a shows the basic structure of the first generation D1 channel bank. Note that, although there is separate interface and sampling circuitry for each user line, the digital hardware for companding, multiplexing, and coding is time-shared among all users. The typical T1 line contains a D-channel bank on each end and has regenerative repeaters every 6000 feet as shown in Fig. 4.4b. The corresponding group of 24 voice channels is sometimes referred to as a *digroup*.

The TDM frame for the T1 level is *word interleaved*, i.e., it contains a complete eight-bit word from each channel. In addition there is one final

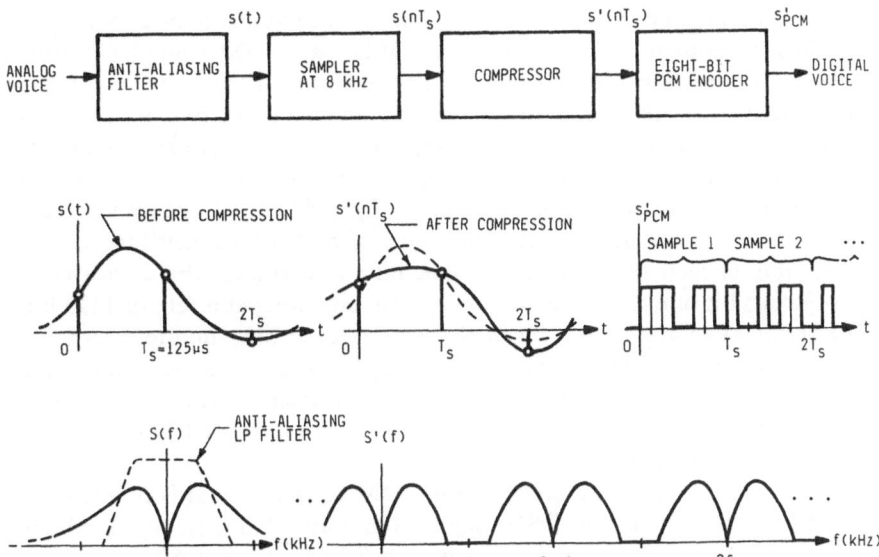

Figure 4.3. Analog-to-digital conversion for speech.

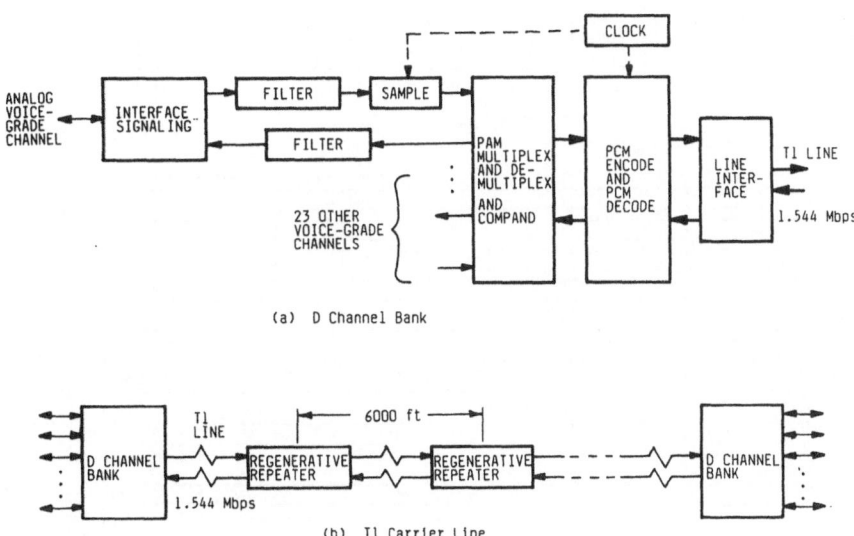

(a) D Channel Bank

(b) T1 Carrier Line

Figure 4.4. Basic T1 transmission line.

(S) bit for frame synchronization resulting in a total frame of $24 \times 8 + 1 =$ 193 bits as shown in Fig. 4.5a. Since there are 8000 frames/sec, then 193 bits/frame $\times$ 8000 frames/sec = 1,544,000 bps = 1.544 Mbps, which is the standard DS1 data rate. In the original scheme seven bits of each word were data, with the eighth for signaling, and the frame synchronization bit alternated between 0 and 1. The later scheme introduced a *superframe* composed of 12 T1 frames as depicted in Fig. 4.5b. In all but frames 6 and 12 the PCM coding uses the full eight bits of each word, resulting in more acceptable speech quality. In the remaining two frames the eighth (least significant) bit of each word is "robbed" for signaling at a rate of 1333 bps. The frame synchronization bit is also modified so that the old alternating pattern is retained on the odd numbered frames, but a new *multiframe alignment pattern* 001110 appears on the even frames. This latter scheme is implemented in all second generation and later channel banks such as D2, D3, D4, and D5.

The multiplexing hierarchy of the TDM carrier system is indicated by Fig. 4.6. Four DS1 or two DS1C signals (96 voice channels) are combined into a single 6.312-Mbps DS2 stream on a *T2 line*, seven DS2 signals (672 voice channels) form a 44.736 Mbps DS3 stream on a *T3 line*, and finally six DS3 signals (4032 voice channels) make up the DS4 stream on the 274.167 Mbps *T4 line*. Note that it takes two T3 circuits, or almost 90 Mbps,

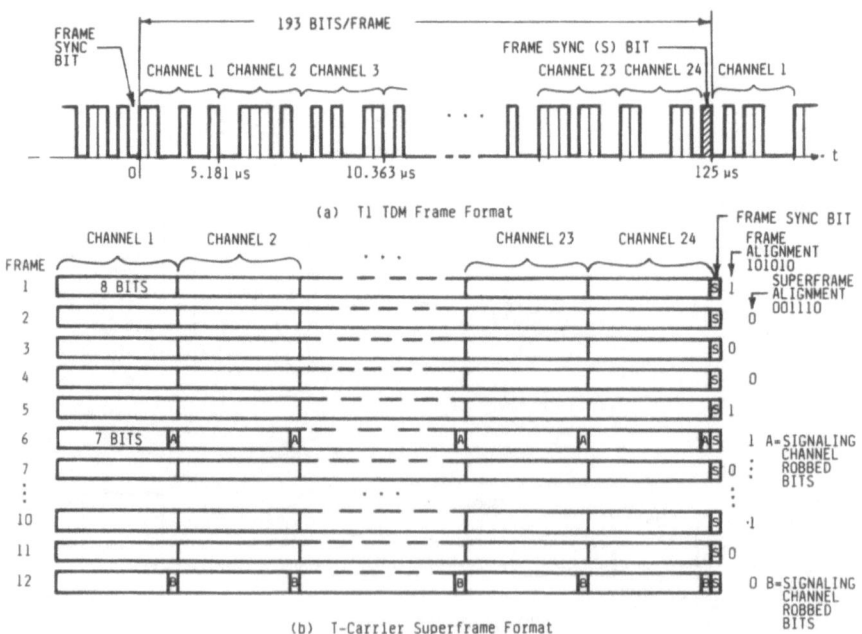

Figure 4.5. T-Carrier frame and superframe formats.

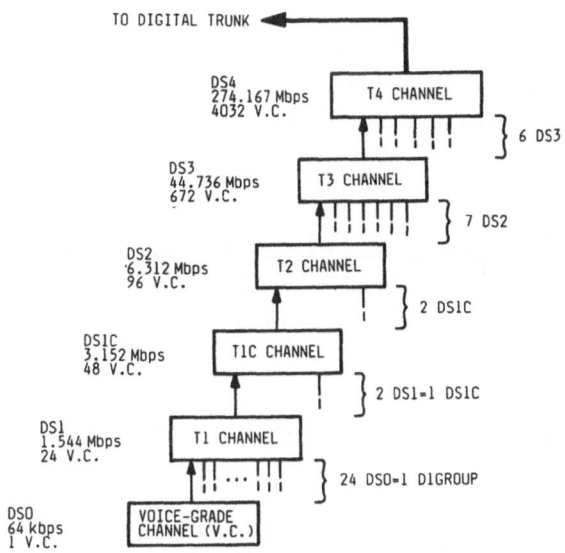

Figure 4.6. Digital T-Carrier TDM hierarchy.

to carry a conventional 6 MHz color TV signal digitally. There is also one other intermediate level in the T-carrier hierarchy, namely, *T1C*, which combines two DS1 signals to get a 3.152-Mbps DS1C signal. There is an obvious parallel between the digital TDM and analog FDM carrier systems. In fact, a voice channel may be carried by both through the use of an internal interface device called a transmultiplexer that converts between TDM and FDM.

All T-carrier multiplexing is synchronous, with timing for the entire system derived from a single atomic clock (the Bell Standard Reference Frequency) through a master/slave tree network. A line coding technique known as *bipolar*, or *alternate mark inversion* (AMI), is used on all but T4 lines, which use polar coding. For an arbitrary data pattern Fig. 4.7 shows the polar code with 100% duty cycle, and also the corresponding bipolar coding with the 50% duty cycle used on T1, T1C, T2, and T3, and the 100% duty cycle for 64-kbps DS0 signals. Note that for bipolar coding the absolute polarity of the data ones is immaterial just so long as they alternate. Bipolar signals have an efficient spectral shape which minimizes crosstalk between adjacent channels, while the absence of dc response allows transformer coupling and isolation to be used. They also allow for straightforward clock recovery from the pulse transitions. The alternating property provides inherent error detection, since most error patterns result in a violation of this coding convention. Since long strings of zeros result in the absence of any line voltage and eventually a loss of timing, intentional code violation

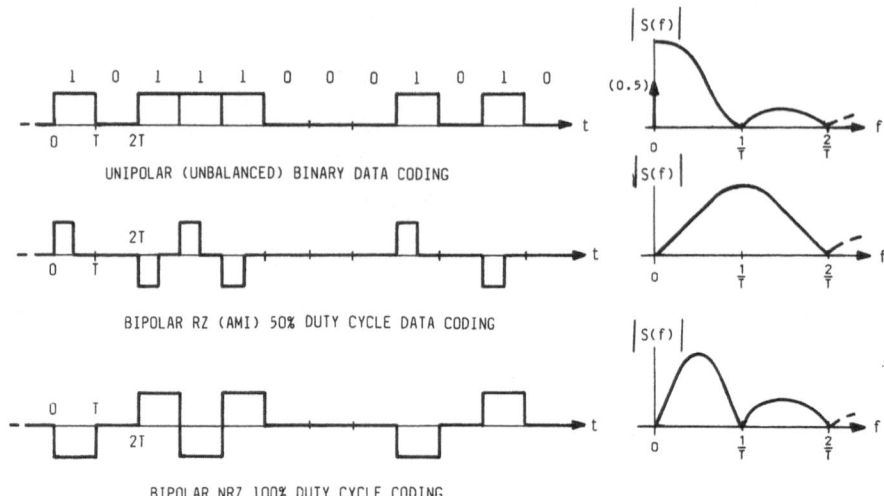

Figure 4.7. Bipolar line coding and spectra.

patterns are substituted on the T2 and T3 lines for strings of six and three zeros, respectively.†

Although the T-carrier system carries digital voice well, our real interest here is the transmission of data. There is a steadily growing demand for such capability, and this has resulted in several data services based on the T-carrier system. The DDS network [1] carries data at 2.4, 4.8, 9.6, or 56 kbps using T-carrier system lines. It now serves well over 100 cities in the United States, offering direct digital input over FDX four-wire leased lines. In place of a modem the user terminal equipment normally interfaces through a *Data Service Unit* (DSU). This device provides the user interface (RS-232C or V.35), bipolar baseband coding with a 50% duty cycle, line driving and receiving, clock recovery, and loopback testing. It is also possible for the user to do his own line coding and clock recovery, in which case the DSU is replaced with a simple line terminator box called a *Channel Service Unit* (CSU). Off-network access to DDS with modems and analog lines is also possible.

The D4 channel bank was introduced by AT&T in 1976, and is considerably more flexible than its predecessors. A selection of features is provided by modular design and the use of integrated electronics. These features include the capability of interfacing with various ESS switches, compatibility with digital pair gain local loop systems, and a "dataport" module with inherent error correction for 56-kbps DDS channels. The basic D4 configur-

† This is known as *binary N zero substitution*: in this case $N = 6$ (B6ZS) and 3 (B3ZS) for T2 and T3, respectively.

ation has a capacity of 48 DS0 channels, hence can carry one T1C channel at 3.152 Mbps or multiplex two T1 channels. It can also be configured with 96 channel capacity (four digroups) for either a conventional T2 metallic line or corresponding lightguide links. The D4 channel bank was designed to handle anticipated future digital transmission requirements while retaining compatibility with older ones. The lower cost, microprocessor-based D5 channel bank adds additional features, such as 64 kbps "clear channel" capability and a more versatile interface to those of the D4 channel bank.

At the central office an *Office Channel Unit* (OCU) terminates the local loop much like the CSU. In addition the OCU regenerates and retimes the user data. Inputs at 2.4, 4.8, and 9.6 kbps are combined by subrate multiplexers in groups of 20, 10, and 5, respectively, to get an actual data rate of 48 kbps. Bit 1 of these words contains a subrate synchronization pattern to indicate which rate is being used, and bit 8 is used for network control with a binary 1 indicating user data. At 56 kbps no subrate multiplexing is necessary and bit 1 is used for data. In either case 24 64-kbps DS0 data streams, including for some multiplexers one control channel, are multiplexed to produce a conventional 1.544-Mbps DS1 stream for transmission over a T1 channel. The entire procedure and data formats are shown in Fig. 4.8. For long-haul transmission DDS may use any of several digital microwave, coaxial cable, and optical fiber lightwave techniques available in the digital carrier system.

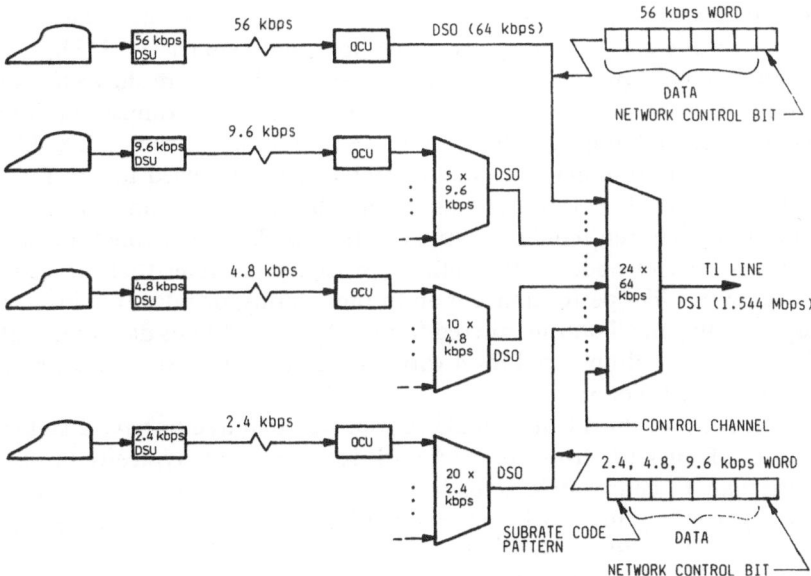

Figure 4.8. DDS input multiplexing scheme.

Although DDS is basically a leased point-to-point synchronous service, there are provisions for multidrop configurations and for asynchronous-to-synchronous conversion. There is also a relatively new capability for a low-speed secondary channel on which the user can transmit his own control or information traffic independent of the main channel. This extracts a significant price in terms of bandwidth. For example, a 9.6-kbps channel requires 12.8 kbps on the loop to provide a 533-kbps secondary channel. At 56 kbps the loop rate is 72 kbps for a 2.167-kbps secondary channel rate.

Besides the DDS system several newer digital services based on the T-carrier system are offered by the domestic telephone companies. At the DS1 rate of 1.544 Mbps there is an FDX dedicated service employing terrestrial links and an FDX switched service over satellites. There is also a 56-kbps switched service that operates over existing local loops by sending short fast bursts of data in alternating directions at 144 Mbps, a technique known as time-compression multiplexing (TCM). Internally the traffic is carried on T-carrier lines. Other similar data services will certainly be offered in the future as the demand for more sophisticated digital transmission capacity continues to grow.

In addition to using the T-carrier system, these digital services make extensive use of digital switching and common channel signaling. The underlying T-carrier system is also evolving towards use of the new 24-frame *extended superframe* format, illustrated in Fig. 4.9. Robbed-bit signaling is again applied to the eighth bit of each channel on every sixth frame, although there is now the option to use the corresponding bits for user data. There are also three signaling options, providing one channel at 1333 bps, two channels at 667 bps, or four channels at 333 bps. The figure shows the latter option. As with the superframe format the frame synchronization S-bit is also time-shared, but here there are three channels. On frames 4, 8, 12, 16, 20, and 24 the framing sequence 001011 is sent and used to synchronize both the channels within the frames and the frames within the extended superframe. On the remaining six even frames the S bits contain a six-bit CRC error check code for the entire extended superframe that can be used for functions like performance monitoring, alarms, or false framing detection. Finally, on all odd-numbered frames there is a 4-kbps data link control channel that could be used for a variety of control, monitoring, alarm, and maintenance functions.

In the past the T1 line bipolar signal was required to have a sufficient density of ones to ensure proper clock recovery. This typically meant no more than 15 consecutive zeros in the user's data and on average at least one 1 per byte. This restriction can be removed by replacing any string of eight zeros by a B8ZS code. This is shown in Fig. 4.10.

There is a steadily accelerating transition from analog to digital techniques in the domestic, as well as worldwide, telephone systems. It is driven

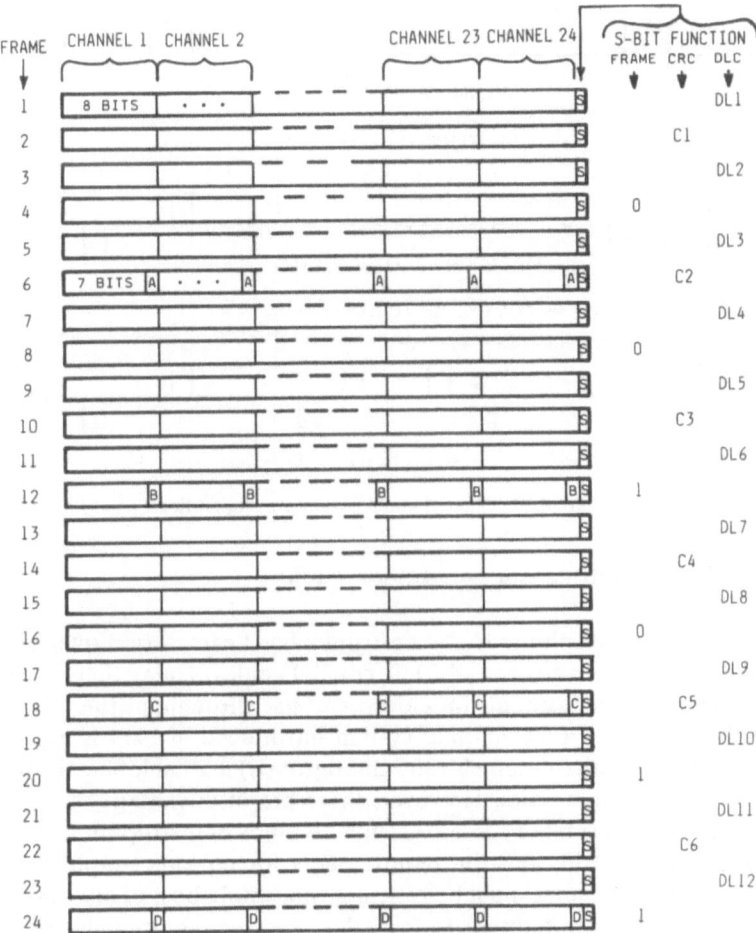

Figure 4.9. Extended superframe format.

by an ongoing user demand for data transmission, by dramatic strides in digital technology, and by increasing competition among common carriers. If current forecasts are correct, in a few decades this transition should result in an all-digital telephone system capable of carrying a mixture of voice, video, and various types of data traffic not just domestically but anywhere in the world. Planning for such an Integrated Services Digital Network (ISDN) is well underway by several organizations and the resulting concepts are beginning to be incorporated into new telephone equipment. Concepts such as the T-carrier system, digital switching, and common channel signaling, which are important today, will be even more crucial for the ISDN. We consider the ISDN concept in more detail in Chapter 13.

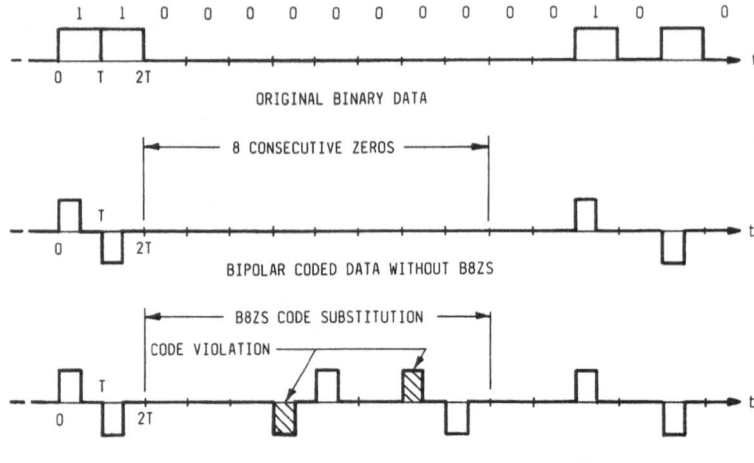

Figure 4.10. Clear channel coding for digital lines.

4.4. Radio, Satellite, and Lightguide Links

Historically telephone traffic on trunks has been carried over open or twisted wires and, since around 1940, coaxial cables. Today there are many coaxial cable carrier systems in widespread use throughout the world. For example, the AT&T L5 long-haul system introduced in 1974 is capable of carrying 108,000 voice channels (30 Jumbogroups) in a cable sheath containing ten coaxial cable pairs. However, there are disadvantages such as the need to physically run the cables through all kinds of metropolitan, residential, and rural terrain, and the requirements for repeater amplifiers every few miles. Buried cables must be resistant to a whole host of environmental hazards ranging from termites to tornados, and the required shielding and alarm systems are complex. These operational costs, along with the costs of easements and construction, make coaxial cable systems relatively expensive.

For cross-country transmission it is generally less expensive to use *microwave radio systems*. Beginning around 1947 the domestic telephone system has used a series of radio systems originally operating in the 4-GHz band but later at 6 GHz. The 11-GHz band is also now widely used for short-haul links. Since at these frequencies signal propagation is line-of-sight, it is necessary to place repeater towers every 30 miles or so because of attenuation and the curvature of the earth. Since this requires very little total land, microwave tends to be less expensive than coaxial cable to install and operate, particularly over rough or remote terrain. Microwave radios suffer from a number of atmospheric effects including signal strength fading, multiple propagation paths (multipath), and rain attenuation at higher frequencies. Fortunately, there are various diversity techniques for the mitigation of fading and multipath distortion. The conventional radio modulation technique has been analog FM owing to its inherent noise immunity. However, SSB modulation is now also used to gain better bandwidth utilization. A representative radio system is the AT&T TH-1, which has six 30-MHz channels in an overall 500-MHz bandwidth around 6 GHz. Each channel contains three mastergroups plus one supergroup for a total of 1860 voice circuits.

Data transmission over analog radio system requires the use of voice-band modems, a rather inefficient and expensive approach. When the volume of digital traffic warrants, it is more practical to use digital radio modulation directly. An early effort to "piggyback" a single digital DS1 channel on existing analog radios was *Data Under Voice* (DUV). Here the 1.544-Mbps bipolar data stream was recoded into seven levels, thereby reducing the transmission bandwidth by a factor of 4, and then was placed at the lower end of the conventional analog FDM spectrum. For the TH-1 system a pilot (synchronization) tone at 308 kHz was relocated to 512 kHz, and the data spectrum located from 0 to 386 kHz. This entire FDM spectrum was then FM modulated up to 6 GHz in the conventional manner.

Completely digital radio systems are also now in widespread use, particularly for short-haul applications. Typical digital modulation in each 30-MHz radio channel is *eight-phase PSK* at 30 MBd or *16-point QASK* at 22.5 MBd, resulting in 90 Mbps, which is sufficient for two 44.736 Mbps DS3 signals. Data is scrambled to facilitate synchronization and prevent distinct spectral lines in the transmitted signal. Adaptive equalization and AGC are invariably used to combat fading, multipath, and frequency response distortion. Multipath is also greatly reduced by using two receiving antennas at slightly different distances from the transmitter and always taking the stronger received signal, a technique known as *space diversity*. Digital radio systems are now most efficient for distances less than a few hundred miles, since most telephone traffic is still analog voice, which requires much less bandwidth than the conventional 64-kbps PCM digital voice. However, digital multiplexing and terminal equipment is much cheaper than analog, and the economic range of digital radio should steadily increase as the percentage of digital traffic grows and as more bandwidth-efficient speech digitization standards are developed.

Geosynchronous communication satellites have been well established for telecommunications since the early SYNCOM III transmitted pictures of the 1964 Tokyo Olympics to the United States. The next year the first commercial communications satellite, INTELSAT I (or Early Bird), was launched by the *International Telecommunications Satellite Organization*. INTELSAT is a worldwide consortium of over 100 nations that share the

communications services of the growing series of INTELSAT satellites. The
United States is represented in INTELSAT by COMSAT, a private corporation created by the 1962 Communications Satellite Act to develop the then
new NASA satellite technology for commercial uses.

Although satellites can be launched into practically any orbit around
the earth, only a few orbits are attractive for continuous communications
links. The most important by far is the *geosynchronous orbit*,† at which the
satellite appears stationary above a fixed point on the earth. This requires
that the satellite be located over the equator to avoid north/south drift, and
that it make one rotation every 24 h just like the earth itself. Since the period
of rotation depends directly on the altitude, geosynchronous orbits must
be approximately 22,300 miles high. This latter restriction introduces two
major problems. First, it is impossible to maintain *exact geosynchronous
position* without using an on-board positioning system which will eventually
become depleted. The orbit is also much higher than that of the NASA
Space Shuttle, so direct maintenance is not presently feasible. The second
major problem is the *transmission delay* inherent over a satellite link. For
the link itself the propagation time at 186,000 mi/sec is

$$2 \times 22,300/186,000 = 0.240 \text{ sec}$$

and if the earth station latitude and various terrestrial and satellite signal
processing delays are included the delay is more like 270 ms. Of course,
the same delay is incurred in the opposite direction; therefore the round-trip
delay is over half a second. In voice conversations this is annoying but not
prohibitive. A classical solution is the use of half-hop circuits, where only
one direction is carried over the satellite with the return traffic carried over
a terrestrial link. The resulting quarter-second delay is much more acceptable
to human speakers.

Satellites are, of course, an expensive wideband resource that must
generally be shared among many different users. This involves *multiplexing*
traffic from various user locations over terrestrial links, *modulating* these
multiplexed signals onto the satellite carrier frequency at a satellite earth
station, and *accessing* the satellite from multiple earth stations without
mutual interference.

Historically, an FDM scheme similar to that of the analog telephone
system has been used with satellites, and the resulting spectrum modulated
with FM in the 6/4-GHz C band. The uplink is sent in a 500-MHz bandwidth
around 6 GHz and the downlink in the same bandwidth at 4 GHz. Since

---

† This is also called the *Clarke orbit*, in honor of Arthur C. Clarke, who first proposed
geosynchronous communications satellites in a prophetic 1945 paper.

these are exactly the same frequency ranges used by microwave radios, care must be taken to avoid interference. Multiple access to a satellite can be obtained by assigning different carrier frequencies to different earth stations, thus allowing them all to operate simultaneously. Such arrangements are appropriately called *frequency-division multiple access* (FDMA), and the entire multiplex-modulation-access technique is denoted FDM/FM/FDMA. Such satellites operate essentially as relays, since they just remodulate and retransmit the received signal. The on-board devices that do this are called *transponders*, which for the 4/6-GHz satellites have a typical bandwidth of 36 MHz each. This is sufficient for hundreds of FDX voice circuits or a few television channels. A modern satellite will contain many transponders, and there may be a single or many FDMA carriers per transponder. Data is sent over FDM/FM/FDMA satellite links in the same way as over the terrestrial analog telephone system; it is converted by a conventional modem into voice-band tones for transmission.

Although FDM/FM/FDMA systems have served as the workhorse satellite technique for many years, they are gradually being replaced by newer techniques. Much more sophisticated and reliable digital hardware can now be built with modern integrated electronics, and the Space Shuttle can launch satellites at lower cost than previous rocket launch vehicles. In addition to the 6/4-GHz C band, satellites now use the 14/11- or 14/12-GHz Ku band, which requires considerably smaller antennas and associated hardware. Even this band, however, is becoming crowded, and consequently there is considerable interest in the 30/17-GHz Ka band for future applications. Higher frequencies are also attractive because of a "parking space" problem in the geosynchronous orbit. To prevent interference at 6/4 GHz, satellites have historically been spaced at angles of 4°, which allows only 90 spaces worldwide. This has now been reduced to 2° with sharper antenna focus and spectral shaping, and the inherently more narrow radiated beams at these higher carrier frequencies.

With larger and more powerful satellites and higher carrier frequencies it is possible to use several antennas on the satellite, resulting in another form of multiple access known as *space-division multiple access* (SDMA). It is then possible to use the same carrier frequency simultaneously with more than one directional antenna; in fact, it is even possible to use the same antenna for two separate channels of the same frequency by the technique of orthogonal field polarization. Antenna beams may broadcast over large (global and hemispheric) areas of the earth, or they may be focused onto relatively small (spot) areas for point-to-point operation.

Digital transmission is growing rapidly over satellite links, and there are many techniques for sending digital traffic. The fundamental multiplexing technique is of course TDM, and many domestic satellites use the T-carrier multiplexing conventions, with 56 and 1544 kbps being the basic data rates. Satellite users can also operate point-to-point without multiplexing in the *single-channel-per-carrier* (SCPC) mode. Carrier modulation may be FM since this allows bandwidth to be traded off for better SNR, but PSK and QASK are more efficient in modern satellites where transmitter power levels are high enough for adequate SNR. Digital access to the satellite is normally via *time-division multiple access* (TDMA). Here each transponder is time-shared among many earth stations. The normal operation is for some controlling station to send via the satellite a synchronizing reference burst which then defines a unique time slot for each station. During its assigned slot each station has use of the entire transponder bandwidth. A representative frame format is shown in Fig. 4.11. The corresponding multiplex-modulation-access technique is denoted by TDM/PSK/TDMA.

TDMA has a number of advantages over FDMA, including better bandwidth utilization, no intermodulation distortion, higher power levels, and the ease of changing the capacity of a station by simply increasing or decreasing its assigned time slot. TDMA also has many variations. For example, instead of having the time slots preassigned, they can be assigned only on demand. TDMA can also be used with SDMA by employing on-board switching to continually switch each time slot to the correct antenna for the intended destination. In this *satellite-switched* (SS) mode, SCPC channels can be individually switched so the system acts like a star network of point-to-point links. Clearly SS/TDMA requires highly sophisticated satellites with extreme reliability. Demand assignment and satellite switching are also possible with FDMA carriers rather than TDMA, but each channel must be demodulated before switching and then remodulated on board the satellite; the analog switching is considerably more complex than digital.

Since TDMA is inherently digital, it allows both *encryption* and *forward error correction* (FEC) to be implemented in a straightforward manner. Satellite voice-grade channels are normally of very high quality compared to their terrestrial counterparts, and satellite links are much less subject to the atmospheric distortions that plague terrestrial radios. On such channels FEC can essentially eliminate the need for retransmissions because of bit

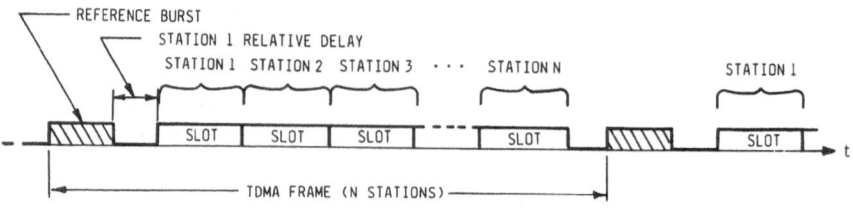

Figure 4.11. Satellite TDMA frame format.

errors, which otherwise could seriously affect throughput due to the extremely long propagation delays.

One significant factor with satellite links is that of echoes. Just as in two-wire dial-up terrestrial circuits, imperfect hybrids can cause echoes from the far end. The inherent 270-ms delay of satellite links makes conventional echo suppressors rather ineffective owing to excessive turnaround times. If not properly set for satellite operation, it is also possible for modems to mistake their own transmission echoes for a valid received signal. The current solution is to use an *echo canceler* to cancel the far-end echo as it occurs. Figure 4.12 shows the basic configuration at the far end. The actual canceler is an adaptive digital filter that adjusts itself to attenuate and delay the received signal $v_R(t)$ so as to produce a replica $\alpha v_R(t - t_d)$ of the echo. This canceler output is then subtracted from the transmitted return signal $v_T(t)$ to remove the echo before it gets to the satellite link. Although developed mainly for voice, cancellation has proved effective for data as well; therefore echoes are seldom a problem on well designed satellite links today.

As an example of an operational satellite we shall consider that of Satellite Business Systems (SBS), the first all-digital satellite system. In late 1980 SBS launched the first of three digital satellites to provide voice and data transmission services to primarily large users. Operating in the 14/12-GHz band with a single on-board antenna, it uses four-phase PSK modems running at the 70-MHz intermediate frequency to send TDMA bursts to and from a large number of earth stations. These can be small roof-top units with antenna diameters of only five meters because of the high carrier frequencies.

The SBS satellite can accommodate data rates from 2.4 kbps to 6.3 Mbps, and has capacity for 13,900 voice circuits or 8000 56-kbps data circuits. There are ten transponders each with a bandwidth of 43 MHz. One station per transponder is configured to provide the TDMA synchronizing

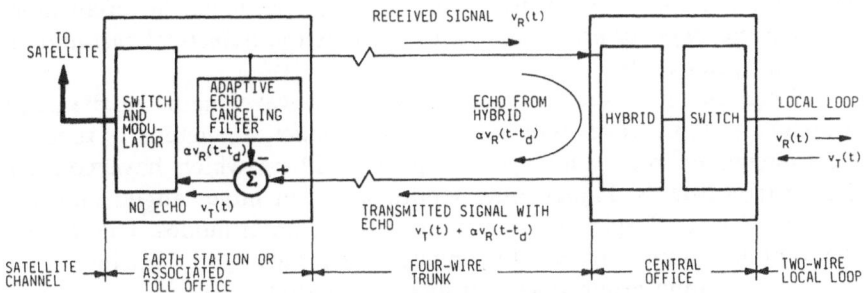

Figure 4.12. Satellite echo cancelation.

bursts, and also to dynamically adjust the capacity assigned to each station by increasing or decreasing the corresponding time slot duration. There are several categories of capacity allocation ranging from full-time service to demand assignment based on available capacity. Voice is digitized at 32 kbps and, to reduce the required bandwidth, no capacity is assigned to silent intervals.

SBS offers a variety of special features to users. FEC is available to reduce the error rate from 10^{-4} to 10^{-7}, and traffic can be encrypted. There are provisions for teleconferencing, facsimile, and electronic mail transfer. The entire system is monitored and controlled from two network control stations. The SBS system illustrates many practical techniques for digital data transmission by satellites. In addition to its historical corporate traffic, in 1985 SBS began carrying long-distance telephone traffic for MCI.

Lightguide transmission is a relatively new technology, dating back to the original laser patents in 1958 and the initial work on dielectric waveguides in the early 1960s. The technology progressed steadily through the decade, and in 1970 the first fiber with low enough attenuation for telecommunications use was announced. By 1976 there were several metropolitan fiber optics trunks in use for telephone traffic. Today, with digital switching and the T-carrier system well established, fiber optics is fast becoming the preferred method for both short-haul and long-haul trunks requiring high capacity. As the cost continues to drop and performance improves still further, there is every indication that fiber optics will eventually be the dominant data communications transmission medium.

Fiber optics cables are actually *dielectric waveguides* made of ultratransparent glass or sometimes plastic. If this round fiber core is surrounded by a cladding with carefully selected dielectric properties, light can be made to reflect almost completely at the interface. Thus a ray of light starting down the fiber will remain in it and propagate along a zig-zag path, attenuated by light scattering and absorption by the fiber. However, at the source the light must be applied within some maximum cone, or else it will be transmitted and lost rather than reflected at the core-cladding interface. The corresponding angle, which depends on the particular materials used, is called the *angle of acceptance* and its sine is the *numerical aperature* of the fiber optics cable.

When the fiber core is more than a few wavelengths in diameter, propagating light waves will have multiple *modes*, analogous to those of electromagnetic waveguides. For *step index* fibers which have constant dielectric constants, higher-order modes reflect at larger angles and thus propagate over a larger overall path than lower-order modes. The result is that different modes arrive at the optical detector at slightly different times, with the resulting group delay causing transmitted pulses to be dispersed in time. This *modal dispersion* increases with distance, causing intersymbol interference (ISI) that directly limits the data rate as illustrated in Fig. 4.13. Modal dispersion can be reduced in two ways. First, *multimode* fibers can be made with diametrically variable dielectric constants in such a way that the higher-order modes travel relatively faster. If the profile of such *graded index* fiber is parabolic then all modes can theoretically be made to propagate at the same velocity, hence eliminating the dispersion. The second approach is to make the fiber core so small that only one mode can propagate. Such *single-mode* fiber has a diameter of only a few wavelengths, typically 5 μm. Hence, it is much more critical to drive and requires a laser as the light source, but modal distortion is totally eliminated.

Unfortunately modal distortion is not the only impairment to fiber optics transmission. Light energy is absorbed by various impurities in the physical fiber, and the degree of absorption varies with the wavelength λ (or frequency $f = v/\lambda$, where v is the light propagation velocity down the fiber). Because of this frequency selectively different wavelengths propagate at different rates, with the result again being a smearing or *material dispersion* of the transmitted pulse. Early systems used a wavelength of around 0.85 μm since this gave the minimum attenuation, but improvements in fiber purity have greatly reduced the attenuation with minima around 1.3–1.6 μm depending on the particular fiber. Other sources of distortion are Rayleigh scattering of the light due to small inhomogeneities, and small energy radiation from bends in the fiber.

Fiber optics light sources may be either *light-emitting diodes* (LED) or *injection laser diodes* (ILD). Although analog modulation is possible in some cases, we shall confine our treatment to the much more prevalent binary on–off light intensity modulation. As a general guide, it is instructive to consider three rough categories of fiber optics systems while keeping in mind that many other categories are also possible. First, for short distances

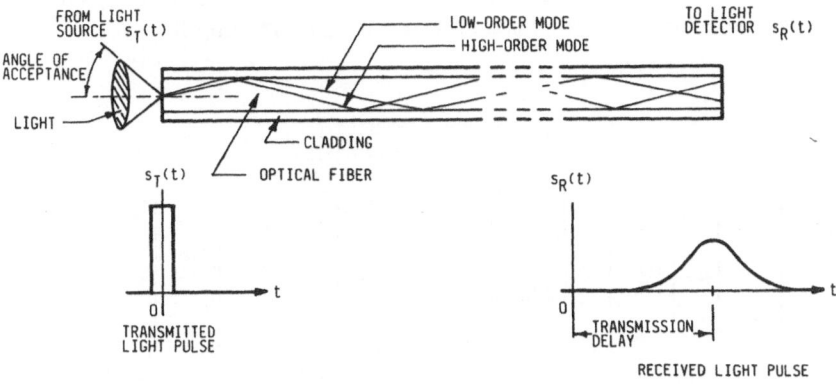

Figure 4.13. Modal dispersion in multimode fiber optic channels.

(under a mile) and low rates (below 50 Mbps) simple LED sources and step-index multimode is adequate. Second, for medium distances (1–5 miles) and rates (50–100 Mbps) multimode ILD sources and graded-index multimode fiber is appropriate. Third, long distances (over 5 miles) and high rates (into the gigahertz range) necessitate the use of single mode lasers with single-mode fiber operating in the long wavelength (high-frequency) range of 1.3–1.6 μm. In this latter case it is possible to send long-haul traffic with repeater spacing of many tens of miles. Thus single-mode systems are increasingly attractive for both inter- and intracity trunks, as well as for long-haul trunks. There are also two general types of photodiode detectors in common use to convert light energy back to an electrical signal. Both *PIN diodes* and *avalanche photodiodes* (APD) are widely used. PIN diodes are easier to work with, but the high inherent gain of the APD makes it very sensitive to weak signals. Thus the APD is particularly suitable for long-haul trunks where SNR is a key consideration. Figure 4.14 indicates the major components of a representative single-mode fiber optics data link for telecommunications applications.

To tie some of these concepts together with a practical example, we shall now look briefly at the relatively early AT&T FT3 Lightwave System. Introduced in 1980, this digital trunk is designed to carry data at the DS3 rate of 44.736 Mbps primarily in metropolitan areas. The graded-index multimode fiber is driven by a semiconductor laser operating at 0.825 μm, and an APD is used for detection. The actual installed cables contain up to 144 fibers, so one cable can carry over 48,000 FDX voice circuits with an error rate of about one every 10^9 bits. The incoming lower rates, such as DS1, DS1C, and DS2, are multiplexed electronically; then the outgoing DS3 signal is scrambled before optical conversion to facilitate clock recovery. Optical regenerative repeaters were spaced every 4 miles in the original FT3 system. Although it performs well, by current standards the FT3 system is already old technologically and several newer systems with single-mode fiber and higher data rates are superseding it.

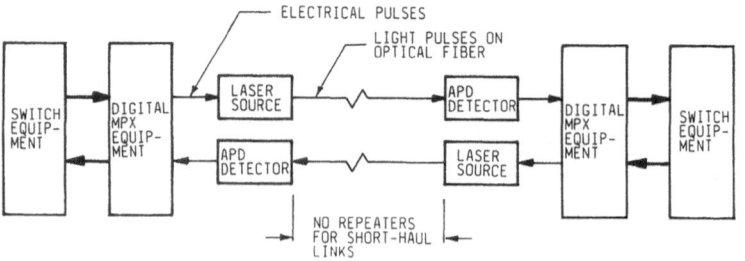

Figure 4.14. Single-mode fiber optic data link.

There are many benefits in using fiber optics. It is easy to install and maintain, and requires few repeaters. This is particularly true for single-mode, where continuous fibers may run well over a hundred miles with no regeneration. It is immune to lightning (unless hit directly!), crosstalk, electromagnetic noise, inductive coupling, and ground loops; and there is no signal radiation, making it difficult to tap. Finally it is inherently digital with huge bandwidth and, perhaps best of all, the entire technology is still young and growing with plenty of potential for new technological developments for many more years. On the other hand, mundane hardware like optical connectors and simple optical switches are still being developed, and of course A/D conversion is required to interface with the analog FDM carrier system.

The full role of lightguide systems relative to microwave radio and satellites is not yet completely clear. There are current plans and programs by many communications carriers and other firms to install huge amounts of long-haul lightguide capacity in the United States as well as in most other industrial nations. A great deal of the replacement and expansion metropolitan trunking will also be fiber. Eventually wideband lightguide local loops may provide a broad range of communications services to homes and businesses. There are also plans for optical undersea cables but they may never completely replace satellites for the larger transoceanic routes. One popular scenario has lightguide systems gradually replacing radio for long-haul links, with radio then being used primarily for regional distribution and alternate routing. However, in many large cities there is already a growing shortage of radio bandwidth at 6 GHz. From a security standpoint radio and satellite signals are far easier to intercept than those of lightguide systems. Satellites will be appropriate for many remote areas and less developed countries that do not require the extensive communications services of the industrial nations. They are also the most suitable way to communicate with mobile locations such as ships at sea and land vehicles outside of metropolitan radio coverage.

Regardless of the carrier technology employed, the user of the domestic telecommunications system essentially only sees the "end of the pipe," and is little concerned with how his traffic is carried through it. Thus it is of great practical importance to understand the input/output characteristics of the particular line being used. In the next section we shall consider the capabilities as well as the limitations and impairment of the common voice-grade telephone line.

4.5. The Basic Voice-Grade Line

This chapter has so far been mainly concerned with interoffice trunk and carrier systems, not with the local loop connecting the subscriber with the central office. Although the local loop is relatively simple and ubiquitous, it is very important for two main reasons: first, it is the only port that the user has into the network, and second, it and its interfaces determine to a large extent the overall characteristics of the voice-grade line through the entire network [2].

The interoffice and intertoll trunks provide a high quality four-wire channel that introduces little distortion, but the two-wire bandlimited local loops at each end are subject to such problems as filter component variation, imperfect hybrid balance, crosstalk, and a host of environmental conditions like moisture, lightning, gophers and industrial interference. There are also common problems due to different wire sizes on feeder, distribution, and drop cables; and, of course, with subscribers who attempt to do their own telephone wiring on their premises.

The nominal 300–3000-Hz frequency response of the voice-grade line is also subject to considerable variation. As Fig. 4.15a shows, the magnitude response tends to be quite good in the lower range where the human vocal chord frequencies are located since otherwise speech becomes difficult to understand. At the upper end, however, there is a good deal more variation, and the response may be well below the conventional 3-dB range at 2500 Hz or even lower. Instead of indicating the phase response, it is conventional practice to use *relative group delay*, with 1700 Hz taken as the reference point. Group or envelope delay was previously defined as the negative

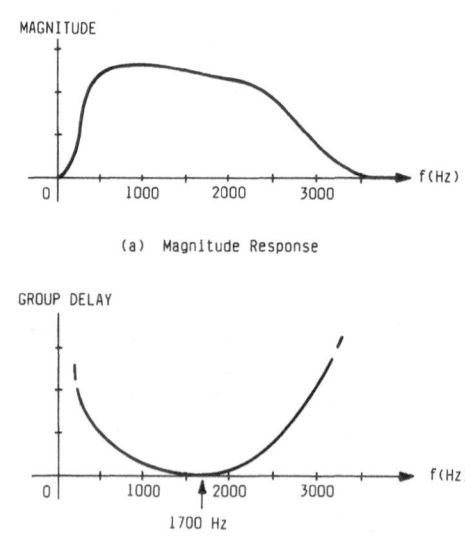

(a) Magnitude Response

(b) Relative Group Delay Response

Figure 4.15. Voice-grade telephone line frequency response.

derivative of phase with respect to angular frequency, i.e., $-d\theta/d\omega = (-d\theta/df)/2\pi$. Although it is not particularly intuitive, group delay is much easier to measure than phase, and provides all the essential engineering information. Figure 4.15b illustrates the typical rapid increase in group delay near the band edges, with a few milliseconds being typical, and also the typically flat midband delay. The group delay curve is often modeled as being parabolic in shape.

These frequency characteristics are due primarily to the bandpass filters used to limit the bandwidth on analog FDM carrier systems or for antialiasing on digital TDM carrier systems. Frequency response may also be affected on some older and longer loops by *loading coils*, which are fixed inductances inserted along the line to improve the magnitude characteristics for voice up to about 3500 Hz. For nearby central offices connected by wire trunks and without pair gain or digital switching, there is no need for bandlimiting filters and the bandwidth of unloaded metallic lines is much greater than with loading.

As we know, local loops can be either dial-up (public) or leased (private). *Leased lines* are nearly always four-wire and are permanently wired around the switches. On carrier systems a specific channel is always reserved for each leased connection. This results in the elimination of switching noise and, of course, the immediate connection of all calls. The channel characteristics are still determined by the bandlimiting filters; but since the end-to-end line does not change it is possible to add additional filtering to produce as good a response as desired. This correction filtering is called *equalization*, and the resulting leased lines are said to be *conditioned*. There are several grades of conditioning available to users; namely, C1, C2, and C4 providing increasing quality of the frequency response. Tolerance limits on noise and harmonic distortion can also be obtained with D1 conditioning. For multidrop configurations the drops are usually attached by bridging electronically across the main line. The connection, called a *bridged tap*, is also a common source of distortion due to reflections and attenuation at the tap. Leased lines are appropriate where high quality and continuous connections are desired. Since they must be paid for whether used or not, they are obviously more cost effective in high use applications.

In contrast, *dial-up lines* are normally two-wire, go through switches which are often noisy, and cannot be conditioned on request. Each dialed call results in a different local loop on the called end, as well as a different internal path through the carrier system. Hence there may be considerable variation in line characteristics from call to call. Figure 4.16 shows connections for a two-wire dial-up voice circuit and a four-wire leased circuit with an FDX modem for data.

Unlike leased lines, there is a variety of management and signaling functions that must be done over the dial-up line. They have historically

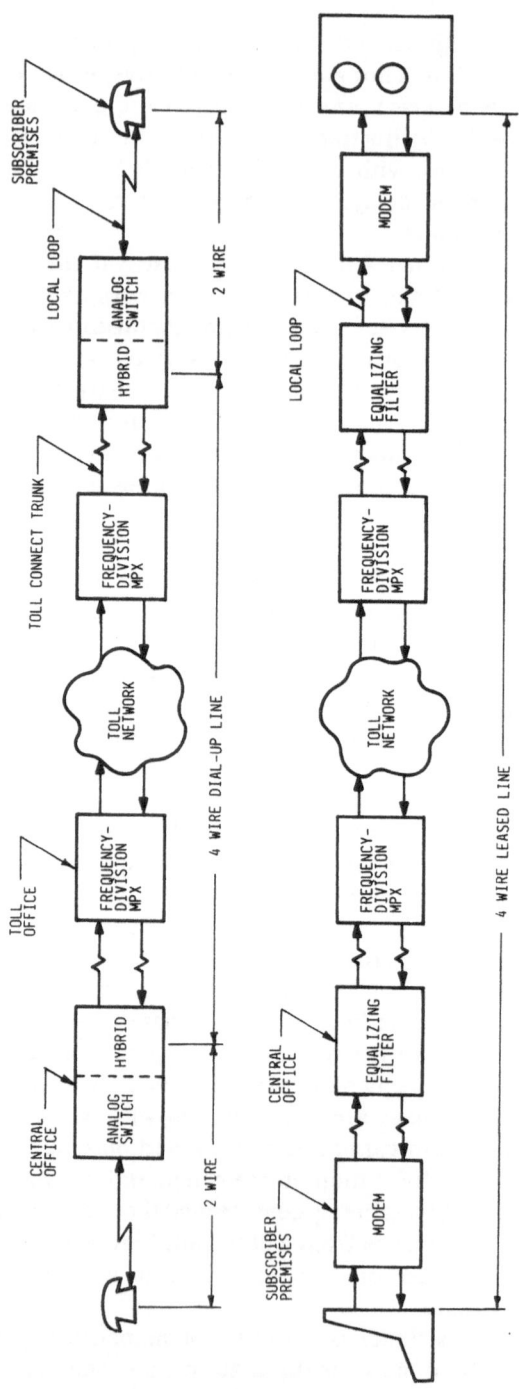

Figure 4.16. Voice-grade line types.

been known by the acronym BORSCHT.† *Battery feed* (B) refers to the application of the −48-V office battery voltage to the *Ring* lead, with the *Tip* lead kept at 0 V. When the line goes off-hook, such as when the handset is picked up, the voltage causes a current between 20 and 80 mA to flow in the loop. This current is then used for switchhook‡ signaling. *Overvoltage protection* (O) refers to protection against lightning, 60-Hz power line coupling, etc.

Ringing (R) is done by applying an ac voltage of around 100 V at nominally 20 Hz,§ but ring frequencies can vary on party lines from 15 to 68 Hz. The typical ringing cadence is 2 sec on and 4 sec off, but again this may vary with the installation. A telephone ringer must always be across the line, and thus contains a blocking capacitor to prevent continuous battery current flow through it. For data equipment there is usually no bell ringer; rather, the automatic answering equipment simply contains a band-pass filter centered at 20 Hz to detect ringing, and then connects the terminal equipment by closing the switchhook relay.

The *supervision* (S) function uses the loop current to detect the off-hook condition, dial pulses, answering, and hanging up on both ends. Pulse dialing is done by interrupting the loop current for 60 msec every 100 msec until the number of pulses equals the digit dialed. Then there is an *interdigit interval* of at least 600 msec before the next digit can begin. For DTMF tone dialing the tone pairs are detected by filters and the interdigit time is only 45 msec, making it much faster than pulse dialing. At the central office all of these local loop signals are, of course, converted to the appropriate trunk signals in transmission through the established end-to-end channel.

Many local loops now terminate on digital central office switches, and in such cases the analog signal must be bandlimited, sampled, and *coded* (C), typically with PCM coding. This analog-to-digital, and the reverse digital-to-analog, process is now almost universally done by *codec* integrated circuits, which may also perform filtering, companding, and some signaling functions.

Since dial-up lines are invariably two-wire, a *hybrid* (H) is necessary to convert to four-wire trunks. Although older transformer hybrids were located on the trunk side of analog switches for economy, digital switching requires one per local loop prior to the codec. This huge potential volume has resulted in a number of electronic hybrid schemes, typically as part of an integrated circuit known as the *Subscriber Line Interface Circuit* (SLIC). SLICs generally also provide some gain adjustment, switchhook signaling,

---

† Borscht (or Borsch) is cold Russian beet soup, and the acronym is not totally inappropriate.
‡ The name "switchhook" comes from the earpiece hook on the early "candlestick" telephones, but it is now operated by the handset cradle on modern instruments.
§ The old hand-cranked magneto ringers produced about 20 Hz and 100 V when cranked vigorously, and hence our modern convention.

and loopback capability. Battery feed, line isolation, ringing, and over-voltage protection are now done externally to the SLIC due to the high potential voltages involved. *Testing* (T) functions include loopback and diagnostic capabilities that are usually executed from the central office. All of these BORSCHT functions are shown for a representative digital interface in Fig. 4.17.

There are many impairments that can afflict the voice-grade line. Besides being out of tolerance, frequency response characteristics can vary suddenly with weather changes and very slowly with component aging. There may be *crosstalk* from coupling between adjacent wire pairs or FDM carrier channels. At the central office there may be spikes of *impulse noise* from various sources, particularly older mechanical switches. Momentary interruptions are arbitrarily classified as *hits* if less than 300 msec and *dropouts* if greater. Linear amplifiers and companders that are out of adjustment can produce *nonlinear distortion* and *clipping*, and *echoes* are caused by impedance mismatches at bridged taps, connection points, and unbalanced hybrids. Echoes from the local hybrid are occasionally strong enough to cause serious distortion of very weak received signals, and those from the remote hybrid can be annoying in long-distance voice calls when not properly suppressed.

The carrier system may also introduce distortions in addition to crosstalk. Slight mismatches in local oscillator frequencies of analog carriers and timing errors in chains of repeaters in digital carriers result in an instability in signal phase called *jitter*. *Harmonic distortion* results when harmonics of one channel appear in the bandwidth of another, while *intermodulation distortion* results when different frequency components are passed through various nonlinearities which produce undesired product frequencies. Finally, serious distortion can result from improper operation of echo suppressors. These voice-operated switches in effect eliminate echoes by forcing long-haul channels to operate HDX. They sometimes malfunction so that speech sounds badly "chopped," or the turnaround time for modems is impaired.

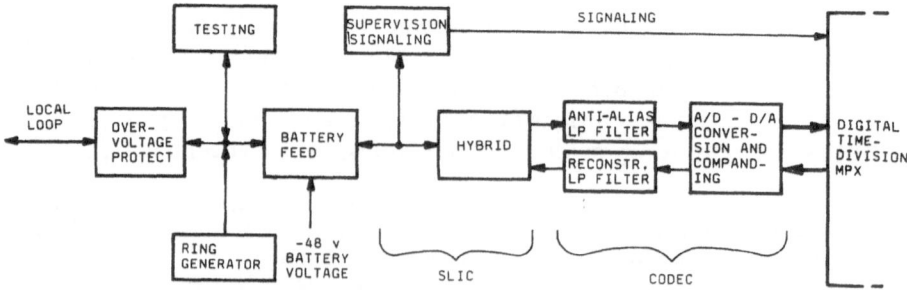

Figure 4.17. BORSCHT function for dial-up lines.

So far we have restricted our attention to the most common 3000-series (or Type 1) voice-grade line, but there are also numerous wideband lines available. Historically, the widely used Telpac service provided bundles of analog lines in such bandwidths as one supergroup (60 voice channels) and four supergroups (240 voice channels), but these have been replaced with a number of other offerings from different common carriers. A similar spectrum of digital offerings is available, including the DDS rates and 1.544-Mbps T-carrier service. Since the divestiture of AT&T the number of both wideband services and vendors has increased, and still more selection will be offered in the future.

As this chapter indicates, the highly reliable and widely available voice-grade line is really a very complex circuit in an incredibly complex communications system. Although most dial-up lines are suitable for voice or data, occasionally a very poor quality line is randomly obtained. This is a more serious problem for data than voice, since many data terminals will not hang up and redial. Even on good lines there will always be some impairments that reduce the efficiency of data transmission.

Eventually the local loop will be a digital channel capable of high data rates. However, it will be the last leg of the ISDN transition due both to the huge number of local loops and to the obvious fact that an analog telephone is presently sufficient for the majority of user's needs. In the meantime much digital traffic will be carried on existing analog facilities. In the next chapter we will study the many techniques and details for transmitting digital data over analog lines. They should remain important for many more years, particularly in more remote areas where all-digital service will be last to arrive.

References

1. *Bell Syst. Tech. J.* (Special Issue on the DDS) **54**(5) (1975).
2. J. E. Nuwer, Why the line acts that way—and what can be done, *Data Commun.* **7**, 63–73 (1978).

Suggested Readings

J. C. Bellamy, *Digital Telephony*, Wiley, New York (1982).

   This is an excellent description of how the digital telephone system operates at a good technical level for the practitioner or student.

J. J. Spikler, Jr., *Digital Communications by Satellite*, Prentice-Hall, Englewood Cliffs, New Jersey (1977).

   This is a classical standard work on all technical aspects of digital satellite communications—a highly useful reference.

S. D. Personick, *Fiber Optics: Technology and Applications*, Plenum Press, New York (1985).

The first thorough, readable treatment of modern fiber optic communications systems, providing an excellent introduction as well as more advanced material.

J. L. Fike and G. E. Friend, *Understanding Telephone Electronics*, Texas Instruments, Dallas Texas (1983).

This is a nice, clear, simple introduction to the digital and analog telephone system from a hardware viewpoint.

D. Talley, *Basic Carrier Telephony* (3rd ed.), Hayden, Rochelle Park, New Jersey (1977).

Another clear and easy-to-read introduction to the classical analog carrier telephone system at the technician level.

R. F. Rey (ed.), *Engineering and Operations in the Bell System* (2nd ed.), AT&T Bell Laboratories, New Jersey (1983).

This tome by the Bell Labs technical staff contains an enormous amount of rather specialized technical and managerial information about the telephone systems at a surprisingly easy to understand level.

R. L. Freeman, *Telecommunications System Engineering: Analog and Digital Design*, Wiley, New York (1980).

A detailed coverage of the many technical aspects of the telephone system in a relatively nonmathematical manner.

Check Your Understanding of Chapter 4—True or False?

1. Leased telephone lines are nearly always four-wire and bypass the telephone switches.
2. A T1 frame contains exactly $24 \times 8 = 192$ bits per frame.
3. DDS is a packet-switched digital network operating at 56 kbps.
4. Multimode fiber optic channels can have ISI much like that in coaxial cable channels.
5. The 12/14-GHz satellites require larger antennas than the older 4/6-GHz ones.
6. Telephone network signaling can be done either in-band or out-of-band.
7. All T-carrier clock signals are derived from one standard frequency reference.
8. The hybrid forces the local loop into two-wire HDX operation.
9. Since divestiture AT&T no longer handles all toll calls.
10. Electronic and digital telephone switching are essentially the same thing.
11. A DSU on a digital network acts much like a modem on an analog network.
12. The battery voltage on a local loop is normally -48 V on the Ring lead.
13. The standard DDS customer access rates are 2.4, 4.8, 9.6, 56, and 1544 kbps in the U.S.
14. A TSTS switch involves both circuit and message switching.
15. The D channel bank multiplexes and demultiplexes voice and data channels.
16. A superframe contains exactly 288 voice-grade channels.
17. The NASA Space Shuttle operates in geosynchronous orbit to service communication satellites.
18. High quality dial-up lines can be obtained by requesting conditioning on them.
19. Companding is used to reduce the number of bits needed to code digital speech.

20. Transmultiplexers convert between HDX digital and FDX analog telephone signals.
21. Crosstalk is often heard in telephone conversations with your spouse's mother.
22. The standard satellite access methods are FSK, PSK, and QASK.
23. The voice-grade line can often be modeled adequately as an LTI analog filter.
24. Historically, microwave radio links have used frequency modulation at 6 GHz.
25. The acronym POTS stands for the telephone creed: Pay Often for Telephone Service.

II

Point-to-Point Data Links

5

Modems and Dialers

5.1. Basic Data Link Considerations

Over the last two decades there has been a dramatic increase in the amount of data transmitted between computers and remote terminals. Although in some cases digital channels are specially constructed to carry data, it is generally much simpler and more economical to use the existing telephone plant. The "telco" is familiar and reliable, and connects the vast majority of locations where computers and terminals might be located. Consequently, both the volume and the percentage of digital telecommunications traffic is steadily increasing and will soon overshadow the voice traffic. This is one of the main reasons that telephone companies worldwide are converting to digital transmission as fast as is economically feasible.

Digital telecommunications, however, is not as simple as the above discussion might suggest. The telephone plant was originally designed for analog voice and will not carry digital data signals in their natural baseband form. Efficient data transmission requires that the digital signals be converted to passband form that more closely resembles voice so as to fit into the nominal 300–3000-Hz bandwidth of the voice-grade line. As we have seen this A/D and D/A conversion process is done by an essential little device known as a *modem.* There are many variations of modems with a variety of special features and options available, but they all do the same basic job of converting digital baseband data signals into transmitted tones that pass efficiently through the voice telephone network, and then converting the tones back to data at the receiving station. In effect, a modem on each end of an analog channel converts it into a digital channel as shown in Fig. 5.1.

Modems provide a standard interface to the digital equipment to which they connect and use this interface to control the flow of data in an orderly manner. Terminal devices such as computers and data terminals are referred to in data communications convention as the *data terminal equipment* (DTE), while the communications devices such as modems are called *data circuit-terminating equipment* or sometimes just *data communications equipment* (DCE). Hence the interface between them is the *DTE/DCE interface.* Figure 5.1 illustrates a simple four-wire leased data communication link with a simple modem and the basic DTE/DCE interface connections. The *Transmit* (Tx or TxD) and *Receive* (Rx or RxD) *Data* lines carry the digital signals for transmission and reception, while *Data Carrier Detect* (DCD or CD)† indicates the modem is receiving a valid carrier and *Clear to Send* (CTS) tells the DTE that any data sent to the modem will be faithfully transmitted. There are numerous other DTE/DCE interface lines which will be covered in the next chapter, but these four, plus ground, constitute a minimal configuration.

There are many different ways of classifying modems: *stand-alone* modems are separate units with their own housings and power supplies, while *OEM* modems are supplied on printed circuit cards for installation inside other equipment; *direct-connect* modems may be connected directly to the telephone line, while other types must connect through a protective

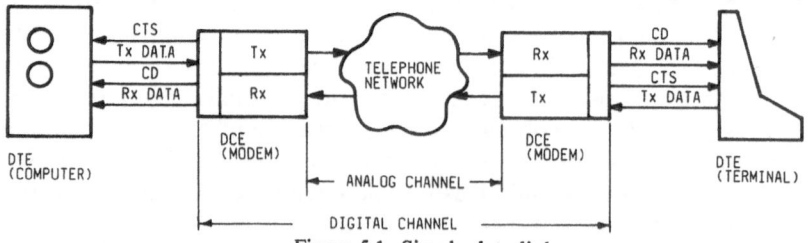

Figure 5.1. Simple data link.

† DCD is usually preferred to CD in actual practice to avoid confusion with the RS-232C interface designation for Data Terminal Ready, which is also CD. DCD also avoids confusion with Carrier Detect for a secondary channel (SCD) when one is used. For simplicity, however, we shall generally use just CD when no confusion will result.

device known as a *data access arrangement* (DAA) to ensure that they cannot damage the telephone system. *Full-duplex* (FDX) modems transmit in both directions simultaneously on a single two-wire line, while *half-duplex* (HDX) devices transmit in only one direction at a time on a two-wire line and require a second two-wire line, i.e., one four-wire line, for full-duplex operation. *Automatic answer* modems are able to detect the ring signal and indicate this condition to the DTE, and *autodial* modems can automatically dial telephone numbers on command from the DTE. *Diagnostic features* are available on modems to indicate poor signal quality and initiate loopback tests for isolating problems with the channel. Finally, modems are frequently classified according to their *speed* (low, medium, or high), *modulation technique* (FSK, PSK, or QASK/QAM), and *bandwidth* (voice-grade or wideband).

Fortunately for users, there are universally accepted specifications for modems to which almost all vendors subscribe. Thus two modems of the same standard type will be compatible, even though they are supplied by different manufacturers. One set of standards is from the CCITT and is widely used throughout most of the world. For modems the relevant CCITT standards are designated by the *V.xx series*, e.g., V.27 for the 4800 bps modem. In the United States, however, there is also a set of de facto Bell standards due to the historical pervasiveness of AT&T. There are also a few vendors with their own unique standards, but the market is usually so limited that this approach is not feasible unless there is a large customer such as the government. We will be concerned only with the CCITT and Bell standards.

This chapter first considers the basic types of modems in use on telecommunications links, including the principles of operation and the procedures (protocols) by which they interface to the telephone lines and to the various DTE devices to which they connect. A detailed understanding of modem types and operation can be invaluable in both the specification and troubleshooting of common data links. Thus, attention to detail here and in the next chapter will be well worth the effort and will also provide a solid foundation for later chapters on data networks, where modems and data links are basic elements. The final section discusses the increasingly important subject of automatic dialers, which are often used integrally with modems to allow a computer to query remote locations to obtain status or data base information.

5.2. Asynchronous FSK Modems

The simplest and, historically, the most widely used telecommunications modems employ FSK modulation. Digital frequency modulation is rather simple and inexpensive to implement, but it is also limited by the modulation method to relatively low data rates. For telephone lines this practical limit is around 1800 bps; however, even this rate is generally an option to the standard rate of 1200 bps and requires conditioned leased lines. FSK modems are also asynchronous, since no clock or carrier synchronization is needed. The FSK modem receiver simply passes the received data to the DTE input circuitry, and it is the responsibility of the DTE to determine what the data means. In this sense, all modems are "dumb"; they just pass bits (and a bit clock if synchronous) to the DTE and make no attempt to interpret these bits.† There are two conventional FSK modem data rates for telephone lines: 1200 and 300 bps. These are also the corresponding baud rates, since the bit and baud rates are equal for FSK.

At 1200 bps the CCITT designation is *V.23* while the corresponding Bell modem is the *202*, but the two are not compatible. Figure 5.2 shows the transmission bandwidths for the 202 and V.23, including an optional *secondary* (or side)‡ channel which can operate anywhere from 5 to 150 Bd. It is also FSK except at 5 Bd, where it is just operated on or off. Secondary channels are used for network control and testing, or sometimes as a feedback link from the destination. The 202 and V.23 mark frequencies are

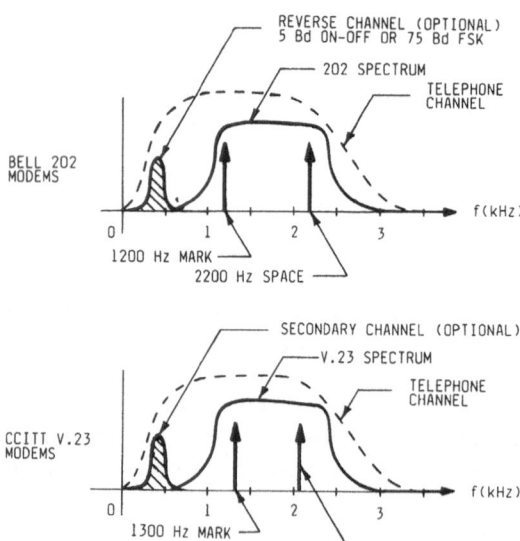

Figure 5.2. 1200 bps FSK modem spectra.

---

† The so-called smart modems for personal computers still just pass bits but also have additional protocol capability for automatic dialing and file transfers, e.g., X.PC.

‡ This is also called a *reverse* channel when used exclusively for return traffic such as acknowledgments.

1200 and 1300 Hz, while the space frequencies are 2200 and 2100 Hz, respectively. The closer spacing of the CCITT frequencies reflects the poor quality of some European telephone lines back when the standard was developed. The effective bandwidth is 1200 Hz in either case, with the main spectrum extending from 1100 to 2300 Hz. These 1200-Bd modems require a full two-wire line to transmit or receive, so full-duplex operation necessitates a four-wire line.

The 300 bps FSK modem, on the other hand, operates FDX on a single two-wire line. This is accomplished by frequency multiplexing the line bandwidth into two separate bands, one for transmitting and the other for receiving. However, like beauty, transmitting and receiving are in the eye of the beholder—the transmitting band on one end is the receiving band on the other, and vice versa! Thus it is necessary to establish a convention for use of the FDM bands. The following is universally used: the modem that *originates* the transmission always transmits on the lower band and receives on the higher, while the reverse is true for the *answering* modem on the other end.† As a consequence there are three basic categories of 300-Bd modems: *originate-only*, which always transmit on the lower band; *answer-only*, which always transmit on the upper band; and the general *answer/originate* modem, which can be set to transmit on either band. For example, a computer center containing a data base might just need answer-only modems since everyone normally would call into the computer, while a computer hobbyist with a home computer might buy an originate-only modem to call his company computer center or one of the commercial data banks. The distinction between originating and answering is shown in Fig. 5.3a.

The Bell standard 300-Bd modem is the *103* (and 113) while the CCITT equivalent is *V.21*, but the two types are again incompatible. In fact, the relative positions of a mark and a space are reversed, as can be seen from the spectra in Fig. 5.3b. For the 103 originating band, the 1270-Hz mark frequency is higher than the 1070-Hz space, but for the V.21 the 980-Hz mark is lower than the 1180-Hz space. A similar relationship holds in the answer band for the 103 mark and space frequencies of 2225 and 2025 Hz, respectively, compared to the V.21 mark at 1650 Hz and space at 1850 Hz. The spacing between the mark and space frequencies in either case is 200 Hz and the bandwidths are always 300 Hz. Since the transmission is full-duplex two-wire, the modem must contain an internal *hybrid* to convert between the two-wire line and the modem transmitter and receiver sections. It must also contain high-quality bandpass filters to separate the two bands; otherwise the receiver will hear its own transmitter as interference to the incoming

† A helpful memory trick for remembering this convention is to recall that "one always starts at the bottom," unless of course, one is digging a well!

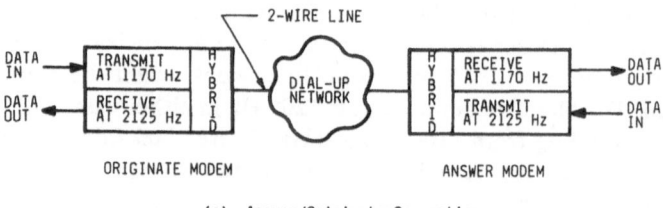

(a) Answer/Originate Convention

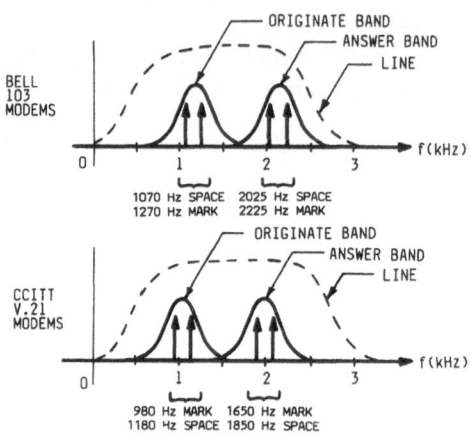

(b) Modem Spectral Band Locations

Figure 5.3. 300 bps FSK modems.

signal, or perhaps even mistake its own transmission for a legitimate received signal.

Since FSK is asynchronous these modems can transmit data at rates below the specified rate, and this is commonly done. However, it is difficult to increase the date rate of FSK above 1800 bps on voice-grade lines due to channel distortion effects on the signals. For example, at 1800 Bd with a 202 modem a 1200-Hz mark tone results in only $1200/1800 = 0.667$ cycle per baud, which is not much time to identify the frequency of the tone in the presence of noise and line distortion. In order to attain higher data rates over voice-grade lines it is necessary to use more complex and efficient modulation techniques which are synchronous. This almost always means either PSK or QASK, which will thus be considered in the next section.

5.3. Synchronous PSK and QASK Modems

In order to achieve higher speeds than with the preceding FSK modulation, it is necessary to increase the number of data bits per line symbol; i.e., we need more than just two different line symbols. In theory this could have been done by using more than two frequencies with FSK, but there are technical problems in implementing such an approach to achieve acceptable performance on voice-grade lines. A more practical approach is to use amplitude and/or phase modulation where the carrier frequency remains constant (on average). By allowing the amplitude and phase combinations to have more than two possible values we can let multiple bits select each distinct line symbol, hence attaining multiple bits per baud and a corresponding increase in data rate. For practical reasons it is desirable to encode an integer number of bits for each baud, which in turn requires that the number of amplitude and/or phase combinations be a power of 2.

With phase encoding it is necessary for the receiver to recover a replica of the carrier that is synchronized in some sense and also a clock signal synchronous with the baud rate which can then be used to generate the bit clock for the DTE. A simple but widely used technique for *baud clock recovery* is to detect the received signal envelope and use this to synchronize a resonant circuit (a narrow bandpass filter) centered at the corresponding frequency. This output is then processed with a PLL for smoothing and used to control a counter that produces the bit-rate clock. The recovered clock is essential for the decoder shift register and is nearly always supplied by the modem to the DTE along with the received data. This approach works because there is a momentary increase in the original signal spectrum bandwidth when the phase is changed (higher frequencies are produced). These frequencies are filtered out by the channel passband, thereby reducing the received signal envelope level at the transitions between baud intervals. In order for this to work properly, of course, there must be continual changes in the transmitted signal phase; long strings of symbols with no changes look like pure tones and have no envelope variation, and such patterns often correspond to the transmitted data. To avoid this situation most synchronous modems randomize the data through a reversible process called *scrambling* before phase encoding so that periodic patterns cannot occur on the line. At the receiver the recovered data is *unscrambled* before delivery to the DTE. Scrambling techniques are discussed further in the Appendix.

Carrier reference recovery is usually done with some type of PLL which locks the received signal to an internal oscillator in the receiver. An intuitive way to understand carrier recovery is to consider the classical squaring loop approach. For simplicity, consider a four-phase PSK signal

$$s(t) = A \cos (\omega_c t + n90°), \qquad n = 0, 1, 2, 3$$

where the phase is either 0°, 90°, 180°, or 270°. If we square $s(t)$ we obtain the sum and difference frequencies. The latter is just a constant dc value, but the sum component is easily shown to be

$$s_{\text{sum}}^2(t) = \tfrac{1}{2} A^2 \cos (2\omega_c t + 2n90°) = \tfrac{1}{2} A^2 \cos (2\omega_c t + n180°)$$

A second similar squaring operation produces a steady fourth harmonic of the carrier frequency without any phase variation:

$$s_{\text{sum}}^2(t) = \tfrac{1}{4} A^4 \cos\left(4\omega_c t + 0°\right)$$

This result can easily be divided down to obtain a stable sinusoid at the carrier frequency. There is, however, a problem with this approach. Although the carrier frequency is obtained exactly, the phase is referenced to only one of the allowable phases but the receiver does not know which. It depends on where the divider begins, which is arbitrary to within the allowable phases. The universal way of resolving this problem in modems is to use *differential phase encoding*, with the corresponding modulation technique called *differential PSK* (DPSK). Here the input data bits to the transmitter are used to select a *change* in carrier phase rather than the absolute phase; hence all information is contained in the phase *difference*. It is not necessary for the receiver to ever know the absolute carrier phase. In fact, the current line symbol can be delayed by one baud at the receiver and then used as a "pseudocarrier" to demodulate the next baud, thereby eliminating the need to recover any carrier at all.

There are three common data rates that use pure PSK: namely, 1200, 2400, and 4800 bps, and four that use QASK: 2400, 4800, 9600, and 14,400 bps. The old workhorse of the synchronous telecommunications modems is the 2400 bps Bell *201* and the CCITT equivalent *V.26*. This is a four-phase HDX two-wire modem that uses DPSK modulation to encode two bits into each baud, using Gray coding† for the phases to ensure that an adjacent phase error (the most common case) results in only one bit error. Figure 5.4a shows the signal spectrum along with the two alternative phase encoding methods. An optional 75-Bd FSK secondary channel can be located below the main channel bandwidth. Note that two differential phase conventions are shown. Since these modems do not employ scramblers, with alternative *A* a long pattern of data zeros results in no phase changes, i.e., a pure carrier. This will result in a loss of clock synchronization. Thus it is conventional to add the additional 45° phase shift (or precession) as in alternative *B* to ensure at least a minimal envelope for clock recovery.

The second type of PSK modem is the Bell *208* and CCITT *V.27*, which operate at 4800 bps. Differential eight-phase PSK is used for HDX two-wire or FDX four-wire operation at 1600 Bd. Three-bit Gray coding is used, so there are three bits per baud. Figure 5.4b shows this phase encoding and the corresponding spectrum for both the 208 and V.27. These modems contain scramblers, so no precession is needed. An additional element that

---

† The Gray code has the specific property that adjacent codes differ by only one bit; e.g., for eight symbols the codes are: 000, 001, 011, 010, 110, 111, 101, and 100. It is widely used for all types of rotational encoding.

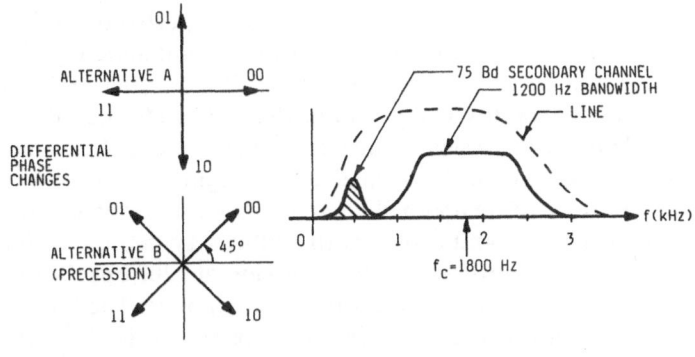

(a)   201 and V.26 Signal Phases and Spectrum

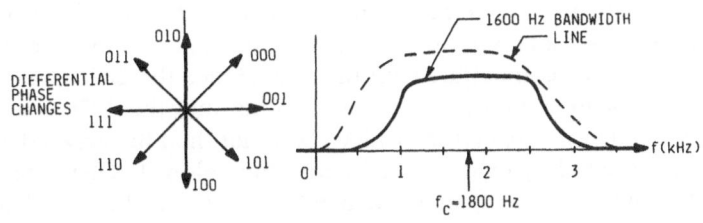

(b)   208 and V.27 Signal Phases and Spectrum

Figure 5.4. 2400 and 4800 bps HDX PSK modems.

is needed for reliable data transmission at 4800 bps (1600 Bd) is an adaptive equalizer. Since a phase error of only 22.5° will cause an error in the received data, it is important to carefully equalize the line group delay. Because of the considerable variation among lines, it is inadequate to use fixed (compromise) equalization based on a nominal group delay. This is particularly true of dial-up operation.

An *adaptive*, or *automatic, equalizer* adjusts itself to each particular line used and then tracks small line variations during data transmission. It is initially set by sending a known *training pattern* before the data begins, which allows these transversal digital filters to adjust their tap coefficients so as to minimize the received signal error. A scrambler is also required for proper equalizer operation, and scrambling is used on all modern synchronous modems (except the Bell 201). The theory and implementation of adaptive equalizers is given in the Appendix.

The third type of PSK modem operates at only 1200 bps, but like the 103 it operates FDX two-wire by frequency multiplexing the line into two bands. The Bell *212* and CCITT *V.22* use four-phase DPSK in both the originate and answer bands, the former always being lower in frequency.

These bands are centered at 1200 and 2400 Hz, respectively, and are wider than the 103 bands, so filtering is more critical. Automatic equalization is not necessary, but scramblers are used to facilitate synchronization. The 212 also has the capability to operate in the 103 mode; i.e., it contains a complete 300 bps FSK modem which can be selected in lieu of 1200 bps. In fact, the 212 receiver will automatically recognize the type of signal it is receiving (FSK or PSK) and will answer in the appropriate manner. The 212 also contains an integral converter that allows asynchronous input data to be converted to synchronous for transmission at 1200 bps. With all these features the 212 is widely used with personal computers. The PSK spectrum and differential phase encoding altenatives are shown in Fig. 5.5a, along with the convention for originating and answering in part (b).

QASK modems are somewhat less standardized than PSK, but at 9600 bps there is the CCITT *V.29* standard, which is widely used even in the U.S. Historically the standard Bell modem at this speed was the *209*, but in recent years a new series has been introduced that includes the Bell 2096, which is not compatible with the 209. Hence the 209 never became a serious de facto standard. It is worth noting that the 2096 spectrum is shifted up so that it is centered at 1800 Hz rather than the original 1650 Hz carrier, so a secondary channel can be inserted below it. There are now 16 different choices of line symbols, hence there are four bits per baud and the data rate is four times the baud rate of 2400 Bd. Figure 5.6a shows the 16 differential phase/amplitude combinations of the 209 and V.29. The corresponding 209 and 2096 spectra are similar but centered at different carrier frequencies. The 209 signal vectors are seen to be arranged in a square pattern of four rows and columns. This particular pattern is a special case of QASK, which is formally called *quadrature amplitude modulation* (QAM), although QAM is often used interchangeably with QASK in practice. In addition to scrambling, AGC, and automatic equalization, it is also common practice to include a multiplexing (split-stream) capability with modems of 9600 bps and higher. This practice allows several of the more common lower-speed inputs to be combined over a single line.

In recent years most major modem vendors have developed higher-speed modems, primarily by using 64 phase/amplitude combinations at 2400 Bd to get 14,400 bps. There are also a few very complex modems that can operate up to 19,200 bps using sophisticated trellis forward-error-connection coding. Additional effective data rates can also be obtained with data compression coding, but this is not normally considered a modem function. There is also the CCITT *V.32* standard for full-duplex two-wire operation up to 9600 bps using a technique called *echo canceling*, but these devices are still relatively new in the United States.

In the other direction, there have been a few FDX two-wire modems operating at 2400 bps for a number of years, and the CCITT has

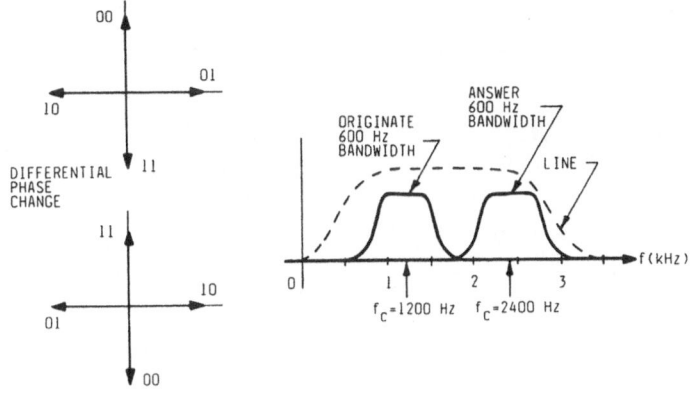

(a)   212 and V.22 Signal Phases and Spectrum

(b)   Answer/Originate Conventions

Figure 5.5. 1200 bps FDX PSK modems.

standardized this type of QASK modem as *V.22bis.*† The comparable U.S. designation is the *224.* These modems use a 16-point set of line signal vectors similar to the 209, including Gray coding and differential phase encoding. They also require adaptive equalization and scrambling and, like the 9600 bps modems, must have an *AGC* circuit to ensure that the multiple amplitudes can be accurately identified. The two FDM bands are centered at 1200 and 2400 Hz just like the 212/V.22, and the 224/V.22bis can fall back to these 1200 bps modes if necessary. Of course, the answer and originate conventions are the same as for the preceding FDX modems. Sometimes 2400 bps FDX modems also contain integral asynchronous-to-synchronous conversion, and may provide various combinations of 212, V.22, 103, or V.21 compatible operation at lower speeds. Figure 5.6b shows the transmit and receive spectra locations on the voice-grade channel.

Table 5.1 summarizes the various voice band modems that we have considered in this section, including a *rough* indication of the common hardware/software implementation and an *even rougher* estimate of the

---

† The CCITT uses the designation *bis* for the second (b) version of its standards, and *ter* for the third (c) version.

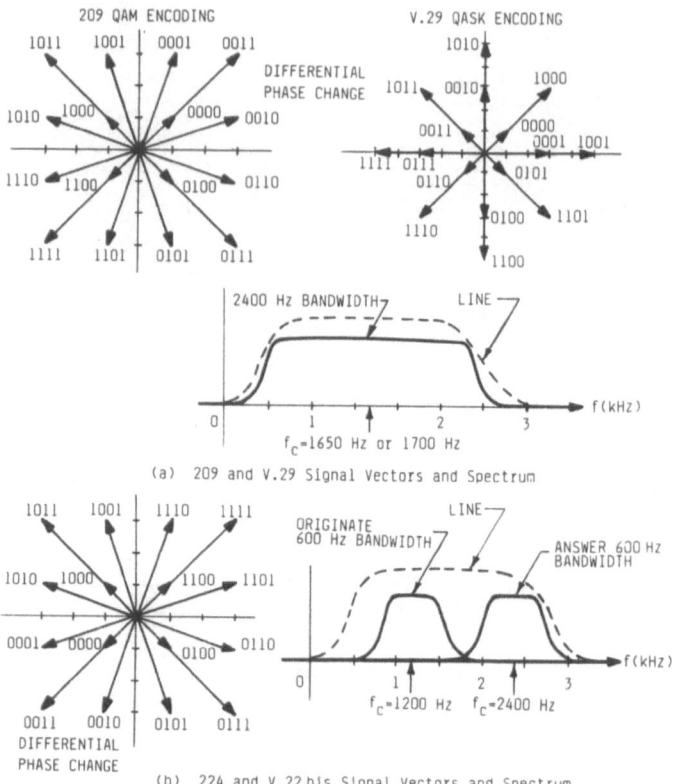

Figure 5.6. 9600 HDX and 2400 FDX modems.

representative prices circa 1985. These different standard modems are used for essentially all of the voice-grade data telecommunications throughout the world, with the notable exception of some government and military organizations. Furthermore, once there are strong standards as there are in modems, then it is very difficult to introduce successfully new nonstandard products. They will be incompatible with all other products and there will be no alternate supplier and little price competition. Thus it is reasonable to expect few changes in the future in these basic types of voice-grade modems.

There are also a variety of *wideband modems* which are worth mentioning briefly. The CCITT *V.35* standard is for sending data at 48 or 40.8 kbps over an analog channel group band (60–108 kHz) and is widely used where RS-232C is too slow. It is also used for DDS operation at 56 kbps. The related *V.36* recommendation also defines a group band modem, but single-sideband (SSB) modulation is used. Data rates of 48, 56, 64, or 72 kbps

Table 5.1. Modem Summary

	103 V.21	202 V.23	212 V.22	201 V.26	224 V.22bis	208 V.27	209 V.29	— V.32
Bell type CCITT type								
Bit rate	300 300	1200	1200 1200	2400	2400 2400	4800	9600	9600 9600
Synchronization	ASYNC	ASYNC	SYNC ASYNC	SYNC	SYNC ASYNC	SYNC	SYNC	SYNC
Modulation	FSK	FSK	PSK	PSK	QASK	PSK QASK	QASK	QASK
Line type operation	2W/FDX	2W/HDX 4W/FDX	2W/FDX	2W/HDX 4W/FDX	2W/FDX	2W/HDX 4W/FDX	2W/HDX 4W/FDX	2W/FDX
Design technology	VLSI Chip	VLSI Chip	MSI/LSI Chip set	MSI/LSI Chip set	μP/LSI	μP/LSI Chip set	μP/LSI Chip set	μP/VLSI
Nominal cost	$100	$200	$300	$400	$800	$1400	$1800	$3500

are available depending on the particular application. There are other Bell wideband modems, of which the 303 series is the most common. Several models of this modem exist, with data rates from 19.2 to 460.8 kbps. For example, the 303C sends 50 kbps over a channel group band and can be extended for 56 kbps DDS operation.

A final category of modems is the so-called short-haul or *limited distance modem* (LDM), which is designed for local operation over telephone links that do not involve the carrier systems and its associated band-limiting filtering. They are much simpler (and cheaper) than long-haul modems, and may operate at baseband over leased lines. Rates as high as 56 kbps are quite feasible, but the data rate is highly dependent on the distance and the modulation complexity. Typical tradeoffs are from around 5 miles at 19,200 bps to 20 miles at 2400 bps. There is essentially no standardization of LDMs and thus little or no compatibility among vendors. However, they are an increasingly important aspect of the overall modem and local area network markets.

Having studied the telephone channel in the last chapter and modems in the last two sections of this chapter, we will turn in the next section to the important question of how the modem is connected to the telephone line. Connection to the other side of the modem (the DTE/DCE interface) will be the subject of the next chapter.

5.4. The Modem Line Connection

The connection of modems and other data communication devices to the voice-grade line is not the trivial procedure that it might at first seem. There are rather strict requirements for such connections by the telephone company and the Federal Communications Commission (FCC). To appreciate these requirements, it is worthwhile to review briefly the history of the interconnect industry. Prior to the FCC Carterfone decision in 1968, AT&T claimed that no private† devices could be connected to the telephone system without its permission, and it often enforced this claim by the threat or actual disconnection of customer service. This policy was, of course, just the point of the Carterfone case, and the outcome was its reversal. However, as part of the decision there was the provision that a protective device would be required in order to ensure that private equipment could not cause damage or malfunction of the telephone plant. Such a device is called a *data access arrangement* (DAA), or occasionally a data coupler. The basic functions of all DAAs are to prevent excessive signal levels from reaching

---

† AT&T calls any equipment that is not provided by itself *foreign* equipment, which corresponds to the use of *private* equipment herein.

the local telephone switching office, to ensure that the transmitted signals have the proper in-band power levels and are attenuated sufficiently out-of-band, to protect from voltage surges such as lightning, and to provide electrical isolation of the private equipment. Some DAAs also have an integral 2800-Hz test tone generator, and certain types provide automatic ring detection (answering) and hook status control.

In 1976 the FCC began the *direct-connect registration* program that allows private equipment to be connected directly to the telephone system if it complies with the FCC requirements for direct-connect. This essentially means that the equipment must pass certain tests set forth by the FCC to ensure that it cannot damage the telephone system. Once these test results are approved, the device is assigned a *registration number* by the FCC and can then be connected without a DAA. Almost all modems manufactured today are direct-connect, since this eliminates the cost of the DAA, which runs from $100 up. In fact, AT&T no longer even furnishes DAAs.

Figure 5.7 shows the various alternatives for connecting modems to voice-grade lines. Note that telephone company furnished equipment never requires a DAA, and neither does equipment for leased lines since no switching equipment is involved. Also, *acoustically coupled* modems contain a microphone and speaker and are connected through the telephone handset itself, so the telephone instrument provides the necessary protection. This type of modem is particularly convenient for portable equipment like lap terminals, since it can be connected over any telephone available, including pay phones. Despite the fact that most new modems are direct-connect, there are still a great many older unregistered ones in use which require a

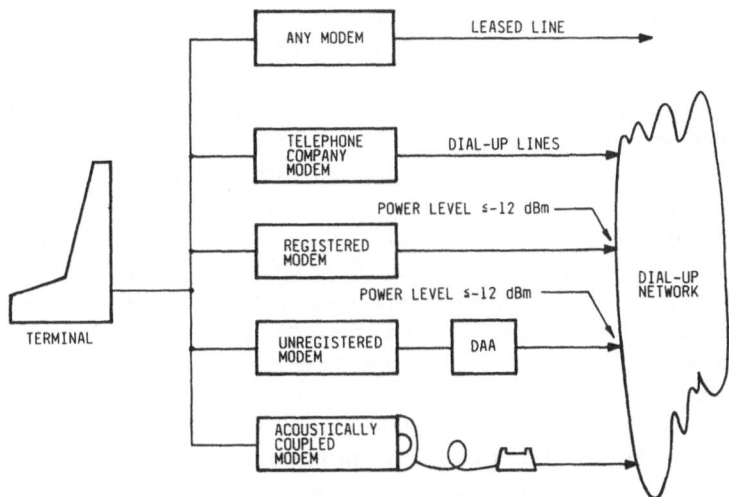

Figure 5.7. Modem line connection alternatives.

DAA. Furthermore, many DAA functions such as impedance matching, lightning protection, and automatic ring detection are now incorporated into direct-connect modem designs. Finally, DAAs are convenient for electronic designers and home computer hobbyists who are developing prototype modems in small quantities for which the expensive registration process is impractical. Thus an understanding of DAA operation is essential for anyone seriously involved in practical computer data communications. Consequently, this section will introduce the three basic types of DAAs and consider some of the more common applications.

The simplest DAA is the Bell *CDT*, or 1000A, which is powered directly from the phone line battery voltage. It basically provides signal limiting and isolation, and perhaps the test tone. Dialing and answering are done by a separate telephone having an exclusion key which requires the user to switch manually from *talk mode* to *data mode*, i.e., to manually transfer control of the line from the telephone to the DAA. This situation is shown in Fig. 5.8a, where lines *Tip* (T) and *Ring* (R) are the phone line, and *Data Tip* (DT) and *Data Ring* (DR) go to the modem. The switchook and exclusion key are manually operated; hence an operator must be present at the modem location to transfer the line from the telephone to the modem. When the DAA is first installed, the modem transmitter power level is adjusted with a programming resistor at the DAA so that at the central office it is at least −12 Bm.† If this adjustment procedure is not feasible, the *permissive* setting of −9 dBm can be employed, since the local loop and DAA will always have a least 3 dB of loss, thus ensuring the required −12 dBm at the switch. Of course, if the line has more than 3 dB loss, then the transmission level and SNR are weaker than they would otherwise be. The basic operational procedure for using the CDT DAA is to dial the number, listen for the answer-back tone from the other end, then raise the exclusion key and begin data transmission. To terminate the call it only is necessary to hang up the telephone.

There are also two other types of DAA, both of which provide *automatic answering* features and may also operate with an automatic dialer. They differ mainly in the way they are powered and the type of interface they require. The *CBS*, or 1001A or 1001F, is powered from a standard 120-V ac source, and uses the common EIA RS-232C interface voltage levels. The *CBT*, or 1001B or 1001D, takes power from the user equipment, e.g., the modem, in the form of 24 V dc. The interface signals are for current driven contact-closures on mechanical devices such as teletypes, and hence are slow and noisy.

Both the CBS and CBT contain filters to detect the 20-Hz ring tone and to pass the signal on to the modem and terminal (DTE). This is done

---

† *dBm* indicates dB relative to one milliwatt; i.e., $10 \log(P/0.001)$, where the signal power P is in watts.

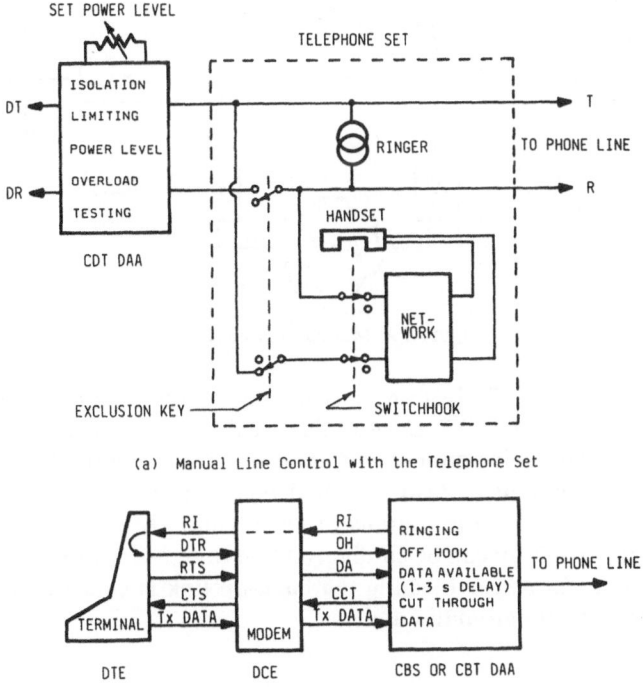

(a) Manual Line Control with the Telephone Set

(b) Automatic Answering Procedure

Figure 5.8. DAA connection.

by raising the *Ring Indicator* (RI) line to the modem, which passes it on to the DTE (terminal). If the terminal wants to answer the call, it sends the signal *Data Terminal Ready* (DTR) to the modem, which in turn tells the DAA to go *Off-Hook* (OH) and that *Data is Available* (DA) for transmission. Operationally, OH corresponds to manually lifting a telephone handset, and DA shifts to data mode. After a few seconds of delay for the telephone billing equipment to operate, the DAA transfers (connects) the line to the modem and sends a signal indicating *Coupler Cut-Through* (CCT), after which data transmission (TxDATA) can begin normally. To terminate the call the modem lowers OH, indicating that the DAA should hang up, which it then does. Note that no telephone instrument is required. The entire sequence is illustrated in the order of occurrence by Fig. 5.8b. This is the basic automatic answering protocol. Automatic answering direct-connect modems perform the same functions internally. In the next section we shall consider the protocol for automatic dialing.

An important application of the DAA is for dial back-up of leased line links. When the main leased line fails, the data transmission can continue, often at a lower rate, over a dial-up link that is then set up by the operators

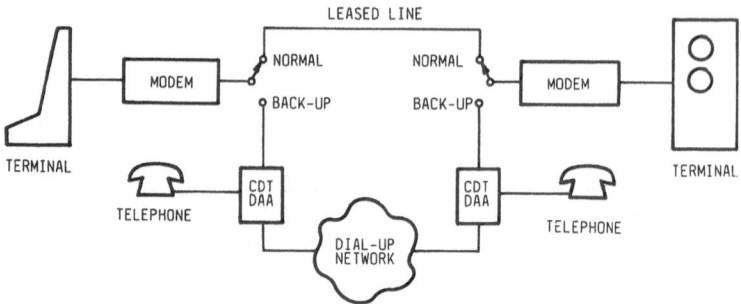

Figure 5.9. Manual dial back-up.

at either end. Figure 5.9 illustrates a typical dial back-up arrangement using manual dialing with CDT DAAs. The back-up line is set up between operators like any voice connection, then the line is manually switched to the DAA. The modem is also switched manually to the DAA. By using automatic dialers with the automatic answering CBS or CBT DAAs, an automatic dial back-up can be configured that does not require any operator intervention as far as establishing the back-up link is concerned. We shall now consider such automatic dialers.

5.5. Automatic Dialers

There are many applications for automatic dialing by office machines and computers today, both for digital and analog transmission. An emerging telemarketing technique, for example, is to automatically dial a presupplied list of prospective consumers and then play a prerecorded advertising message to each. However, we shall only be concerned here with data communications applications, of which there are many. The dial back-up example in the preceding section can be automated by replacing the manual telephone with an automatic dialer that is controlled by the DTE, thus eliminating the need for human operators to be present in the event of line failure. Other common applications include the periodic updating of central data bases from remote locations, the polling of various types of alarm systems that must be frequently interrogated, and the automatic selection of various destinations for message delivery such as the delivery of electronic mail to a distribution list at predetermined times. There is a variety of different approaches and configurations for using automatic dialers in data networks, but before considering them we will review the sequence of events involved in any dial-up connection.

To successfully dial a telephone connection we first go *off-hook* (OH). Then we must wait for and detect a dial tone, since any digits dialed before the dial tone arrives are ignored by the central office. Next *each digit must be dialed* in sequence, with the proper frequencies for touch-tone, or the proper pulse format for pulse dialing. If there is a *busy tone,* or if there is *no answer* for an excessive period of time, then we should hang up and perhaps try again later. If the other end does answer, then we must detect and verify the appropriate response, e.g., an answer-back tone for modems, and then begin information transfer. This involves *transferring control of the line* to the appropriate user DTE. Finally, we must detect the end of transmission and hang up on both ends. With an automatic dialer, or *automatic calling unit* (ACU), all of these operations must be done automatically and provision should be made for recovery from error conditions should an error occur. Clearly, the ACU operation is closely related to the accompanying modem, and dialers are often supplied integral to auto-dialing modems.

There are several basic ACU classifications. Obviously, they may provide either touch-tone (DTMF) or pulse dialing. They may also have a fixed internal list of numbers, or the numbers may be supplied through the interface. They are commonly connected to two-wire dial-up lines, operating either FDX or HDX, but remotely located ACUs may be accessed via leased or multiplexed lines. For example, a leased long-distance line might be used at night to access a remote dialer for updating a data base from numerous terminals located in the same city as the remote dialer. However, the most fundamental classification from the user's viewpoint is based on the type of port used to supply the digits of the desired number. The *parallel port* ACU accepts each digit as a four-bit BCD-coded number, and is well standardized throughout the world. The alternative *serial port* ACU is not well standardized and varies with each different vendor. Most, however, use ASCII control characters for the various functions required in dialing, and send the digits as numerical characters having from four to seven bits each. We shall now consider the details of each method, with the latter illustrated by a viable example.

Parallel port dialers are defined by several standards, all of which are essentially compatible. The EIA RS-366, CCITT V.25, and Bell 801 standards differ in nomenclature and in the amount of detail they include, but we shall not attempt to make such distinctions and will focus our discussion on the RS-366 version. Telephone numbers to be dialed can either be stored in some versions of this type of ACU and then addressed later by the DTE, or the numbers may be passed one at a time across the interface in BCD-coded form. Figure 5.10a shows the latter version, which will be described here. At the interface the DTE initiates an automatically dialed call by raising the control line *Call Request* (CRQ), which causes the dialer to go OH and wait to detect a dial tone (typically simultaneous tones of 350 and 440 Hz). When the dial tone is detected the ACU raises *Present Next Digit*

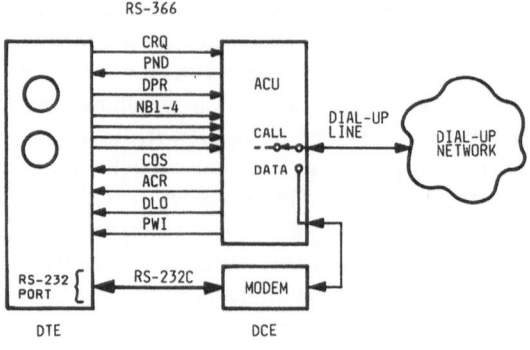

(a)   RS-366 DTE-ACU Interface

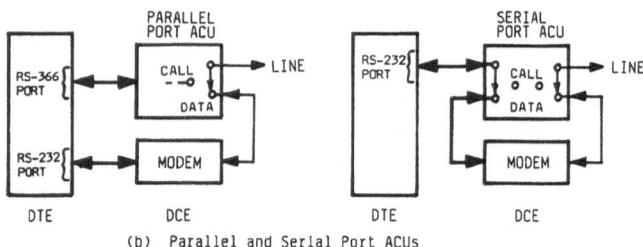

(b)   Parallel and Serial Port ACUs

Figure 5.10. Automatic calling units.

(PND), which indicates that it is ready to accept the next BCD digit over the four parallel *Number Bit* lines NB1, NB2, NB4, and NB8. The DTE then puts the next digit on the lines and raises *Digit Present* (DPR). Once the digit has been accepted, the ACU drops PND, the DTE drops DPR, and this handshaking sequence repeats for each successive digit.

Since phone numbers can vary in length from three to around 20 digits, it is necessary to indicate to the ACU when the number is complete. This can be done in two ways. The straightforward method is to send an *End of Numbers* (EON) code across the digits interface. It is defined to be 1100 binary, or 12 in decimal. Alternatively, the ACU may time out while waiting for another digit, in which case it will assume there are no more digits to follow. It then dials the number. With EON the line is immediately transferred to the modem. Otherwise the ACU dials the number and then keeps control of the line until an answer-back tone is detected from the other end. This ensures that the dialing computer or terminal is not connected to an incorrect or inoperative number. Only after this last step is completed does the ACU transfer control of the line to the modem, after which it raises *Call Originate Status* (COS) back to the DTE to indicate that the dialing is completed and the modem is connected to the line. Now the usual modem initialization procedures take place without regard for the ACU, and data transmission begins.

If, after dialing the number, one receives either no answer or a busy signal (typically 60 or 120 Hz pulsed on and off), then the ACU raises the *Abandon Call and Retry* (ACR) line causing the DTE to drop CRQ and initiate whatever contingency procedure for which it has been programmed. The DTE can also abort any call in progress by dropping CRQ, although once the line has been transferred to the modem hanging up is usually done through the modem. There are also signal lines for *Data Line Occupied* (DLO), indicating the line is not idle, and *Power* (PWI), indicating power is supplied to the ACU for general use by the DTE if desired. The above procedure is the typical operating sequence for an RS-336 (V.25, Bell 801) ACU, but of course there are many subtle variations and parameter options. These units can also be operated in conjunction with PBXs in a straightforward manner. Most ACUs also contain automatic answering capabilities.

The *serial port* dialers are not standardized, although most vendors follow a similar procedure [1]. Since these devices receive control and digit information serially, typically as ASCII characters, they can be operated from remote locations very easily. Serial dialers can operate over the same RS-232C interface as the modem, thus saving the cost of a separate and rather complex RS-366 interface for the DTE. They also allow much simpler DTE equipment to be used for automatic calling, such as microcomputers with at least one serial port or intelligent multiplexers. Such arrangements can greatly facilitate remote applications. The serial ACU is illustrated in Fig. 5.10b.

The general serial procedure may be understood by considering a typical example of a serial dialer procedure, i.e., protocol. The DTE initiates the dialing by sending a block of ASCII control characters that uniquely indicates that a phone number is to follow, plus the modem address if necessary. Then any additional control characters are sent followed by the number digits as conventional ASCII characters, e.g., 011 0101 for the digit 5. These are then followed by a character for EON such as ASCII " < " which is 011 1100 and thus a logical extention of the previous BDC EON code 12. The ACU then accepts and buffers the entire phone number and locks the modem out of the shared interface and line while placing the call. It next takes control of the line (goes OH), then listens for the dial tone, dials the number, and waits for the answer-back tone before transferring the line and the interface to the modem. Provisions for busy, no answer, and aborting calls already in progress can be handled in a variety of ways, but a common method is to basically mimic the corresponding RS-366 controls with selected ASCII characters. Overall control of the dialer by the DTE may be effected with the RS-232C *Request to Send* (RST) control, with the dialer being disabled whenever RTS is turned off. This can, for example, be used to abort a call. Control can also be effected with conventional ASCII control characters such as *Start of Text* (STX) and *End of Transmission* (EOT). It is also desirable for the modem to be able to indicate its status to the ACU which is easily done with the conventional RS-232C data mode indication line *Data Set Ready* (DSR).

Besides conserving DTE ports, the main operational advantage of serial dialers lies in their ability to be located remotely from the DTE. They can be downloaded and controlled with serial ASCII characters, and can be co-located with rather simple remote devices like modems, multiplexers, line selectors, or PBXs. The big disadvantage, of course, is the lack of standardization, which may lock the user into a single vendor for hardware, and perhaps also for the associated communication software. On the other hand, parallel dialers are well standardized and widely available from many vendors. They are more complex and expensive, and require a special DTE port configured for RS-366 (although RS-232C conversions are possible). As with most data communications decisions, the best choice depends on the circumstances of each individual situation.

In this chapter we have considered in detail the basic building block of most data communications channels, along with some associated equipment. A good understanding of the modem will prove fundamental to much of the later material. For the point-to-point data links which today constitute the bulk of existing data communications systems, the modem is both the key communications component and the starting point for most testing and troubleshooting. Thus a good modem background can greatly reduce the time, cost, and frustration normally associated with line failures. In the next chapter we shall extend our study of modems to the physical interface between the modem and the DTE.

Appendix: Adaptive Digital Equalizers and Scramblers

Adaptive equalization was one of the most important early applications of digital signal processing theory and hardware, and was the key development leading to high-speed modems for voice-grade lines. The basic idea of equalization is to correct for the ISI caused by amplitude and phase (group delay) distortion of analog lines. Although there is considerable variation in these parameters, particularly with dial-up lines, it can generally be satisfactorily mitigated at low data rates with fixed compromise equalizing filters. However, at data rates above 2400 bps the margin against error is increasingly smaller, and each line must be equalized individually. This is done efficiently in practice with a digital filter in which the coefficients are adjusted iteratively according to an optimal algorithm. A number of such algorithms and filter structures have been proposed, and used, over the past two decades or so. However, the most widely used by far is the *transversal filter structure* of Fig. 3.23a, with the tap-gains adjusted to minimize the *mean-squared error* (MSE) between the actual filter output and the corresponding decision value. We shall confine our presentation here to this fundamental version. Figure 5.11 illustrates the basic modem blocks with an adaptive equalizer, including the equalizer *input and output signal samples* $x(n)$ and $y(n)$, respectively. The *error sample* $e(n)$ is calculated by simply subtracting the *decision* $\hat{a}(n)$ from $y(n)$, and is then used to drive the tap adjustment algorithm. This is appropriately called *decision-directed* operation.

To develop the mathematical algorithm it will be convenient to use vector notation. For simplicity we shall assume only one sample per baud, although it is common practice to use more than one. If the filter has N taps, then at the nth sampling time the samples at each tap will be the set $\{x(n), x(n-1), \ldots, x(n-N+1)\}$. These numerical values are just the contents of the filter delay line, and the corresponding variable tap gains are the set $\{c_k(n), k = 0, 1, \ldots, N-1\}$. In column vector form we define

$$\mathbf{x}(n) = (x(n), x(n-1), \ldots, x(n-N+1))^T$$
$$\mathbf{c}(n) = (c_0(n), c_1(n), \ldots, c_{N-1}(n))^T$$

with the superscript T denoting the vector transpose. The filter output $y(n)$ can then be consisely written as

$$y(n) = \mathbf{c}^T(n)\mathbf{x}(n) = \sum_{k=0}^{N-1} c_k(n)x(n-k)$$

We now define the MSE criterion, or *cost function*, as the statistical expected value† (E) of the (nonnegative) square of the error $e(n)$ between the equalizer output $y(n)$ and the actual transmitted symbol $a(n)$, thus

$$J(\mathbf{c}(n)) = E[e^2(n)] = E[y(n) - a(n)]^2$$

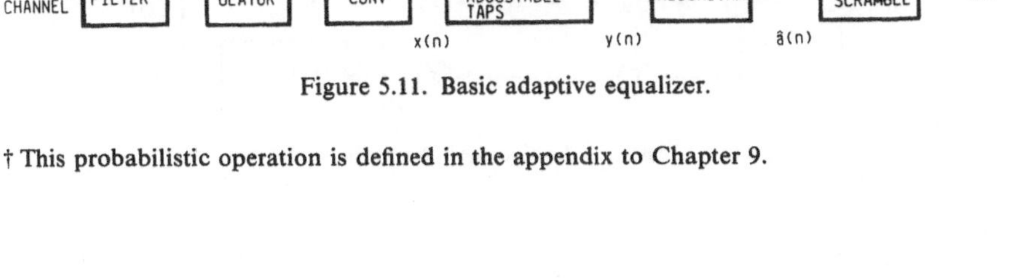

Figure 5.11. Basic adaptive equalizer.

† This probabilistic operation is defined in the appendix to Chapter 9.

For practical channels the function $J(\cdot)$ is strictly convex, which means it has a unique minimum value that can be determined by simply setting the vector derivative, or gradient ∇, with respect to $c(n)$ to zero,

$$\tfrac{1}{2}\nabla J(c(n)) = E[x(n)x^T(n)]c(n) - E[x(n)a(n)] = 0$$

Defining the $N \times N$ matrix $A = E[x(n)x^T(n)]$ and the N-vector $b = E[x(n)a(n)]$, the optimal solution results from solving the classical Wiener-Hopf equation, which in our case with A invertible is simply

$$c_{opt} = A^{-1}b$$

In practice the matrix inversion is usually computationally inefficient (but not impossible!). The more common procedure is to use the *steepest descent algorithm* to iteratively find c_{opt},

$$c(n+1) = c(n) - \tfrac{1}{2}\gamma\nabla J(c(n)) = c(n) - \gamma E[x(n)e(n)] = c(n) + \Delta c(n)$$

For convergence the scalar γ must be small enough to ensure stability, and it is necessary to estimate the vector $E[x(n)e(n)]$. A common solution [2] is simply to use the argument, which is clearly unbiased, resulting in

$$c(n+1) = c(n) - \gamma x(n)e(n)$$

Figure 5.12 shows the equalizer implementation of this algorithm. Once the filter has properly converged the error rate will be very small and the decisions $\hat{a}(n)$ can be used to estimate the true values of $a(n)$ in the calculation of $e(n)$. To obtain initial convergence, a known training sequence

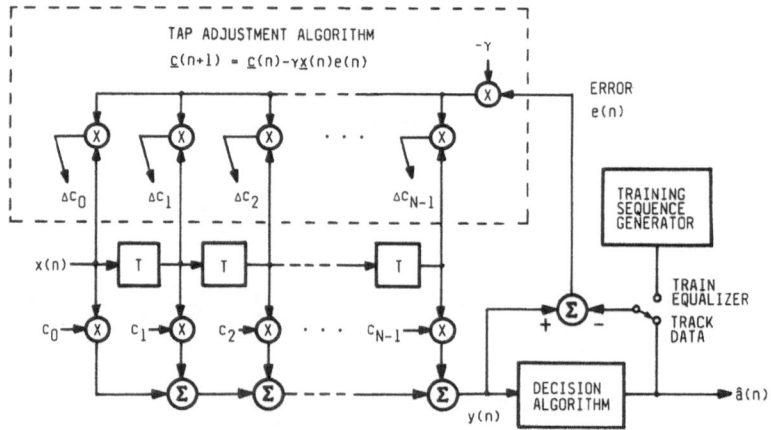

Figure 5.12. MSE adaptive digital equalizer structure.

is sent and an exact replica stored in the receiver is used for $a(n)$, as shown in the figure. Intuitively, the equalizer correlates the error $e(n)$ with the input sample $x(n - k)$ for each tap, and iteratively adjusts the tap gain $c_k(n)$ so as to always reduce this correlation. In the frequency domain the equalizer tries to adjust itself so that overall transmitter-channel-receiver frequency response is approximately flat in magnitude and has constant group delay and hence attempts to produce the Nyquist shaping discussed in Chapter 3.

The preceding adaptive equalizer can be extended in a straightforward manner to the *two-dimensional*, or complex, form necessary to equalize the in-phase (p) and quadrature (q) components of PSK and QASK signals. The result is a set of updating equations for tap gains $c_k(n)$ and $d_k(n)$ that are "cross-coupled,"

$$c_k(n + 1) = c_k(n) - \gamma[x_p(n - k)e_p(n) + x_q(n - k)e_q(n)]$$
$$d_k(n + 1) = d_k(n) - \gamma[x_p(n - k)e_q(n) - x_q(n - k)e_p(n)],$$

$$k = 0, 1, \ldots, N - 1$$

The resulting equalizer is shown in Fig. 5.13.

In order for an adaptive equalizer to operate properly, it should see a random signal pattern without any periodic characteristics. Otherwise, the signal spectrum will have harmonic peaks with nulls between which the

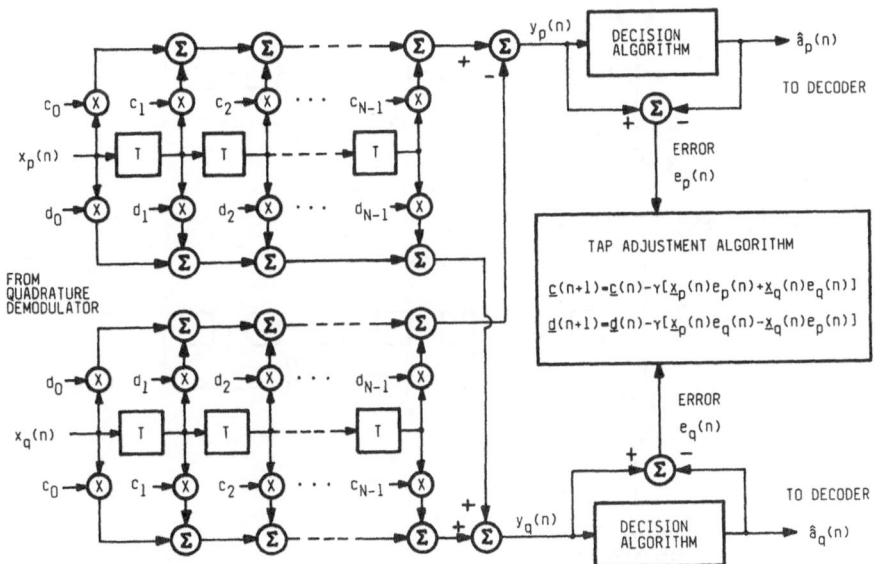

Figure 5.13. Two-dimensional MSE adaptive equalizer.

filter cannot flatten the spectrum, and it will attempt to equalize the signal as well as the channel. Randomness of the received signal is also desirable for clock synchronization, since otherwise the clock recovery circuit could track data harmonics. To ensure this randomness without placing any constraints on the user data, a digital *scrambler* is used in the modem transmitter before modulation and a *descrambler* in the receiver after demodulation and equalization. These are simple binary shift registers with appropriately chosen feedback taps that use modulo 2 (exclusive-or) addition. Figure 5.14 shows the typical seven-stage scrambler and descrambler as used in V.27 modems. The essential function of a scrambler is to ensure that periodic patterns do not occur in the binary output stream, regardless of the input data. Very long periodic output patterns are always possible, although improbable, but are of little concern since they contain very low harmonics with little potential for distortion. The other possibility is an overall system block diagram of Fig. 5.15 gives the sequential relationship among the various modem components that have been considered separately in this chapter.

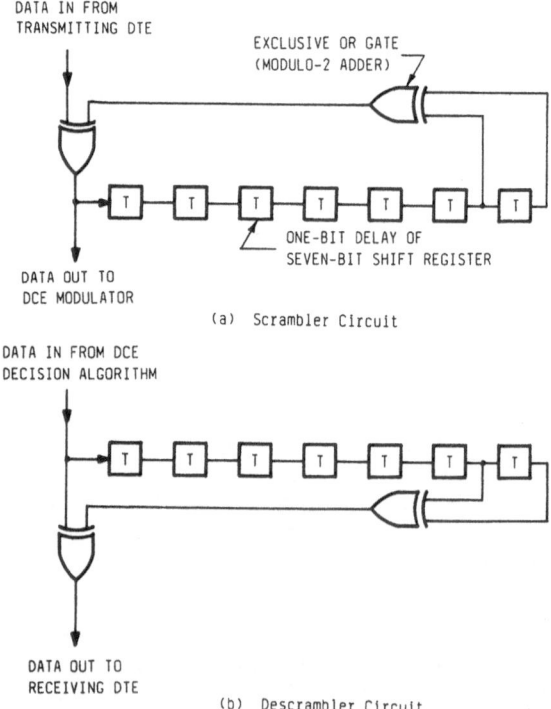

Figure 5.14. Data scrambler/descrambler pair.

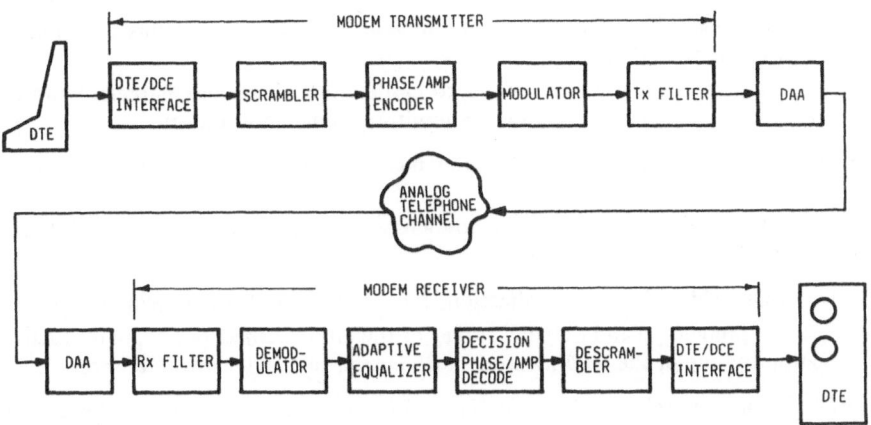

Figure 5.15. Complete modem data link.

Modern scramblers are also *self-synchronous*. This means that the descrambler will acquire or recover synchronization with the corresponding scrambler without any outside intervention. Initially this requires that a few bits of the training pattern be allocated to setting the descrambler. Once it is set, any erroneous bit will propagate through the scrambler with a propagation time of just the register length. In Fig. 5.14, for example, an isolated bit error will cause two more bit errors when the erroneous bit reaches stages 6 and 7, but will then be discarded with no further effect.

Scramblers are used in all standard synchronous telephone line modems except the 201 and V.26, which were the first PSK modems developed. The overall system block diagram of Fig. 5.15 gives the sequential relationship among the various modem components that have been considered separately in this chapter.

References

1. A. L. Wedgman, Serial interface autodialer protocol, *Comput. Design* **20**, 17–31 (1981).
2. B. Widrow, J. M. McCool, M. G. Larimore, and C. R. Johnson, Jr., Stationary and nonstationary learning characteristics of the LMS adaptive filter, *Proc. IEEE* **46**(8), 1151–1162 (1976).

Suggested Readings

J. R. Davey, Modems, *Proc. IEEE* **60**(11), 1284–1292 (1972).

A concise, simple description of basic modem operation, components, and earlier types.

R. Glasgal, *Basic Techniques in Data Communications*, Artech House, Dedham, Massachusetts (1977).

A comprehensive survey of standard modem techniques in Chapters 4 and 5, and dialers in Chapter 6, all at an intuitive, nontechnical level.

W. J. Barksdale, Understanding the fundamentals of line equalization, *Data Commun.* **8**(5), 107–117 (1979).

A rather intuitive introduction to how and why equalization is done on telephone lines.

S. U. H. Quershi, Adaptive equalization, *Proc. IEEE* **73**(9), 1349–1378 (1985).

A comprehensive and detailed survey of the entire subject of adaptive equalization by one of the leading modem designers, plus extensive references.

IEEE Journal on Selected Areas in Communications (Special Issue on Voiceband Telephone Data Transmission) **2**(5) (1984).

A nice collection of tutorial articles covering the more recent developments in modems, with particular emphasis on high-speed transmission using echo canceling and trellis error coding techniques.

Check Your Understanding of Chapter 5—True or False?

1. All modems should cease transmitting when CD (DCD) goes off.
2. The end of a telephone number is indicated in RS-366 by either END or a time-out.
3. Initial scrambled marks in a 212 modem indicates that operation is to be at 1200 bps.
4. To test for an open telephone line one should use the remote analog loopback test.
5. The modem is basically a very specialized A/D and D/A converter.
6. All PSK modems in use today on telephone lines contain a scrambler.
7. Some serial dialers can be controlled by using the RTS interface line of RS-232C.
8. An HDX modem transmitter is normally controlled by RTS/CTS in RS-232C interfaces.
9. A modem secondary channel is normally used as a backup if the main channel fails.
10. CCITT modem types can be readily identified by the V.-series designation.
11. Adaptive equalizers for high-speed modems are invariably digital today.
12. Most personal microcomputers can readily accommodate an RS-366 dialer.
13. The DAA is intended primarily to ensure that the user's equipment is not damaged.
14. To tell a DAA to answer an incoming call, a DTE will raise the RS-232C ANS control line.
15. Dial backup must always be initiated manually.
16. Modem scramblers are very complex digital filters requiring fast multiplications.
17. An FSK modem can accept either asynchronous or synchronous input data.
18. FDX two-wire modems always transmit on the lower band and receive on the higher.
19. Scrambling facilitates clock and carrier recovery as well as equalizer training.
20. With QASK and QAM modems some type of AGC is essential for proper operation.
21. To obtain higher data rates, modems must eventually pack more bits into each baud.
22. Gray coding is used in PSK modems to reduce the number of line symbols required.
23. Differential phase coding is used in PSK modems to reduce the number of line symbols.
24. Answer-only modems are always called and cannot transmit to the caller.
25. The analog modem should be eventually replaced by all-digital networks.

The DTE/DCE Physical Interface

6.1. Interface Functions

The interface between the DTE (data terminal equipment, e.g., terminal or computer) and the DCE (data circuit terminating equipment, e.g., modem) is of fundamental importance in data telecommunications since it not only transfers data, but also provides electrical grounding and a host of control and status indications in both directions. On the DTE side of the interface the input/output (I/O) port must be able to take data from the high-speed internal parallel bus of the user's terminal and convert it to serial form at much lower speed in the proper format for transmission by the modem. It must be able to indicate when data is ready to be sent and then to send the data at the proper time and synchronization. Conversely, the I/O port must be able to take data from the modem at the proper time and synchronization, indicate when it is ready and when there are errors, and then convert this data from serial to parallel form for the internal data bus. The heart of most I/O hardware is an LSI chip called a USART or, if only asynchronous capability is provided, a UART. These devices provide the necessary serial/parallel conversion, clocking, synchronization, error detection, and modem control signals for an efficient interface. They are typically programmable, or at least strappable, and are controlled by the I/O driver routines of the DTE operating system or monitor. These routines normally either poll the USART to determine its status periodically, or else respond to software or hardware interrupts that the USART may produce.

On the modem side of the interface there are numerous control and status signals, but fortunately they are almost universally standardized by the EIA RS-232C interface standard and its CCITT equivalent, V.24. Over the last few years there has been increasing use of the higher performance EIA RS-449. It was developed in 1977 for interface applications requiring performance beyond that of RS-232C. In this chapter we shall study both of these important interfaces, along with the much simpler but more limited current loop. The chapter also considers the USART and I/O ports. Finally, there are some representative examples of how these interfaces operate in a variety of common applications such as HDX transmission, synchronization and training periods before data transmission, polled network configurations, and line loopback testing.

The simplest interface, and the one most widely used locally, is the *20-mA current loop*. Historically, the current loop was used in telegraphy to drive the old mechanical teletype machines, but today it is used for totally electronic terminal devices, particularly the so-called *"dumb" terminals*.† The current loop requires only two wire pairs: transmit send and return, and receive send and return. The current source can be located at either end of the loop (source or destination), and voltages may range over 100 V, depending on the particular interface design. If the I/O hardware contains the current source it is called *active*; otherwise it is *passive*, as shown in Fig. 6.1. Optical couplers are commonly used for electrical isolation and for current-to-voltage conversion with current loops.

There are two different conventions for binary signal representations. In both conventions a mark (binary 1) is represented by a current of +20 mA; for the *polar* signal convention a space is −20 mA, while for the *neutral* signaling mode the mark is the same but a space is represented by no current. In some older equipment +60 mA is required, but this is seldom used today. There is no restriction on how far the 20-mA current loop can run. In practice it is not uncommon for there to be runs of several thousand feet, depending on the transmission speed and the driving voltage. It is important to note that no DCE equipment is needed, since the loops connect terminals and computers directly. Because there is no separate provision

† In the *data communications context*, a dumb terminal is one that cannot be addressed, i.e., it "doesn't know its name!"

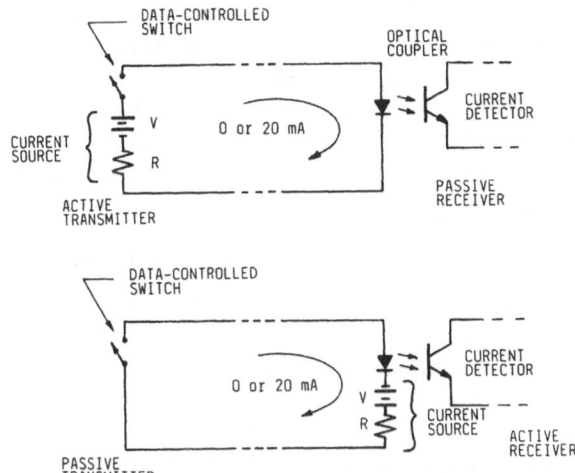

Figure 6.1. 20-mA current loop interface.

for a clock or other synchronizing signal, the normal mode of operation is asynchronous. Current loops are not normally usable over telephone lines, although some 103 modems will accept an asynchronous current loop input when using leased lines. Besides the nonstandard nature of the current loop, the main drawback is that there is no provision for control and status signals. Hence it cannot be used for such applications as automatic dialing and answering, transmitter control in HDX transmission, synchronizing clocks for synchronous communications, and secondary channel control. Clearly a more sophisticated interface is needed for modern digital telecommunications. The present standard for voice-grade transmission is RS-232C, which we consider next.

6.2. The EIA RS-232C Interface

The EIA RS-232C and the closely related CCITT V.24 standards [1, 2] are of fundamental importance in data communications today, since they are used for nearly all voice-grade modem interfaces. This is true despite the development of the higher performance RS-449 and V.24rev standards, and will probably remain so for many years hence. In this section we will concentrate on RS-232C, but will use the descriptive circuit names so that the description should be easily extendable to V.24 as well as the similar U.S. government MIL-STD-188C standard.†

---

† Care must be taken to adhere to the polarity conventions, e.g., the 0 and 1 polarities are opposite in RS-232C and MIL-STD-188C.

The physical RS-232C connector is a standard 25-pin cable connector in which each pin is assigned a particular function. There are four basic function categories, namely, *grounding and signal return*, *data transfer*, *control*, and *timing* (clocks). All lines are binary, although the binary voltage levels are different from the usual 0 and +5 V because the standard was developed before the now familiar TTL logic levels were widely used. For RS-232C the voltage ranges are as follows:

-3 to -25 V   mark (1) for data   OFF for control

$+3$ to $+25$ V   space (0) for data   ON for control

Note that the polarity for data is inverted from conventional usage; the low state represents a binary 1 while high represents a binary 0.† The range from -3 to $+3$ V is undefined, and RS-232C lines should never be stable at levels within this range. In fact, the signal transition through the range should never take more than 1 msec.

The EIA standard recommends limiting the RS-232C lines to 50 ft, but it is both feasible and common practice to exceed this limit with the use of good quality line drivers and receivers. However, care must be exercised with equipment from different vendors since only 50 ft need be guaranteed. Signaling rates can vary from zero to 20,000 bps, and the data can be sent synchronously or asynchronously. RS-232C applies to leased and dial-up two- or four-wire lines, and includes provisions for automatic dialing (which is covered in the EIA RS-366 standard). All line drivers and receivers are *single-ended*, i.e., there is only one sending wire, with all the signal returns over a common ground line. Such connections are a limitation on performance since they are more susceptible to crosstalk and noise pickup than the alternative of using balanced lines. Figure 6.2 shows both types of line drivers and receivers. RS-232C is very similar to MIL-STD-188C, but the latter is limited to only 4000 bps, and the logic levels are ±6 V. It is also similar to the V.24 standard, which is somewhat more extensive. V.24 includes several additional control and clock circuits, and even a pair of audio lines for voice answering devices. It also includes a complete section specifying the automatic calling procedure. Finally, CCITT standard V.28 gives the electrical characteristics for the unbalanced functions of V.24.

The RS-232C standard defines 21 functions, which are normally connected to the DTE and DCE equipment with a standard 25-pin connector. It is conventional for the DTE and DCE always to have female connectors. By combining two functions on one of the lines only 20 connector pins carry actual signals, while two are reserved for testing and the remaining

† However, data on buses and transmission lines is often inverted because historically the line drivers and receivers were simpler and less expensive to build as inverting amplifiers.

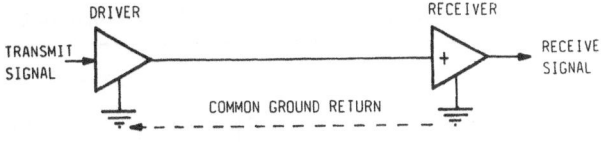

(a) Single-Ended Line Driver and Receiver

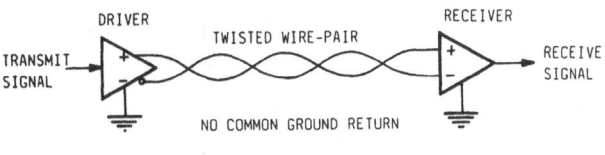

(b) Balanced Differential Line Driver and Receiver

Figure 6.2. Interface line drivers and receivers.

three are unassigned. The 20 active lines are called *interchange circuits*, and are designated by a two-letter code†; e.g., the signal ground and common return line is designated as circuit AB. In practice it is awkward to keep up with these arbitrary designations, and thus simple and easily remembered mnemonics, such as GND, will be used instead. Table 6.1 gives each RS-232C circuit, including the pin numbers, the EIA and CCITT designators, and the commonly used functional name. Typically only a few of the 20 possible lines are used in any one modem application. Having introduced the RS-232C standard, we now shall consider some of the essential control circuits in a rather intuitive manner. The exact functions vary with the particular type of modem as well as with the way it is used. Consequently, a number of representative DTE/DCE handshaking examples will be considered in the last section of this chapter.

Besides *Ground* (GND) the two main data interchange circuits are *Transmit* and *Receive Data* (TxDATA or TXD and RxDATA or RXD respectively). The directions are, of course, *relative to the DTE* and not the DCE; hence TxDATA is sent on pin 2 by the DTE and "received" by the DCE, again on pin 2, for transmission over the telephone line. Modems are always wired according to this convention, but it can cause confusion when two RS-232C DTE are connected directly. The circuit *Received Line Signal Detector* (RLSD), more commonly called *Data Carrier Detect* (DCD or CD), is sent to the DTE to indicate that it is receiving a valid carrier at some predetermined signal level from the other end of the line. This is important since, if the carrier is lost, the line or destination equipment may

† It is worth noting that CCITT V.24 uses a three-digit number for designating interchange circuits; for example, signal ground/common return is circuit 102.

Table 6.1. RS-232C Circuit Functions and Designations

Pin	Function	Mnemonic	EIA	V.24
			Designations	
1	Protective (Chassis) Ground	—	AA	101
2	Transmitted Data	TxDATA or TXD	BA	103
3	Received Data	RxDATA or RXD	BB	104
4	Request to Send	RTS	CA	105
5	Clear to Send	CTS	CB	106
6	Data Set Ready	DSR	CC	107
7	Signal Ground	GND	AB	102
8	Received Line Signal Detect	RLSD, DCD, or CD[a]	CF	109
9	Reserved for Testing	—	—	—
10	Reserved for Testing	—	—	—
11	Unassigned	—	—	—
12	Secondary Received Line Signal Detect	SRLSD or SCD	SCF	122
13	Secondary Clear to Send	SCTS	SCB	121
14	Secondary Transmitted Data	STxDATA or STXD	SCA	118
15	Transmit Bit Clock (from DCE)	TxC	DB	114
16	Secondary Received Data	SRxDATA or SRXD	SBB	119
17	Receive Bit Clock (from DCE)	RxC	DD	115
18	Unassigned	—	—	—
19	Secondary Request to Send	SRTS	SCA	120
20	Data Terminal Ready	DTR	CD	108.2
21	Signal Quality	SQ	GC	110
22	Ring Indicator	RI	CE	125
23	Data Signal Rate Selector	RATE	CH/CI	111
24	External Transmit Bit Clock (from DTE)	XTxC	DA	113
25	Unassigned	—	—	—

[a] Care should be taken to avoid confusing this mnemonic with the EIA designation for Data Terminal Ready, which is also CD!

be inoperable. In such a case the modem usually shuts off its receiver and just sends a *mark hold* (continuous binary ones) back to the DTE over RxDATA. A minimum configuration for any FDX modem connection requires GND, TxDATA, RxDATA, and CD.

There are four additional control circuits that are of major importance to many common DTE/DCE configurations, namely, *Request to Send* (RTS), *Clear to Send* (CTS), *Data Terminal Ready* (DTR), and *Data Set Ready* (DSR). The exact usage of these signals is somewhat dependent on the modem type, but the general functions are simply stated. DTR is used in dial-up situations by a terminal in response to *Ring Indicator* (RI) to tell the modem to automatically answer the phone. Conversely, when DTR goes off, it says to hang up, i.e., disconnect from the line. On leased lines, when used, DTR just indicates that the DTE power is on. Once the modem has gone off-hook, shifted into data mode (disconnected the telephone instrument), and completed any answer-back tones, then it turns DSR on to indicate completion of the answering process. If DSR ever goes off during transmission, it indicates an aborted call. In any case, DSR only indicates the status of the local modem and gives no guarantee about the overall channel. For leased lines, DSR only indicates that the modem is powered and ready to accept further commands from the DTE.

While DTR/DSR is used for establishing a dial-up connection, the pair of lines RTS/CTS is mainly used for controlling the direction of transmission in HDX operation. RTS is used by an HDX terminal to tell the modem that it wants to transmit. Formally, it conditions the local DCE for transmission and maintains the transmit mode as long as it is on. In response to RTS the modem turns on its transmitter and initiates any initialization and training sequences necessary, then raises the signal CTS indicating to the terminal that it can now begin sending actual data. The DTE should never send data across the interface unless CTS is on. For FDX operation there are several different conventions regarding RTS and CTS, but the simplest is for the terminal to keep RTS on all the time and transmit as long as the modem keeps CTS on. There is always a delay between RTS and CTS, appropriately called the *RTS/CTS delay*, which can usually be set on the modem. This delay is an important factor in the *line turnaround time* for HDX links such as polling. As Table 6.1 shows, there are also interchange circuits for transmit and receive clocks (TxC and RxC), signal quality as determined by the modem, and another separate set of control and data lines for the secondary channel when one is used. The modem nearly always furnishes the clocks to the terminal, although there is provision for supplying the transmit clock externally from the DTE.

To illustrate the operation of the RS-232C interface, we will first consider a basic FDX dial-up procedure for a 103-type modem with manual dialing and automatic answering. Figure 6.3a shows the data link with the pertinent interchange circuits in the order that they become active. The order of operation is indicated from top to bottom throughout. Since this example is a dial-up line operating FDX, it will be necessary to disable any echo suppressors that might be installed.† This is done by the 2225-Hz answer-back tone before data transmission begins. Assuming the terminal A has DTR on, the call is dialed manually and the line switched to the modem by the exclusion key on the telephone. The 20-Hz ring signal is detected by a filter in modem B, which then raises RI to the DTE. To have

---

† Echo suppressors are disabled by sending a single tone between 2010 and 2240 Hz for at least 400 msec at 0–5 dB below the maximum data signal level. They will reenable if there is no energy on the line for about 100 msec.

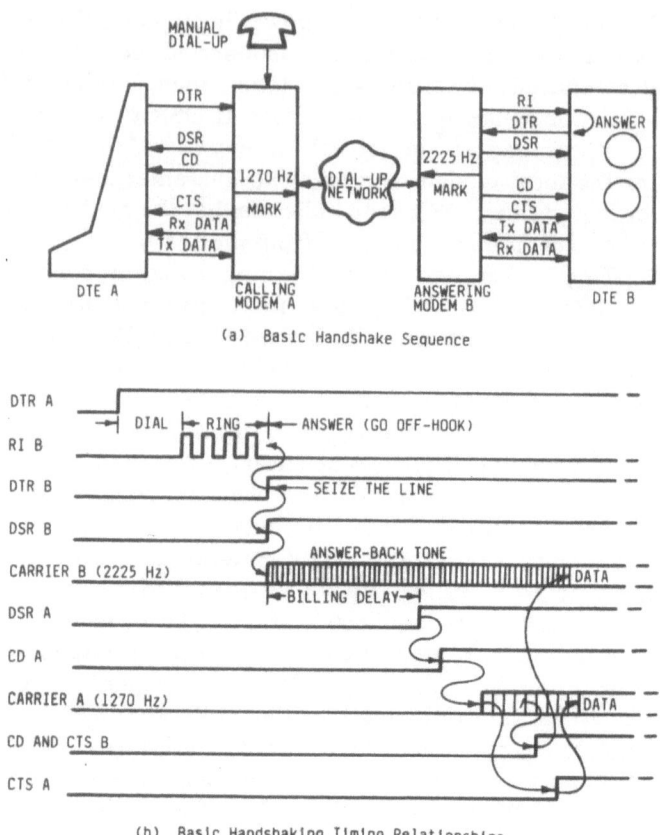

(a) Basic Handshake Sequence

(b) Basic Handshaking Timing Relationships

Figure 6.3. RS-232C dial-up protocol for 103 modem.

the modem answer the call, the DTE raises DTR and the modem then connects to, or *seizes*, the line. This type modem immediately sends its mark frequency (answer-back tone) of 2225 Hz and raises DSR, indicating completion of the answering sequence. Modem A hears the answer-back tone, but delays for a predetermined time (typically 1–3 sec) to allow the telephone system billing equipment to start charging for the call. It next raises DSR and CD, then sends its own mark frequency of 1270 Hz. When modem B detects this tone, it raises CD and CTS, indicating that the DTE can now begin to send data. Finally, after a preset interval of sending its carrier, modem A also turns on CTS and the corresponding DTE begins sending data. This last interval should be set so that both ends get CTS and can thus begin transmitting data at roughly the same time. The entire sequence is shown in the timing diagram of Fig. 6.3b, in which the arrows indicate causal effects. Such a sequence is often referred to in practice as the *handshake protocol* between the DTE and DCE.

This example illustrates the basic operation of the RS-232C DTE/DCE interface, but it should be noted that there are many subtle variations in the timing even for this simple case. And, of course, there are many other comparable handshake sequences for other modems and other configurations. Each will be unique, but we shall consider a selection of highly representative sequences in the last section. Before doing so, however, we shall look at the comparatively new RS-449 interface, which provides much improved performance necessary for many of today's sophisticated applications, and also look at I/O ports briefly.

6.3. The EIA RS-449 Interface

The RS-232C standard has been a reliable and widely accepted means of connecting DTE and DCE equipment for many years, and will continue to be so in the future. However, in recent years the need has developed for an interface having greater speed, range, and control capabilities. This trend is now accelerating. In response, the EIA in the United States and the CCITT worldwide have developed relatively new standards that are meant to overcome many of the shortcomings of RS-232C and V.24. For example RS-232C is limited to 50 ft and 20,000 bps, and this is often inadequate in modern data communications systems. There is also some inconsistency in the RTS/CTS and DTR/DSR protocols under certain conditions. Finally, there is no provision for indicating loopback tests across the interface; rather they must be initiated by a manual switch on the modem. The new standards, EIA RS-449 and CCITT V.24rev, contain provisions to overcome all of these problems. They will, of course, work in all the possible configurations of RS-232C and V.24. In addition, they are designed to operate around the more contemporary logic voltages of -5 and $+5$ V. We shall concentrate here on RS-449 since it is well established in the United States and is quite similar to the comparable CCITT standard.

To resolve these distance and data rate limitations of RS-232C, RS-449 uses two different types of line driver-receiver pairs as shown in Fig. 6.4a. Those interface circuits that do not require extremely high speed are called *Category II* circuits, which are unbalanced in the sense that the line *drivers are single-ended* (one side grounded) and the *receivers are differential.* They are defined in the separate *RS-423* standard which allows transmission rates up to 100 kbps. This is a considerable improvement over the RS-232C arrangement, since the separate return wire for each such circuit greatly reduces noise pickup. Furthermore, the differential receiver cancels any noise common to both wires (common mode noise rejection) and can thus

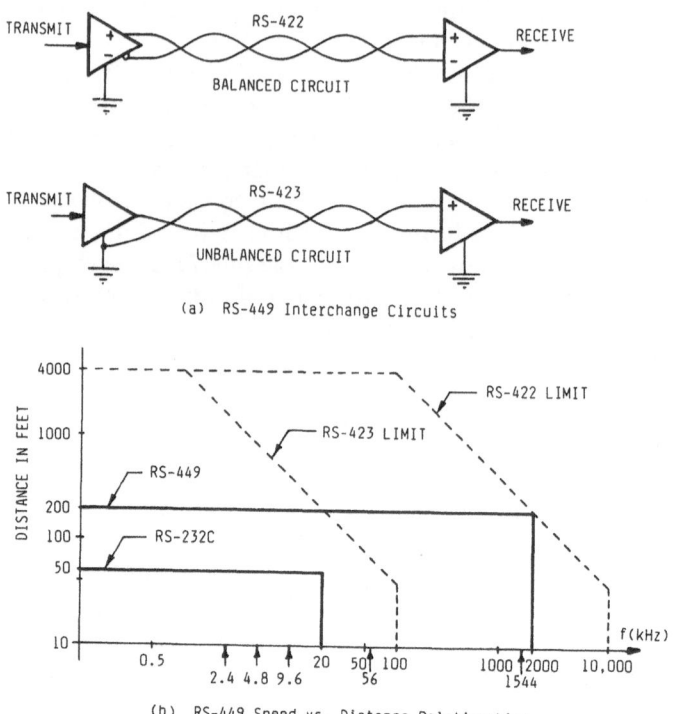

(a) RS-449 Interchange Circuits

(b) RS-449 Speed vs. Distance Relationships

Figure 6.4. RS-449 interface characteristics.

respond accurately to received voltage differentials of as little as ±0.20 V. Specifically, for RS-423 the driver voltage ranges are

   −4 to −6 V   mark (1) for data   OFF for control

   +4 to +6 V   space (0) for data   ON for control

and the corresponding differential receiver voltages are

   < −0.20 V   mark (1) for data   OFF for control

   > +0.20 V   space (0) for data   ON for control

Note that polarities are the same as RS-232C, and that the received signal is undefined in the transition range of −0.20 to +0.20 V.

   The RS-449 circuits that require high-speed capability are called *Category I*. For rates above 20 kbps they are specified electrically by the *RS-422* standard, which defines a completely *balanced line driver-receiver pair*. Again there is the desirable common mode noise rejection, and here the noise pickup is further reduced by a complete isolation from the common ground. As with RS-423, the minimal differential voltage is only ±0.20 V as compared to ±3.0 V for RS-232C. For data rates below 20 kbps RS-449 allows the optional use of the slower RS-423 drivers and receivers for Category I circuits. This option will generally result in a simpler implementation at the expense of reduced noise immunity. It is also desirable when interoperability is required between RS-449 and the unbalanced RS-232C. The RS-422 differential driver voltage ranges are

−2 to −6 V   mark (1) for data   OFF for control

+2 to +6 V   space (0) for data   ON for control

while the differential receiver voltages are the same as for RS-423. Again the received signal is undefined in the transition region between −0.20 and +0.20 V.

Although RS-422 provides for Category I signaling rates of up to 10 Mbps, the basic RS-449 limits this to 2 Mbps at 200 ft (or 60 m) except for timing circuits which are allowed to go as high as 4 Mbps. At lower rates the distance may be increased to 4000 ft. Figure 6.4b shows these nominal RS-449 limits of 200 ft and 2 Mbps along with the 50 ft and 20 kbps comparable limits for RS-232C. It should be noted that while RS-232C includes all voice-grade modems, RS-449 can operate on wideband lines and on the 1.544-Mbps T-carrier digital lines. There are *ten* entirely new RS-449 circuits compared with RS-232C, plus several of the existing circuits have been modified. However, these modifications do not preclude the interoperability with RS-232C provided minor adjustments have been made. Table 6.2 indicates all circuit functions of RS-449, along with the circuit category and the nearest RS-232C equivalent. The new circuit names and two-letter mnemonics are quite similar to the familiar RS-232C ones, except for DSR, which has been renamed *Data Mode*, and RI, which is called *Incoming Call*. There are also three lines for testing: *Local Loopback*, *Remote Loopback*, and *Test Mode*, all of which allow automatic initiation of loopback tests that are used to isolate the cause of failures. These tests must be done with a manual switch in RS-232C. There are separate common lines for each direction (sending and receiving), which reduces noise and crosstalk. Note that two different connectors are specified for the entire RS-449 interface; one 37-pin main connector and a separate 9-pin connector that includes all secondary channel functions. If there is no secondary channel, then this 9-pin connector is not necessary. From Table 6.2 one sees that all of the differential Category I circuits have two pins assigned; e.g., *Receiver Ready* (RR), which corresponds to Carrier Detect in RS-232C, connects to pins 13 and 31 of the 39-pin connector. Note also that pin 16 has a dual function.

EIA RS-449 and the accompanying RS-422 and RS-423 standards are similar to the U.S. Government standard MIL-STD-188-114, but there are some significant differences. This is also true of the CCITT V.24rev standard,

Table 6.2. RS-449 Circuit Functions and Designations

Pins	Function (Category)	EIA	V.24rev	RS-232C Equivalent
		— Designations		

The following circuits are on the main 37-pin connector:

Pins	Function (Category)	EIA	V.24rev	RS-232C Equivalent
19	Signal Ground (II)	SG	102	GND
37	Send Common (II)	SC	102a	—
20	Receive Common (II)	RC	102b	—
28	Terminal in Service (II)	IS	—	—
15	Incoming Call (II)	IC	125	RI
12, 30	Terminal Ready (I)	TR	108.2	DTR
11, 29	Data Mode (I)	DM	107	DSR
4, 22	Send Data (I)	SD	103	TxDATA
6, 24	Receive Data (I)	RD	104	RxDATA
17, 35	Terminal Timing (I)	TT	113	XTxC
5, 23	Send Timing (I)	ST	114	TxC
8, 26	Receive Timing (I)	RT	115	RxC
7, 25	Request to Send (I)	RS	105	RTS
9, 27	Clear to Send (I)	CS	106	CTS
13, 31	Receiver Ready (I)	RR	109	CD
33	Signal Quality (II)	SQ	110	SQ
34	New Signal (II)	NS	—	—
16	Select Frequency (II)	SF	126	—
16	Signaling Rate Selector (II)	SR	111	CH/CI
2	Signaling Rate Indicator (II)	SI	112	—
10	Local Loopback (II)	LL	141	—
14	Remote Loopback (II)	RL	140	—
18	Test Mode (II)	TM	142	—
32	Select Standby (II)	SS	116	—
36	Standby Indicator (II)	SB	117	—
3, 21	Undefined Spares (II)			
1	Shield			

The following circuits are on the optional secondary 9-pin connector:

Pins	Function (Category)	EIA	V.24rev	RS-232C Equivalent
5	Signal Ground (II)	SG	102	GND
9	Send Common (II)	SC	102a	—
6	Receive Common (II)	RC	102b	—
3	Secondary Send Data (II)	SSD	118	STxDATA
7	Secondary Request to Send (II)	SRS	120	SRTS
8	Secondary Clear to Send (II)	SCS	121	SCTS
2	Secondary Receiver Ready (II)	SRR	122	SCD
1	Shield			

and the accompanying electrical specifications V.10 for unbalanced circuits and V.11 for balanced ones. Despite these minor differences, all of the newer standards were carefully designed to eventually replace the older ones in a smooth evolutionary manner over many years. Considerable attention was given to facilitating this transition by making it as simple as possible to interconnect the new interfaces with existing equipment having the older ones, for example between RS-449 and RS-232C equipment.

Although RS-232C and RS-449, and their CCITT and government equivalents, are by far the most widely used interfaces for analog telecommunications lines, there are a number of other wideband interfaces too. Most notable is the interface portion of the CCITT V.35 group band modem standard. Used for 40.8, 48, and 56 kbps wideband and the digital 56 kbps DDS transmission, V.35 is a straightforward interface. In fact the control (RTS, CTS, DSR, CD) and ground conform to V.28 just like V.24 does. However, for the transmit and receive data and corresponding clock circuits, a balanced differential driver and receiver are defined to accommodate the higher data rates. Here a mark (1) is defined as a differential voltage of -0.55 V while a space (0) is $+0.55$ V.

In this section and the preceding one we have studied the DTE/DCE interfaces in considerable detail. This should provide a solid basis for the data link operational examples that conclude this chapter. However, before considering them we shall in the next section briefly consider the nature of the I/O port in the DTE.

6.4. Serial I/O Ports and the UART

Because of the basic two-wire nature of telecommunications links it is generally much more practical to send data serially than in parallel. However, modern DTE equipment invariably is based on parallel high-speed bus structures. Consequently, it is necessary to convert data from parallel to serial form for transmission, and vice versa upon reception. There are also many other "housekeeping" functions that must be performed in the *DTE communications I/O port*: speed conversion, error control, delineation of individual words, provision of interface controls to the DCE, and indication of the status of transmitted and received data. All of these operations require considerable processing capability, but fortunately there are a number of standard sophisticated integrated circuit chips that perform such I/O functions both efficiently and cheaply. These devices, known as *Universal Asynchronous Receiver/Transmitters* (UARTs), are the heart of practically all serial low-speed asynchronous computer and terminal I/O hardware. For synchronous transmission various data link protocols are commonly used, and for this type of communication there are more sophisticated chips known as *Universal Synchronous/Asynchronous Receiver/Transmitters* (USARTs).

Before considering the UART per se, it is important to understand the way asynchronous data is formatted for serial transmission. When there is no data to be sent, it is customary for the transmission line to send a *holding* or *idling pattern*, the simplest of which is typically just a continuous string of binary ones called a *mark hold*. When a character is to be sent, the UART will first send a *start bit*, which is a space or binary zero, then send the bits of the character in serial form usually with the least significant bit first. The last character bit is followed by one or more *stop bits*,† which are binary ones. If there is no immediately following character, the mark-hold resumes and appears as a continuation of the stop bit; otherwise the stop bit is immediately followed by the start bit of the next character. This is illustrated for the common case of one stop bit and eight character bits in Fig. 6.5.

The last character bit shown is a *parity bit*, which is a simple error check technique. For *odd parity* it is selected by the UART transmitter so that there is always an odd number of binary ones in every word; e.g., 1101001P would have the parity P set to $P = 1$ to make five ones, and 0010110P would become 00101100 since $P = 0$. *Even parity* is similar, except that parity is selected for an even number of ones. Since most errors are only one bit per character (unless the transmission line is very bad), then the receiver can detect them by simply counting the ones in each character received; for odd parity the sum should always be an odd number. For a UART to send and receive such data it is necessary to preset several

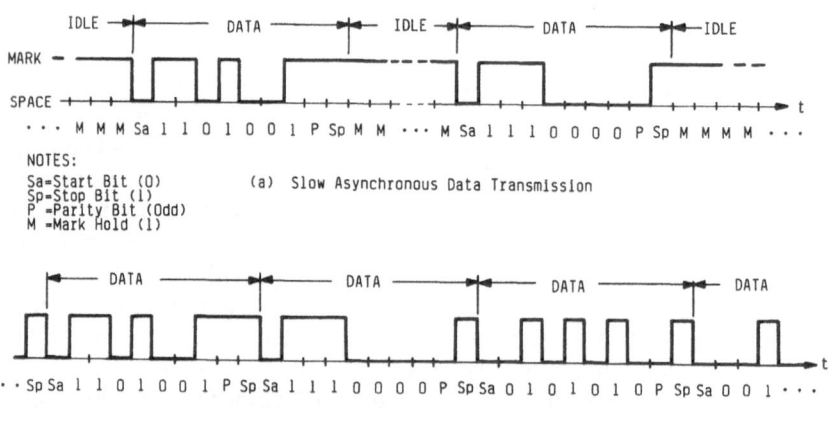

Figure 6.5. Asynchronous data transmission.

† Most asynchronous data transmission uses only one stop bit except for some old 110-Bd teletypes that require two.

parameters, including the word (character) length, the number of stop bits, and the type of parity to be used if parity is to be used at all. These parameters may be set either by logic levels on the appropriate UART pins or by sending control words to the UART port, depending on the particular device. It is, of course, also necessary to provide a clock synchronized with the bit rate and perhaps to assign the appropriate I/O port addresses.

From the DTE point of view, it would like first to initialize the UART and then be able to read the UART status and, if not busy, send a character to it over the parallel data bus for transmission. The UART transmitter should add the start, stop, and parity bits, and then transmit the bits serially at the proper data rate to the DCE, normally observing the RS-232C protocol. Conversely, the UART receiver should be able to detect start bits (always a 1 to 0 transition), then synchronize its sampling clock to sample at the center of each data bit, check the parity, and verify that the stop bit is present. If there are *errors in the parity or framing*, they should be indicated by flags or interrupts. The UART should also indicate to the DTE processor that the data has been received, again typically through status flags or interrupts. The UART should then convert the received data from serial to parallel form for the processor to read on the internal parallel data bus.

If the UART is not read before the next character is received, then the present character will be overwritten and lost; this condition is normally signaled by the UART by indicating a *character overrun error*. There are some other properties that many UARTs have, including FDX capability (the receiver and transmitter operate independently and simultaneously), double buffering (one register for current data character and another to accept new data), and programmability (UART parameters can be set and changed by software instructions in DTE programs). It is also nice to have modem controls such as RTS/CTS for HDX operation and DTR/DSR for automatic dialing and answering.

Figure 6.6 shows the functional block diagram of a simple basic UART (the classical TMS 6011, which is no longer in production). There are separate transmit and receive sections that include the buffering, status, and error control logic circuits. Note particularly the control register inputs and the various status output lines. This single chip clearly replaces a great deal of discrete logic hardware. One of the most popular and enduring USARTs is the Intel 8251. Thus we shall look at the asynchronous function of it here. Figure 6.7 describes this device and includes a functional block diagram. There are again separate transmit and receive sections, each with its own status and clock lines. There is also an eight-bit tristate buffer for the DTE data bus, with reset, clock, read, and write controls. The C/$\overline{\text{D}}$ line selects the *control* or *data* mode. In the former mode the USART status (errors, busy, etc.) can be read, or command words can be written to the USART for enabling and resetting flags and modem controls. The modem

functional block diagram

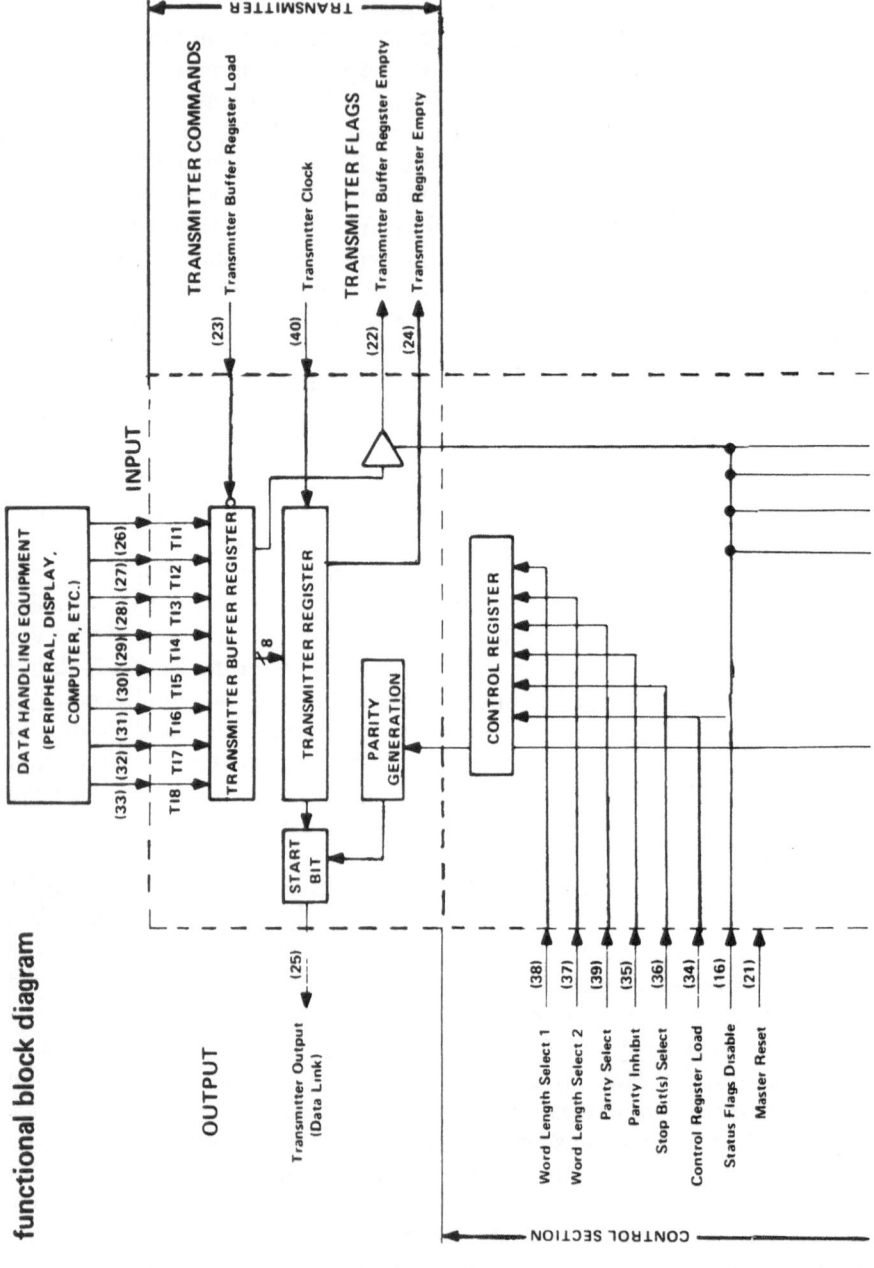

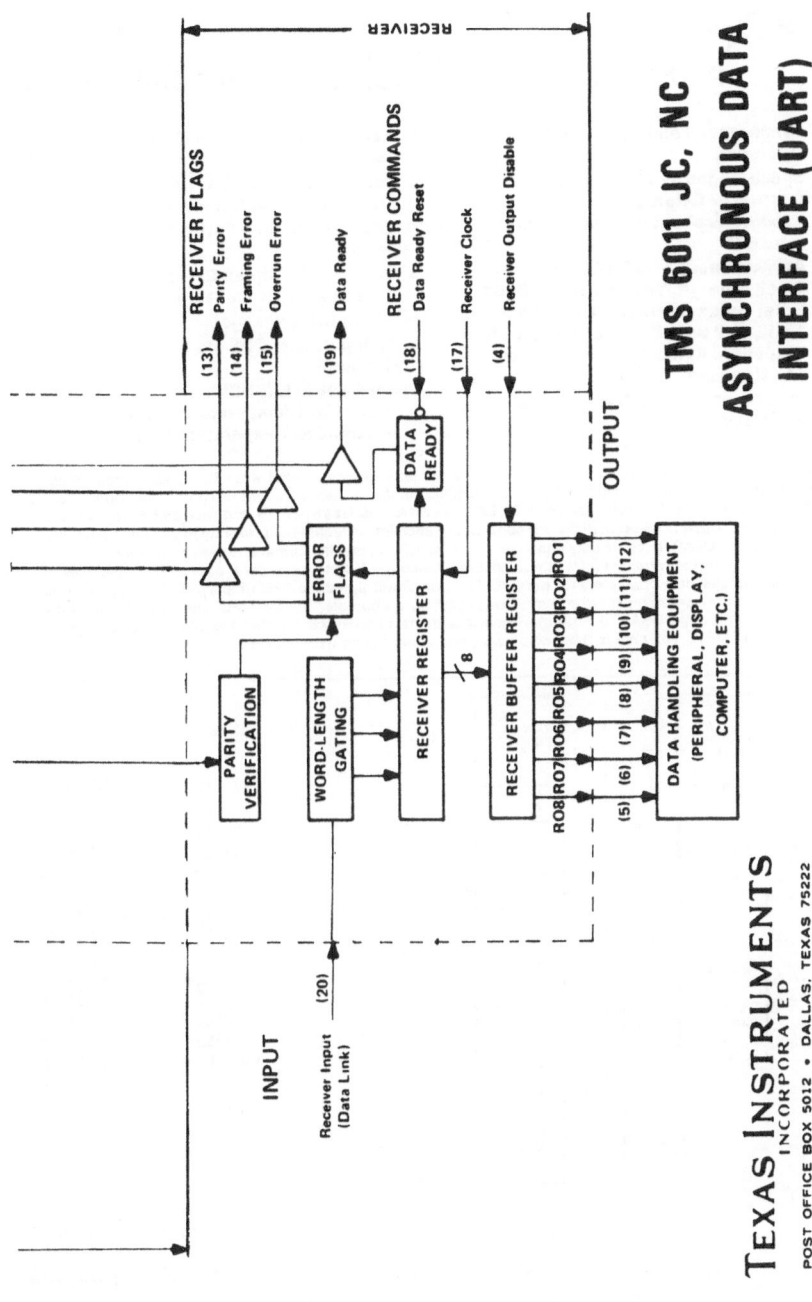

Figure 6.6. Basic UART functional organization. (Courtesy of Texas Instruments Incorporated.)

intel®

8251A
PROGRAMMABLE COMMUNICATION INTERFACE

- Synchronous and Asynchronous Operation
- Synchronous 5–8 Bit Characters; Internal or External Character Synchronization; Automatic Sync Insertion
- Asynchronous 5–8 Bit Characters; Clock Rate—1, 16 or 64 Times Baud Rate; Break Character Generation; 1, 1½, or 2 Stop Bits; False Start Bit Detection; Automatic Break Detect and Handling
- Synchronous Baud Rate—DC to 64K Baud

- Asynchronous Baud Rate—DC to 19.2K Baud
- Full-Duplex, Double-Buffered Transmitter and Receiver
- Error Detection—Parity, Overrun and Framing
- Compatible with an Extended Range of Intel Microprocessors
- 28-Pin DIP Package
- All Inputs and Outputs are TTL Compatible
- Available in EXPRESS
  —Standard Temperature Range
  —Extended Temperature Range

The Intel® 8251A is the enhanced version of the industry standard, Intel 8251 Universal Synchronous/Asynchronous Receiver/Transmitter (USART), designed for data communications with Intel's microprocessor families such as MCS-48, 80, 85, and iAPX-86, 88. The 8251A is used as a peripheral device and is programmed by the CPU to operate using virtually any serial data transmission technique presently in use (including IBM "bi-sync"). The USART accepts data characters from the CPU in parallel format and then converts them into a continuous serial data stream for transmission. Simultaneously, it can receive serial data streams and convert them into parallel data characters for the CPU. The USART will signal the CPU whenever it can accept a new character for transmission or whenever it has received a character for the CPU. The CPU can read the complete status of the USART at any time. These include data transmission errors and control signals such as SYNDET, TxEMPTY. The chip is fabricated using N-channel silicon gate technology.

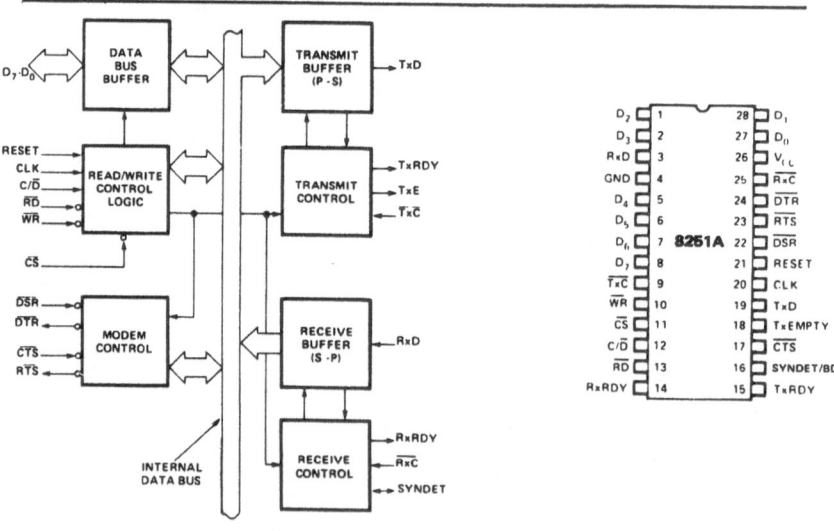

1. Block Diagram   2. Pin Configuration

Figure 6.7. Typical USART features and organization. (Courtesy of Intel Corporation.)

control block provides RTS/CTS and DTR/DSR outputs at normal TTL levels. Thus it is only necessary to shift the voltage levels to those required by the interface standard, e.g., perhaps to +12 and −12 V with reversed polarity for RS-232C. This can be done with discrete transistors, but for this purpose special integrated circuit chips are also available which simplify the design. The USART can also handle a number of common synchronous data link protocols, but we shall defer our study of this capability until our study of these protocols in the next chapter.

Having now both introduced the DTE I/O port and studied the physical DTE/DCE interface in detail, in the next section we will now consolidate and extend our study with some representative examples of interface operations.

6.5. Handshaking Examples

This section is designed primarily to illustrate how the physical interface protocol actually operates in practice. There are many possible configurations and parameters for the DTE/DCE interfaces that we have considered, but there is no concise way to illustrate them all. However, we shall look at six examples, which have been carefully selected to represent a large proportion of those situations encountered in practice. Our orientation will be on RS-232C because of its pervasiveness, but the examples are applicable with slight nomenclature changes to other interfaces as well. It must be kept in mind that each actual installation is unique, and there are usually a number of time delays and optional parameters that must not only be reconciled between the modems on either end (usually by internal strapping or switch settings), but must also be compatible with the terminal I/O ports and software.

Five of our six examples deal with point-to-point data links, with the final one being a multidrop line. In particular, the following cases are included:

1. Two-wire FDX leased point-to-point;
2. Two-wire HDX leased point-to-point;
3. Four-wire FDX leased point-to-point;
4. Two-wire FDX manual dial-up point-to-point;
5. Two-wire HDX ACU dial-up point-to-point;
6. Four-wire FDX leased polled multidrop.

Each example is oriented towards a particular modem type, but can generally be adapted for other modems in a straightforward manner.

The first example, in Fig. 6.8, is the simplest type of 103 connection over a permanent two-wire leased line. Recall that the 103 is inherently

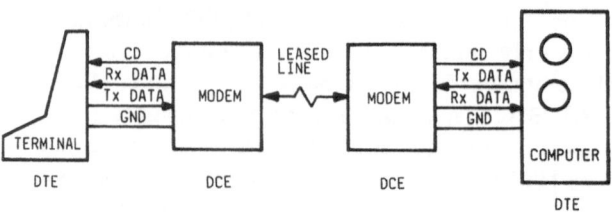

(a) Data Link Interface

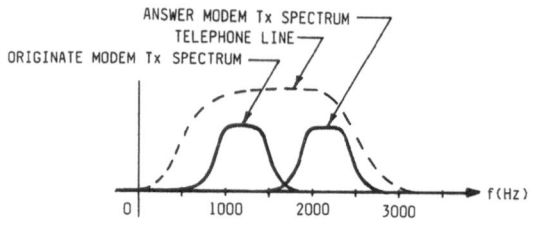

(b) Spectral Bandwidth Assignment

Figure 6.8. Two-wire FDX leased point-to-point link.

FDX and asynchronous, so no line turnaround or clock timing is required. Besides ground there is only TxDATA and RxDATA and the single control line CD. The latter should always be included, since otherwise large amounts of data could be blindly sent to a failed line or an inoperative destination modem. In the initial installation it is, of course, necessary to designate one modem as originating and the other as answering, since this then determines which modem transmits in which frequency band. This data link might be used for an alarm system or random data entry station which operates 24 hours per day.

Example two is also a two-wire leased point-to-point line, but now is operated HDX with a synchronous modem such as the 201. There must be a synchronous bit clock from DCE to DTE, and this requires a short training pattern to enable the receiver to acquire carrier and clock synchronization before data transfer can begin. In HDX the transmitter and line turnaround are normally controlled with RTS/CTS. Thus, in Fig. 6.9a, the originating terminal raises RTS to initiate the local modem training sequence. Upon completion of training the modem returns CTS, indicating that any data will now be faithfully transmitted. The terminal then begins sending its data.

At the destination the modem acquires synchronization and, in due time, raises CD to its DTE and prepares to receive and deliver the data. Should the training fail, there is no provision at the physical level for informing the sending end. Rather, this must be done at a higher protocol

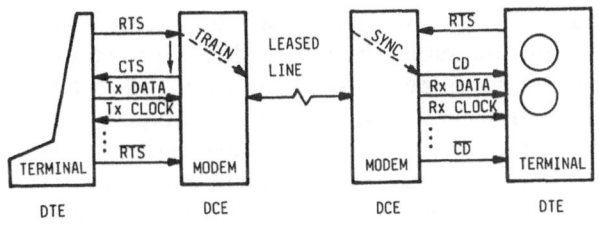

(a) Forward Direction Transmission

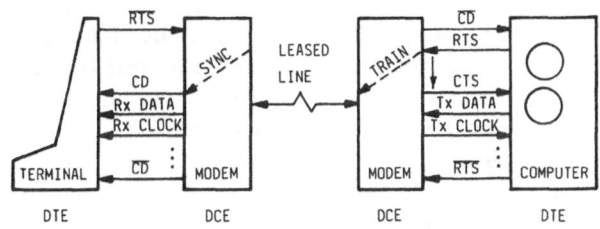

(b) Line Turnaround and Transmission

Figure 6.9. Two-wire HDX leased point-to-point link.

level, typically through the data link protocol acknowledgment procedure. This will be considered in the next chapter. Once successful data transmission is completed the sending terminal simply turns off RTS and the modem promptly shuts off its transmitter. The resulting loss of carrier at the destination is conveyed to the DTE by turning CD off ($\overline{CD}$), which the DTE acknowledges with RTS. The HDX channel is now idle and may either remain so, or the line may be turned around.

To turn the line around the previously receiving DTE raises RTS and the modem initiates training. After the RTS/CTS delay, the CTS line is turned on and data transfer may begin. At the other end the modem raises CD after successfully acquiring synchronization, and then conveys the received data to its respective DTE. Upon completion the transmission is again terminated with $\overline{RTS}$ and the resulting $\overline{CD}$. The same procedure can then be repeated as often as necessary. Note that in all cases the modems supply the clock signals to their respective DTEs.

Although we have focused on the 201 for simplicity, the same protocol is applicable to higher speed modems that require considerably longer training times for additional AGC and adaptive equalizer circuits as well as for scramblers. For applications such as alarm polling or point-of-sale transactions, this modem training time can be longer than the actual messages. Thus, ironically, a higher overall data throughput may actually be obtained with slower speed modems.

Our third example is a four-wire leased line operated FDX with perhaps 208 modems. However, rather than always keeping the transmitters on, they are only activated when there is traffic to send. In Fig. 6.10 note first that DTR/DSR are not needed in this leased-line protocol; so they are looped back to ensure proper DTE operation. This may not be necessary, depending on how the DTEs are configured. To initiate transmission from either end the DTE raises RTS, which initiates the training sequence that sets the AGC, carrier recovery, clock recovery, and equalizer circuits in the receiver. When completed, CTS is returned indicating that the modem is now ready for the data. At the destination CD indicates receipt of a valid signal. Normally the modems supply a bit clock for the data at both ends, as shown. Either end can terminate transmission by just turning RTS off, or the line can be kept active between data transmissions by transmitting a *mark hold* (binary ones) or perhaps some other idling pattern defined by higher protocol levels.

The fourth example, shown in Fig. 6.11, is somewhat more complex. It is a two-wire FDX dial-up link using a 212 modem operating in the 1200-bps PSK mode. Although the 212 is widely used in the United States, its protocol is rather unique and differs considerably from its CCITT V.22 counterpart. For 212 autoanswer operation DTR is always kept on, so the operator begins the sequence by simply dialing the destination. The answering modem detects the ring and passes RI to its DTE. Since DTR is already on, the modem simultaneously goes OH, raises DSR, and begins sending a 2225-Hz answer-back tone. Upon detecting this tone, the calling operator switches the line over to the originating modem, which indicates it is now in data mode by turning on DSR. It then listens for the answer-back tone.

Once this tone is detected, the originating modem begins sending (scrambled) marks on the lower band. This indicates to the answering modem that the data rate is to be 1200 vice 300 bps. Once this pattern is detected and synchronization acquired, the answering modem returns (scrambled) marks for a fixed time to confirm 1200-bps operation. It then

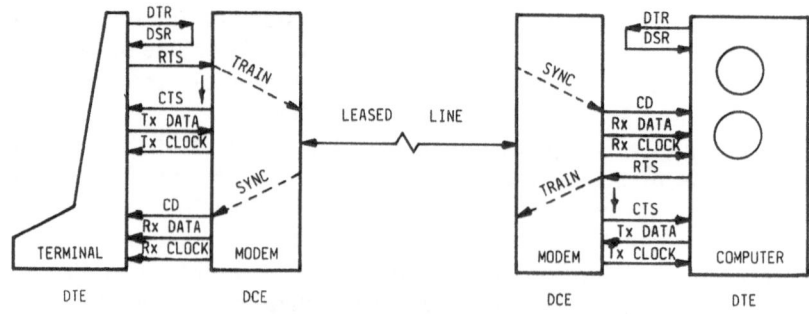

Figure 6.10. Four-wire FDX leased point-to-point link.

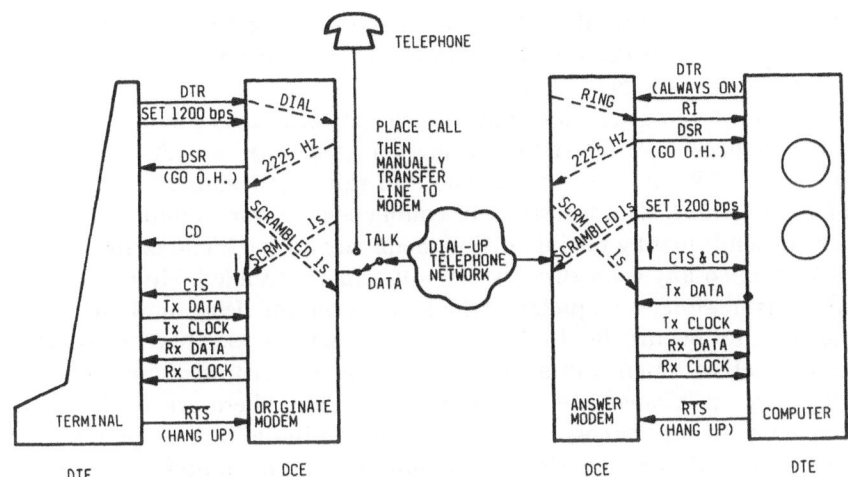

Figure 6.11. Two-wire FDX manual dial-up point-to-point link.

raises CTS and CD simultaneously to the DTE, indicating that data transfer can begin. Back at the originating end the modem detects the scrambled marks and raises CD. When the training period is complete it stops sending scrambled ones and raises CTS to the originating DTE for data to begin. In the synchronous mode clocks are provided by the modems to the DTEs but are, of course, absent for asynchronous operation at 1200 and 300 bps. FDX operation normally continues in both directions until either end hangs up by lowering DTR. Termination will also result if there is a loss of carrier, or in some cases, if there is a sufficiently long spacing interval known as the *long space disconnect.*

Despite the complexity of this example, it is quite instructive. In practice there are numerous little millisecond intervals at every step of the protocol. On the modem side they are designed or strapped into the unit but can present a real challenge to anyone writing I/O software for a computer or terminal.†

Our fifth and last point-to-point example is a two-wire HDX dial-up link with automatic dialing and answering. We shall use the 202 modem for this case. Assuming the originating terminal is equipped with an RS-366 port and dialer, the call is dialed according to the parallel-port ACU procedure of the previous chapter. Once all digits are dialed, the answering modem indicates RI to its DTE across the RS-232C interface. Assuming the normal DTR response, it then seizes the line and begins sending the

† This is particularly true of relatively unsophisticated personal computers with limited I/O handler software.

202 answer-back tone of 2025 Hz, as shown in Fig. 6.12a. The ACU hears this tone, transfers the line to the modem, and indicates this to the DTE by turning the RS-366 COS line on.

Figure 6.12b shows the conventional physical level protocol once the call is connected. At the end of the answer-back tone interval each modem will have DSR on. The first to send will raise RTS and, after the preset RTS/CTS delay, data-transfer will commence. In our example the called modem sends first by raising RTS immediately after the end of the answer-back tone, so any echo suppressors will remain disabled. However, when this transmission is completed and RTS goes off, the delay until the reverse transmission begins should be neither too long nor too short. If too long, then the abort timers will assume there is no futher traffic and both ends will hang up. Conversely, if it is too short and there are inadvertently enabled echo suppressors on the line, then the initial part of the messages may be lost during the 100 msec required for the suppressor to reverse directions. This latter situation is usually guarded against by always setting the RTS/CTS delay to around 200 msec in this configuration. However, if it is certain that no suppressors are present, such as for short distances, or

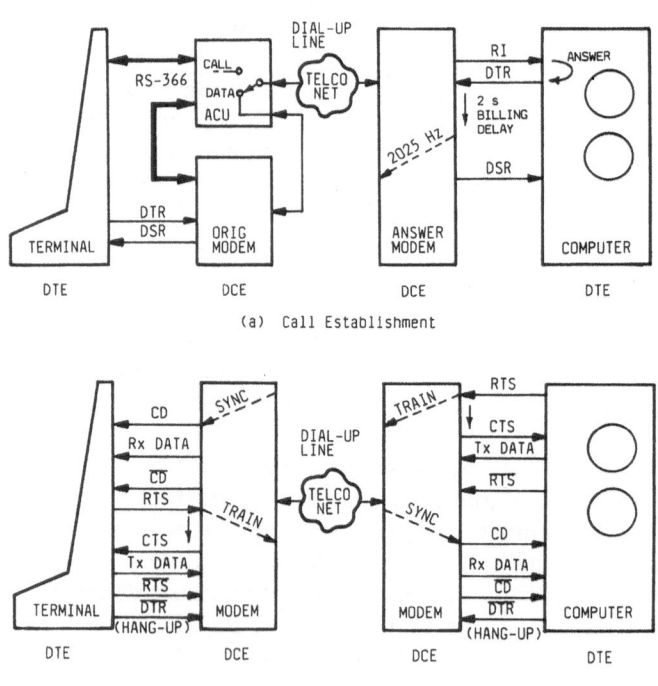

(a) Call Establishment

(b) Called Terminal Sends First and Calling DTE Replies

Figure 6.12. Two-wire HDX ACU dial-up point-to-point link.

if echo suppressors can be kept off during turnaround,† then much shorter RTS/CTS delays are possible.

In the example, once the line is turned around the originating end raises RTS and then sends its data after receiving CTS. This completes the call, which may then be terminated by either end turning DTR off. Alternately, it could also terminate if both RTS and CD are off for the abort timeout period. Although we have specifically considered the asychronous 202 in this example, the corresponding protocol for synchronous modems like the 201 is essentially the same except for the time intervals and the need to reacquire synchronization and training with each line turnaround.

Our last example involves a polled *multidrop*, or multipoint, data link configuration. Before delving into the specific case selected, it will be helpful to reconsider polling briefly. Among the various methods for the control of multidrop lines, polling has historically been the most common. There is a single *master* station, typically a computer or front-end processor (FEP),‡ that communicates in some sequence with a number of remote stations. This sequence is often determined by a *polling list*. As shown in Fig. 6.13, the connections may be on a single multidrop line or there may be an individual line to each remote station. This latter case may be used to simplify the master station control, but clearly does not save on line utilization. In Fig. 6.13a, for example, an FEP at a firm's headquarters might

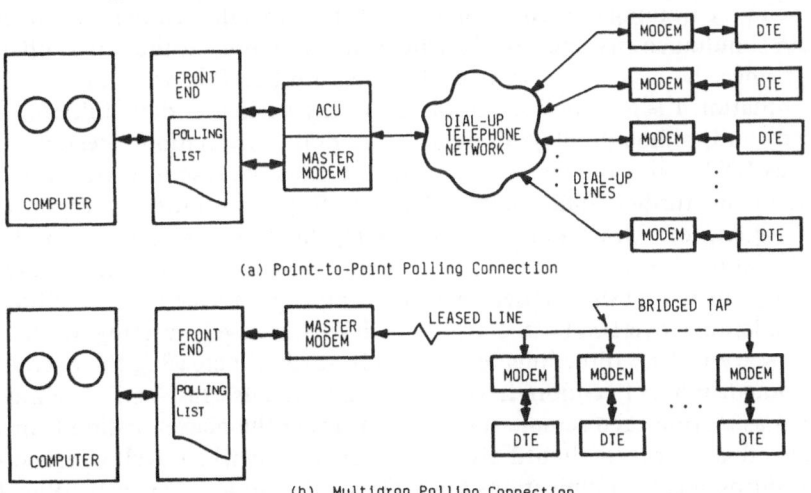

(a) Point-to-Point Polling Connection

(b) Multidrop Polling Connection

Figure 6.13. Polling arrangements.

† This is often done when there is a secondary channel on one of the modems simply by keeping it turned on continuously once the suppressors are initially disabled by the answerback tone. Additionally, the 208 provides a special 600-Hz "out-of-band" tone for this purpose.

‡ Many statistical multiplexers are now capable of polling, too.

use an autodial modem to poll each branch office once each day to get a compilation of the day's transactions. A conventional HDX dial-up protocol could be used with only one of the autoanswer remote modems connected to the master modem at any given time.

The second configuration in Fig. 6.13b is a multidrop line, where each remote station is connected to a single leased line with a bridged tap. Here, unlike the preceding case, whenever the master modem transmits, all remotes hear. Thus the remote DTEs must be addressable (intelligent) so that only the polled station will respond. Operation is generally HDX, even on four-wire lines, because this is the natural mode for polled functions like checking status or receiving daily transaction data. Since conventional modems simply pass bits and are not addressable themselves, the link operation is highly dependent on the particular data link protocol used by the DTEs. This is covered in the next chapter.

Our primary concern here is with modem operation on multidrop lines, as illustrated in Fig. 6.14. This example is a leased four-wire link operated HDX using synchronous dial-up modems, perhaps the 209 or V.29 at 9600 bps. The normal RS-232C HDX protocol is used by the master and by each remote modem with provision made for the rather long training time required for the automatic equalizers.† The master station initializes the entire multidrop link by raising RTS, thus causing the master modem to send the training sequence and return CTS. Note that *all* remote modems train simultaneously and remain synchronized so long as the link is operating, since the master is always transmitting on this four-wire line.

Station 1 is now polled (according to the particular data link protocol used) and, although all stations hear the poll, only station 1 responds. It raises RTS, trains the master modem, and sends its response after CTS. If there is no further traffic, station 1 turns off its transmitter in response to RTS going off. Upon normal loss of CD, the master station proceeds to poll station 2 in exactly the same manner. Note that each remote modem must train the master modem before it can reply to the poll, resulting in considerable overhead. If a two-wire line was used, training would be necessary in *both* directions for each poll, further increasing the overhead. Although modem vendors have introduced various "fast-poll" techniques such as storing each remote station's settings at the master station (jamsetting), this overhead should always be considered when evaluating polled multidrop telecommunications links. Once all stations have been polled the master station may start the polling cycle over with station 1, or it may terminate the link by simply dropping its RTS line.

---

† Training time may be relatively long, e.g., 250 msec for V.29, compared to the average message length, in which case faster overall performance may be obtainable with slower modems having shorter training times!

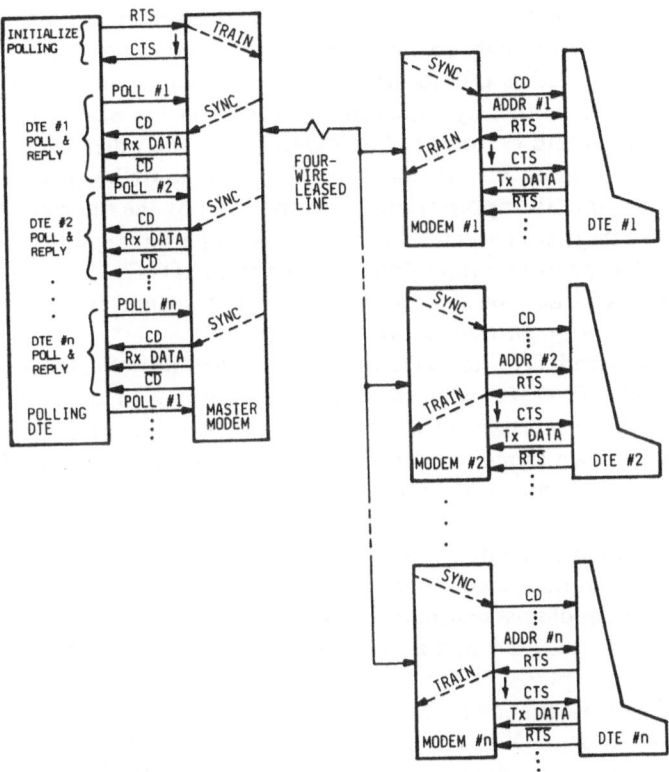

Figure 6.14. Four-wire HDX leased polled multidrop link.

Although by no means exhaustive, the preceding six examples should provide a good intuitive grasp of modem and physical interface protocol operation. In practice it is invariably necessary to consult with the particular equipment manuals or vendors to ensure optimal data link operation, but a good grasp of the fundamentals can greatly facilitate these interactions. This is even more true with troubleshooting, which we consider in the next section.

6.6. Null Modems and Modem Tests

We now turn to two additional practical considerations with interfaces: namely, the direct interconnection of two DTEs without intervening modems and then the increasingly important subject of loopback testing. Direct interconnection of two DTEs, each with an RS-232C port, is extremely common for local connections that do not require telecommunications links and modems. For example, this is the normal way to connect a local terminal to a computer. Although common, it is not trivial, since both DTE I/O ports are configured to talk to modems. It is, therefore, necessary for each DTE to "fool" the other by appearing to be a DCE. For example, TxDATA on one DTE should be rerouted to RxDATA on the other DTE, and vice versa, and any essential modem control signals must be provided either permanently or at the appropriate times specified by the interface protocol. It may in some cases be necessary to provide separate logic level voltages, or to insert appropriate time delays between the DTEs. For synchronous operation it will also be necessary to provide transmit and receive clock signals just as a synchronous modem would.

The required DTE-to-DTE interconnections are effected in practice with a special RS-232C cable having a male connector on each end and containing the necessary transpositions and rewiring for compatibility. Such devices are known as *null-modems*, or sometimes no-modems, or "haywire" connectors. They are also known as *modem eliminators*, although these generally also handle the more complex synchronous case. We shall restrict our introduction to some rather basic configurations. More sophisticated modem eliminators are commercially available as "black boxes."

For the simple asynchronous FDX case it is only necessary to cross TxDATA and RxDATA and tie the CD lines to a high logic level, with all other lines (except GND) unconnected. A little more elaborate version of this FDX haywire is shown in Fig. 6.15a. Since many DTEs will not operate unless CTS and/or DSR are on, RTS is looped back to immediately give CTS, while DTR produces an immediate CD and DSR. This tricks the terminal into providing all of its own FDX modem controls with no delays. Note that only a three-conductor cable is required with a 25-pin connector on each end. Another FDX variation is shown in the (b) part of the figure, where each DTR enables the opposite CD, DSR, and DTR. Hence FDX operation cannot occur unless both DTEs have DTR on, and RTS is not needed or used.

For asynchronous HDX dial-up connections the most popular haywire is shown in Fig. 6.15c. Here DTR produces RI and DSR at the opposite end, thus effectively bypassing the dialing and answer-back tone HDX procedures. The normal DTR response to the RI has exactly the same effect but in the opposite direction. Thus both DTEs now have DSR turned on and either may then initiate transmission via RTS. This produces the local CTS and remote CD that are required for successful HDX transmission. Turning off RTS allows the line either to be reversed or to go idle if there is no return traffic.

Our final subject of this chapter is loopback testing and network control techniques using modems and physical level interfaces. Loopback test capabilities are available on all but the simplest 103 and 202 modem types.

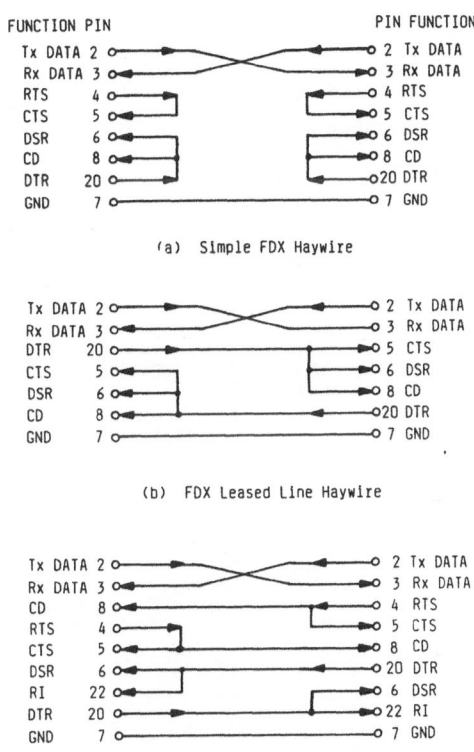

FUNCTION PIN   PIN FUNCTION

Tx DATA 2   2 Tx DATA
Rx DATA 3   3 Rx DATA
RTS   4   4 RTS
CTS   5   5 CTS
DSR   6   6 DSR
CD   8   8 CD
DTR   20   20 DTR
GND   7   7 GND

(a) Simple FDX Haywire

Tx DATA 2   2 Tx DATA
Rx DATA 3   3 Rx DATA
DTR   20   5 CTS
CTS   5   6 DSR
DSR   6   8 CD
CD   8   20 DTR
GND   7   7 GND

(b) FDX Leased Line Haywire

Tx DATA 2   2 Tx DATA
Rx DATA 3   3 Rx DATA
CD   8   4 RTS
RTS   4   5 CTS
CTS   5   8 CD
DSR   6   20 DTR
RI   22   6 DSR
DTR   20   22 RI
GND   7   7 GND

Figure 6.15. RS-232C null-modem con-
nections.   (c) HDX Dial-Up Line Haywire

They are defined formally for CCITT modems by the V.54 recommendation.
With RS-232C the tests are activated manually by switching into *test mode*.
They are usually done by systematically testing outward from the local DTE.

Loopback testing can be invaluable for saving time and money in the
event of a failure, since it systematically isolates the failed component. Such
capability is of increasing importance today as computer data communications networks are steadily growing larger with more sophisticated equipment and diverse telecommunications channels from an expanding selection
of vendors. Clearly those good old days when the telephone company could
quickly supply and maintain an entire network have passed, with the network
responsibility usually now falling on the data processing or communications
manager and his staff.

Local digital loopback, shown in Fig. 6.16a, tests the local DTE/DCE
cable and I/O ports by returning the DTE (digital) data signal as soon as
it enters the modem interface. Next the *local analog loopback* in Fig. 6.16b

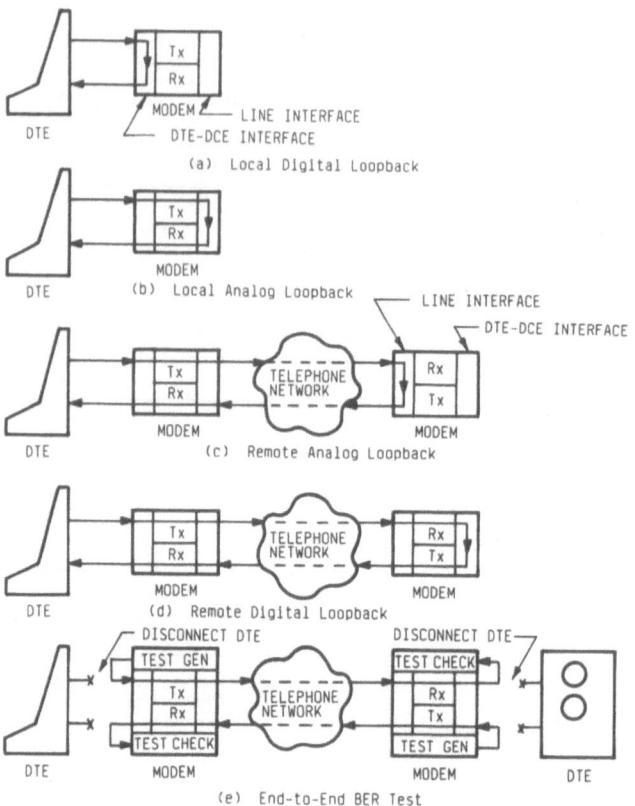

Figure 6.16. Modem loopback tests.

loops the modem transmitter (analog) output right back into the receiver, with a possible frequency translation for FDX two-wire modems. These first two tests are done by the local operator alone, but the remaining tests require coordination with the remote location. Figure 6.16c shows the *remote analog loopback*, where the received (analog) signal is looped back and transmitted without being demodulated by the remote modem, although again frequency translation is provided in FDX two-wire modems. This procedure tests the telephone line itself, assuming the local tests were satisfactory. Finally, the remote modem can be checked with the remote digital loopback test of Fig. 6.16d, where the demodulated (digital) signal is looped back at the remote interface and retransmitted to check the remote modem from the local end.

This systematic testing procedure can be invaluable for locating failed components or at least determining whom to call for repairs. Such tests can prevent a great deal of expensive and time-consuming "finger pointing" among different vendors. However, in some cases data links do not fail outright but rather gradually degrade in performance over an extended time. In such cases the so-called *end-to-end bit error rate (BER) test* is useful. Here each modem is equipped with an internal or external generator that produces a fixed digital test pattern, and also with a receiver for the same pattern with the capability of detecting and counting erroneous bits. When the test is switched on at both modems, the DTEs are consequently disconnected and replaced by the pattern generators and receivers as in Fig. 6.16e. The known test pattern is then sent by each transmitter and, at the opposite ends, the received pattern is checked for bit errors. These results can be used to detect exceptionally poor error performance, or to provide early detection of continually degrading performance before any serious failures occur.

With RS-232C these loopback tests are manual, with operators at each end typically communicating by voice over the telephone. An attractive capability of the newer RS-449 interface is a pair of control lines to the DCE called *Local Loopback* (LL) and *Remote Loopback* (RL) for setting these test modes through the interface, plus a status line from the DCE indicating *Test Mode* (TM). This allows loopback testing to be done automatically under program control without the need for an operator to be present. This is particularly nice for unmanned remote stations which are difficult or inconvenient for an operator to reach whenever testing is required.

There are also a number of standard test devices for monitoring data links. They range from simple *breakout boxes*, which are inserted in the DTE/DCE path and display the on/off status of all lines with LEDs, to highly sophisticated (and expensive) *BER testers* that measure error rates in test patterns, perform polling to check multidrop lines, emulate various terminals and data link protocols, and record the received data preceding and following the occurrence of preselected (trap) data patterns.

A natural evolution of loopback testing on data links is its extension to large networks containing many links. There is typically a central *network control station* that can initiate loopback tests, run BER tests, collect stored statistical information, effect dial backup, and even reconfigure the network to eliminate or replace failed components. This can be done either *in-band* as part of the normal data stream or *out-of-band* via secondary channels [3]. Such capability is becoming vital as networks become larger with components from many diverse vendors (none of whom seem to cooperate with the others). The penalties for networks being inoperative are also increasingly severe as dependence on them increases, and of course, there is always a shortage of good technicians willing to catch a midnight flight across the country to work on a network problem all week-end. Loopback testing will provide a basic foundation for studying complete network management and control systems in later chapters dealing with wide area networks.

In this chapter we have introduced the functional and electrical characteristics of the major DTE/DCE physical interfaces used today. Although fairly complete, our emphasis has been on a practical intuitive understanding rather than on exact technical specifications. To this end the examples of this final section should serve to tie together much of the concepts of the chapters covered so far. A good understanding of this physical level protocol will add the necessary dynamics to the modem material in the preceding chapter. Combined with the data link protocol in the next chapter, this material should provide the comprehensive background for practical design, operation, and maintenance of the point-to-point and multidrop data links that make up the majority of today's digital telecommunications systems. In addition, practically all large public and private digital networks use complex layered protocols in which the physical interface is invariably RS-232C or V.24.

References

1. *Interface Between Data Terminal Equipment and Data Communication Equipment Employing Serial Binary Data Interchange* (EIA Standard RS-232C), EIA, Washington, D.C. (1969).
2. *List of Definitions for Interchange Circuits between Data-Terminal Equipment and Data Circuit-Terminating Equipment* (CCITT Recommendation V.24), ITU/CCITT, Geneva (1981).
3. R. B. Freeman, Net management choices—sidestream or mainstream, *Data Commun.* **11**, 91–108 (1982).

Suggested Readings

M. Sargent III and R. L. Shoemaker, *Interfacing Microprocessors to the Real World*, Addison-Wesley, Reading, Massachusetts (1981).

A nice, application-oriented paperback that discusses current loop and RS-232C modem interfacing with the 8251 USART.

M. D. Seyer, *RS-232 Made Easy*, Prentice-Hall, Englewood Cliffs, New Jersey (1984).

This entire book on RS-232 gives an extremely clear and simple explanation of how the interface operates, including various haywire configurations plus an extensive appendix on personal computer connections.

Check Your Understanding of Chapter 6—True or False?

1. A computer will normally use DTR to tell an autoanswer modem to answer the phone.
2. Most UARTs will just send parity bits if there is no other traffic.

3. Null modems allow two DCE devices to be connected directly without any DTEs.
4. The 2225-Hz answer-back tone should disable any echo suppressors on the line.
5. RS-449 is designed to eventually replace the lower-performance RS-232C interface.
6. In high-speed modems, the training time may be longer than the data transmission time.
7. A major disadvantage of RS-232C today is that the data rate is limited to 50 bps.
8. RS-449 defines a voltage of -2 V as a data mark or a control off.
9. RS-232C defines a voltage of $+2$ V as a data space or a control on.
10. Polling with modems is done only on multidrop lines.
11. In-band network control systems operate over the modem secondary channels.
12. The USART is basically a very specialized A/D and D/A converter.
13. Many older interfaces with data rates above 20 kbps use the V.35 interface.
14. Most synchronous modems require all 20 of the RS-232C interchange circuits.
15. Each remote modem must be retrained at each poll on a multidrop line.
16. To test for a degrading telephone line the remote analog loopback test is indicated.
17. A 208 modem would normally use RTS/CTS to turn an HDX line around.
18. The 20-mA current loop interface is not normally used with modems.
19. RS-449 interfaces will handle the telco DS1 rate for up to 200 feet.
20. The RS-232C signal to hang up is usually to turn RTS off.
21. The mnemonics DCD, SCD, CD, and RLSD all refer to the same interface circuit.
22. With RS-449, a signal on circuit IC is answered by the DTE turning on circuit DM.
23. Some UARTs provide modem control signals such as RTS/CTS.
24. For loopback tests RS-232C and RS-449 must be switched manually by an operator.
25. The breakout box is only used if a high-speed modem catches on fire.

7

Data Link Protocol

7.1. Basic Data Link Protocol Considerations

In the preceding chapter the mechanical and electrical, i.e., physical, interface between DTE and DCE was considered, but such an interface alone is usually insufficient to handle conventional digital telecommunications between remotely located terminals. The reason is that the physical level, along with the accompanying modem, only passes bits blindly without any regard for what the bits mean or how they are grouped together. There is no indication, for example, as to where characters and data fields begin and end unless some convention is agreed on beforehand between the users of the physical link. Clearly, it is desirable to have a well-defined set of rules for communicating over point-to-point links that can be used by a wide range of different terminal users. In general, such a formal set of rules is called a protocol, and so the set of rules for link operation is a *data link protocol*.

For digital data communications, data link protocols define such things as how the links are set up and taken down, how errors are detected and controlled, how information bits are blocked and sequenced for transmission, how acknowledgements and retransmissions are provided, and how various types of failures are handled. For multidrop lines the data link protocol will also determine how the basic functions of polling and selection are handled, including provisions for addressing different drop stations along the line. Controlling the rate of data flow across a link is also a function of the data link protocol, therefore more sophisticated flow control techniques are usually necessary in addition to simple acknowledgements.

There are many different data link protocols in use today throughout the U.S. and the world. Although there are some official standards, particularly by the CCITT, there is also a large number of protocols specific to various computer equipment vendors and applications. Very small systems may even have "homemade" protocols. In the past it was conventional practice to use the particular protocol provided by the computer vendor, since most networks were supplied by a single vendor. Today the vast increase in the number of equipment vendors and in the complexity of data networks has made it necessary for different types of terminal equipment and computers to communicate over common links. Any resulting discrepancy in protocols may be resolved in two ways, short of replacing terminal devices. Either protocol conversion devices can be inserted in the links, or the communication software and hardware can be changed to accommodate the same protocol. In the latter case it is clearly desirable to have a well-known standard protocol which is supported by all, or at least most, terminal and computer suppliers. Fortunately, there is at least one such standard that is in widespread use today, namely, the ISO *High-Level Data Link Control* (HDLC), which was derived from the popular IBM *Synchronous Data Link Control* (SDLC) first introduced by IBM in 1974 for links to large IBM mainframe computers.

There are several important ways to classify data link protocols. First, they can be either *HDX or FDX* depending on whether or not data can be transmitted in both directions simultaneously. It is important to note that this is not the same thing as HDX and FDX physical lines. An HDX physical line must use an HDX data link protocol, with the inherent line turnarounds; however, an FDX physical line may be used with either type of data link protocol in the same way as the HDX four-wire multidrop line example of the preceding chapter. A second important classification is whether the protocol is *synchronous or asynchronous.* An asynchronous data link protocol requires each character to be synchronized separately between the transmitter and receiver, typically through the use of start and stop bits. This procedure is generally not very bandwidth efficient for remote serial data links owing to the large number of overhead bits and the continual timing adjustments at the receiver. Practically all of the more sophisticated data link protocols in use today are synchronous and use a special synchronization character (such as SYN = 0010110 in ASCII) to align the transmitted characters at the receiver. Note that at the data link level synchronization pertains to characters and blocks of characters, while at the physical (modem) level synchronization is limited to only bits. It is common practice, for example, to use an asynchronous FSK modem together with a synchronous data link protocol.

A fundamental difference between the physical and data link protocol is again illustrated: the former just passes bits with no regard for their meaning or grouping, while the latter passes characters that may be interpreted as either data or as link control information. Proper response to this control information requires considerable "intelligence" at the data link level. A third key classification of link protocols concerns the means of distinguishing between *control information and data.* It is essential that there be no confusion here, particularly in the misinterpretation of data patterns for control characters. It is highly desirable that the transmitted data be *transparent* to the data link protocol so that no restrictions need be placed on transmitted text or numerical data. There are three major approaches to this transparency problem, leading to *bit-oriented, byte-oriented,* and *character-oriented* protocols.

In addition to these basic protocol classifications, there are numerous others. Some protocols contain special features that allow more efficient operation on satellite links. There are various provisions for multipoint operation, and also sometimes for operation on loop networks where messages are passed from terminal to terminal around a closed path. This chapter describes four of the most widely used data link protocols. Through these representative examples we shall study the nature of practical data link protocol and its relationship with the physical level protocol and with link components.

7.2. Character-Oriented Protocol

Many of the older and less sophisticated protocols are character-oriented, but they are still widely used where network efficiency is not an overriding consideration, e.g., bank ATM machines or retail POS systems. Most such systems are based on a central computer with terminals radiating out from it over point-to-point or multidrop lines. This leads to the notion of a *primary or control station* that controls the link, and a *secondary or tributary station* that can respond only to the primary. Assuming the link is set up, the primary station can initiate data flow by either polling or selecting. A *poll*, as in the preceding chapter, is an invitation to the secondary to send any data it may have waiting, while a *selection* is a notification that the primary has data for the secondary. This section will first consider the IBM *Binary Synchronous Communication* (BSC or BISYNC) link protocol and then the Burroughs *Basic Line Control Procedure* (Poll/Select) as practical character-oriented data link protocols.

BSC is an older IBM synchronous data link protocol that operates HDX, with an acknowledgment required for each transmitted data block before the next block can be sent. It is designed for use on point-to-point and polled multidrop networks with a single primary station. Figure 7.1a shows typical frames† of the protocol which we will now describe. There are at least two SYN characters that determine the character boundaries, and then a *Start of Header* (SOH) character. A header field may not be necessary on point-to-point links, in which case SOH and the header can be omitted. Otherwise the header is user-supplied and typically contains such information as addressing, message numbering, and routing. It is normally terminated with the next BSC control character *Start of Text* (STX). The text field following STX is the actual information being conveyed from the user directly or perhaps from a higher-level protocol in the form of a data packet. It may also be only a *block* of a longer message being transmitted in multiple frames. Such blocks are terminated by an *End of*

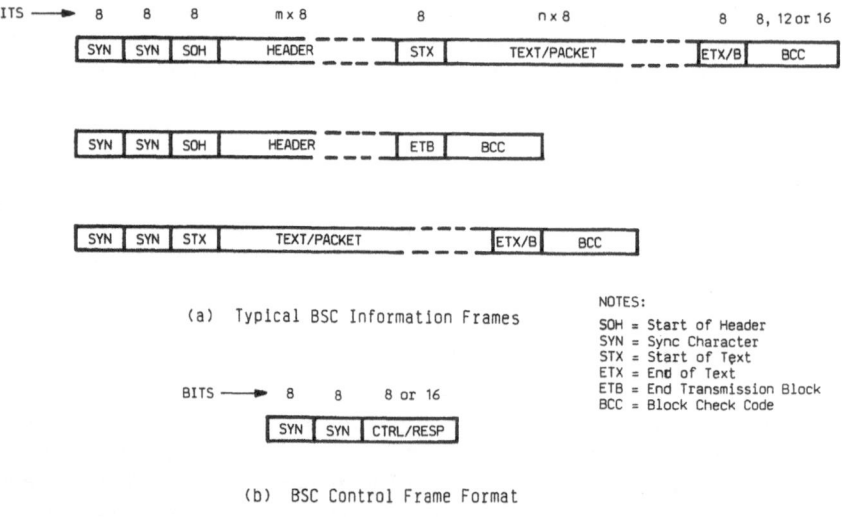

(a) Typical BSC Information Frames

NOTES:
SOH = Start of Header
SYN = Sync Character
STX = Start of Text
ETX = End of Text
ETB = End Transmission Block
BCC = Block Check Code

(b) BSC Control Frame Format

Figure 7.1. BSC frame formats.

† We shall call the block of data bits transmitted across the link a *frame* and will later have occasion to refer to the data or text field of the frame as a *packet*. This distinction will provide consistency of presentation throughout this book but is by no means strictly adhered to in the literature.

Transmission Block (ETB) character until the final block (or a one-frame message) which is terminated by *End of Text* (ETX).

The last field is a *Block Check Code* (BCC) for detecting errors in the preceding text and header fields. Three different BCC algorithms are specified, with the particular one used dependent on the binary code used. For ASCII data the BCC is determined by a combination of odd parity (vertical redundancy checking, or VRC) on each character, and a modulo 2 summation of corresponding bits over all characters (longitudinal redundancy check, or LRC) including the parity bits. Thus this BCC is eight bits long. For the EBCDIC code preferred by IBM the BCC is a 16-bit cyclic redundancy check (CRC) which is considerably more powerful in detecting errors than the VRC/LRC. Finally, there is a provision for a 12-bit CRC BCC with the European Transcode. The operation of error check codes is described in the Appendix to this chapter.

In addition to the basic information frame described above, there are several other frames used for control purposes. These all have the basic form shown in Fig. 7.1b. To initialize a point-to-point link the originating station sends an *Enquiry* (ENQ) control frame, and should receive back an *Affirmative Acknowledgment Zero* (ACK 0) as shown in Fig. 7.2a. Then each information frame is sent and acknowledged in turn until the session is terminated by an *End of Transmission* (EOT) control frame as indicated in the preceding figure. Note that the acknowledgments alternate between ACK 0 and ACK 1. This provides some protection against lost frames at the expense of a two-byte character for ACK 0/1.†

A frame received with errors, i.e., the BCC does not match the corresponding result at the receiver, is acknowledged with *Negative Acknowledgment* (NAK) control frame, and the erroneous frame is then retransmitted. In this way the protocol makes the link look essentially error-free to the terminal devices. There is one remaining type of acknowledgment called *Wait-Before-Transmit Positive Acknowledgment* (WACK), which indicates a frame was correctly received but that the receiver does not want any more frames until a later time. Upon receiving WACK the transmitting station will normally send ENQ; the stations continue to alternate WACK and ENQ at least every two seconds until the normal ACK 0 or ACK 1 is received, at which time normal transmission resumes. This sequence is shown in Fig. 7.2b. There are a few other control frames applicable to point-to-point links, such as *Reverse Interrupt* (RVI) acknowledgment to turn the line around in the middle of a transmission, but they will not be considered here. Complete details are found in the IBM specification [1].

For multipoint operation it is necessary to provide for polling and selecting, including the addressing of tributary stations. There are no

---

† The actual codes used for ACK 0 and ACK 1 are DLE 0 and DLE 1, respectively.

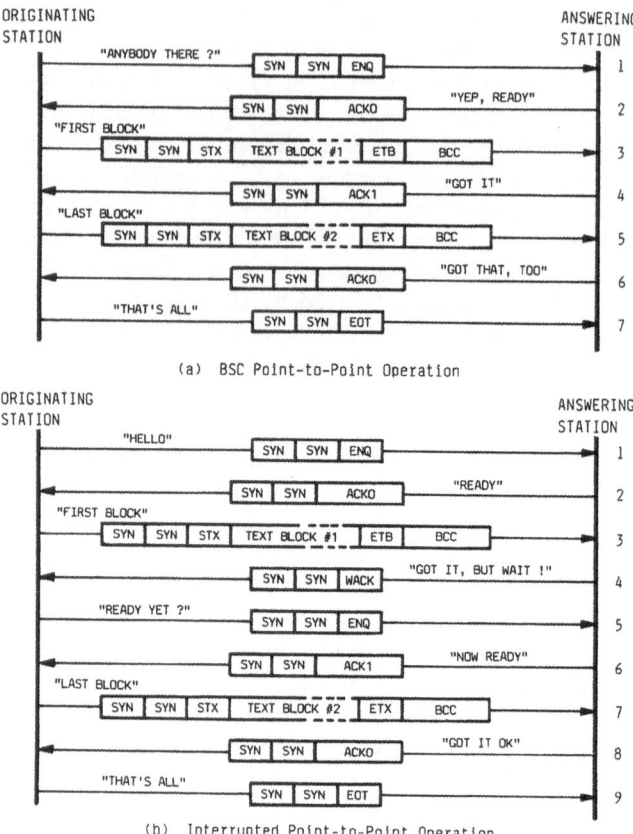

(a) BSC Point-to-Point Operation

(b) Interrupted Point-to-Point Operation

Figure 7.2. BSC protocol operation.

separate poll and select characters in BSC, so the distinction is made through address assignments. For example, a printer would always be selected. The basic polling and selecting frames look the same, namely, "PAD SYN SYN address ENQ PAD," with a standard poll response being either a conventional information frame or, if no traffic, an EOT in the form "PAD SYN SYN EOT PAD." For a select, the proper tributary response is ACK 0 if ready, otherwise NAK or WACK. An initialization sequence for the polling cycle is just the combined frames and has the form "PAD SYN SYN EOT PAD SYN SYN address ENQ PAD." The leading and trailing PAD characters aid in modem operation, and their coding is dependent on the particular installation.† The former serves as a warning to the modem that a frame is coming so that it will not distort the following SYN characters. Continuous

† A common code for the leading PAD is 01010101 (alternating pattern) and for the trailing PAD is 11111111.

SYNs are also used as the idle pattern to maintain character synchronization during periods when there is no other traffic to transmit.

Our final consideration with BSC concerns transparency. If, for example, the text contains the particular bit pattern corresponding to ETX or ETB, then the receiver will assume the frame has ended and will unsuccessfully try to use the next one or two bytes (of data) as a BCC and assume that errors have occurred. Retransmissions will simply repeat the process until various timeouts result in a link failure and perhaps disconnection. Such situations are avoided by use of the *BSC Transparent Mode*, which basically uses the special control character *Data Link Escape*† (DLE) to ensure that valid control characters are unique. The algorithm by which this is done is as follows:

Algorithm 7.1 (*BSC Transparent Mode*)
1. *DLE is "stuffed" in front of all valid control characters in the frame header and trailer.*
2. *Whenever a DLE pattern occurs within the text, another DLE is stuffed directly in front of it to produce DLE DLE.*

This simple character-stuffing operation makes the text transparent, as illustrated by the following example text string in which the stuffed DLEs are underlined:

Original text: A DLE ETX G O T O DLE X ETB
Transparent text: A DLE DLE ETX G O T O DLE DLE X ETB
Resulting frame: SYN SYN DLE SOH header DLE STX A DLE DLE
ETX G O T O DLE DLE X ETB DLE ETX BCC

Although the actual implementation of transparent BSC protocol appears rather complicated, it is straightforward and there are a number of sophisticated USART-like integrated circuits that perform functions for BSC as well as for other common protocols. Figure 7.3 shows such a protocol controller chip for BSC which also handles asynchronous UART functions and the SDLC protocol described later in this chapter. There are a few other details of BSC but the salient points have been considered above. Although BSC is an old protocol going back to 1968, it is still widely used in practice where HDX operation is sufficient.

The *Burroughs Basic Line Control Procedure* [2] is commonly called *Poll/Select*, since there are separate control characters for polling and selecting. It is an HDX link protocol and also makes the distinction between

† It may be helpful to consider DLE as being analogous to the shift key of a typewriter; thus control characters are analogous to upper case letters in the transparent mode and this makes them unique from any data patterns.

8274
MULTI-PROTOCOL SERIAL
CONTROLLER (MPSC)

- **Asynchronous, Byte Synchronous and Bit Synchronous Operation**

- **Two Independent Full Duplex Transmitters and Receivers**

- **Fully Compatible with 8048, 8051, 8085, 8088, 8086, 80188 and 80186 CPU's; 8257 and 8237 DMA Controllers; and 8089 I/O Proc.**

- **4 Independent DMA Channels**

- **Baud Rate: DC to 880K Baud**

- **Asynchronous:**
  - **—5-8 Bit Character; Odd, Even, or No Parity; 1, 1.5 or 2 Stop Bits**
  - **—Error Detection: Framing, Overrun, and Parity**

- **Byte Synchronous:**
  - **– Character Synchronization, Int. or Ext.**
  - **– One or Two Sync Characters**
  - **– Automatic CRC Generation and Checking (CRC-16)**
  - **– IBM Bisync Compatible**

- **Bit Synchronous:**
  - **– SDLC/HDLC Flag Generation and Recognition**
  - **– 8 Bit Address Recognition**
  - **– Automatic Zero Bit Insertion and Deletion**
  - **– Automatic CRC Generation and Checking (CCITT-16)**
  - **– CCITT X.25 Compatible**

- **Available in EXPRESS**
  - **—Standard Temperature Range**

The Intel® 8274 Multi-Protocol Series Controller (MPSC) is designed to interface High Speed Communications Lines using Asynchronous, IBM Bisync, and SDLC/HDLC protocol to Intel microcomputer systems. It can be interfaced with Intel's MCS-48, -85, -51; iAPX-86, -88, -186 and -188 families, the 8237 DMA Controller, or the 8089 I/O Processor in polled, interrupt driven, or DMA driven modes of operation.

The MPSC is a 40 pin device fabricated using Intel's High Performance HMOS Technology.

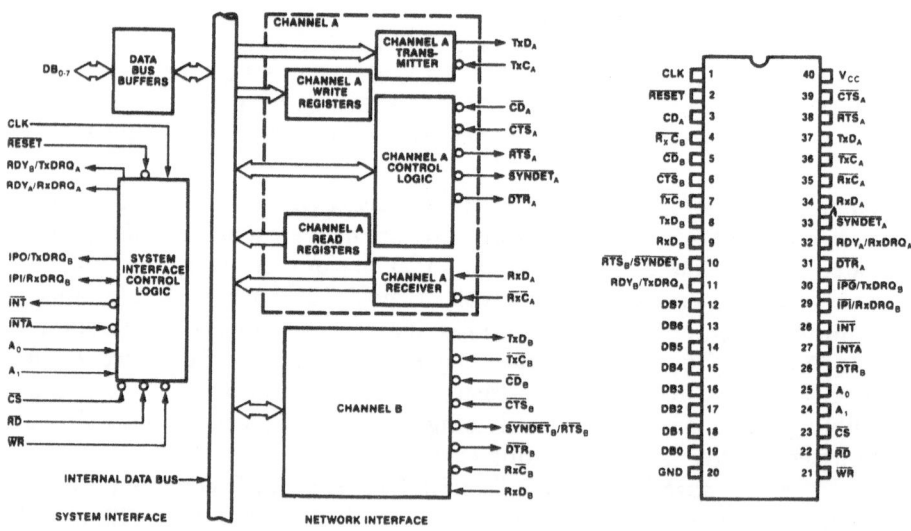

1. **Block Diagram**   2. **Pin Configuration**

Figure 7.3. Controller for BSC data link protocol. (Courtesy of Intel Corporation.)

control and tributary stations. A provision for message numbering is inherent in the protocol, for which a major application is financial data transmissions where missing frames can result in large financial losses.† There is also a dedicated field for a 16-bit address. Poll/Select may operate either synchronously or asynchronously.

Rather than enumerate the complete protocol, we shall indicate the main features and applications of it by example. In Fig. 7.4 a dial-up point-to-point link operation is shown in which a two-block message is sent from the originating to the answering station. Transmission originates with an ENQ, and frames are acknowledged with ACK or, if errors are detected, with NAK. The EOT character indicates no further traffic in a sequence of frames, and it is also used to precede polls and selects to ensure the called terminal is listening. Dial-up links are normally disconnected with DEOT, which is composed of the two characters DLE EOT. The address and

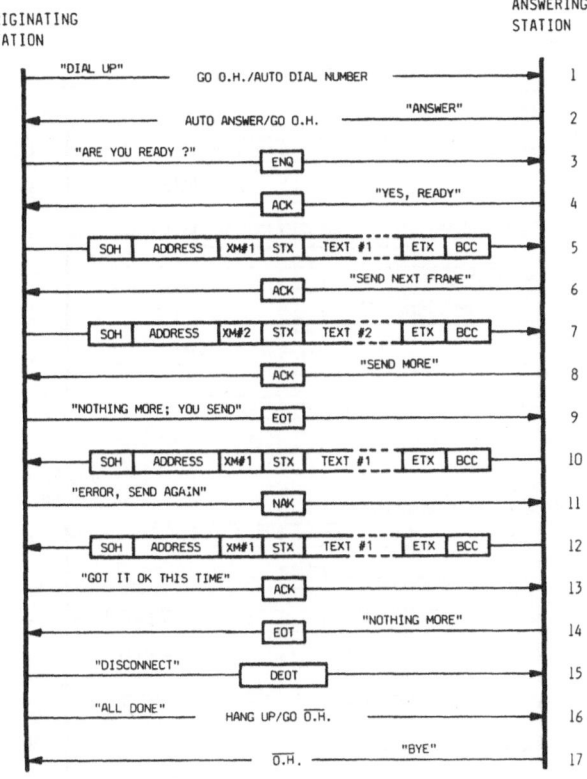

Figure 7.4. Poll/select dial-up operation.

† Or large financial gains for embezzlers!

message numbers are indicated by AD1, AD2 and XM#n, respectively, and the eight-bit BCC uses a form of LRC for error checking. The SYN characters used for synchronous link operation are not shown. For asynchronous operation each character shown is sent with the conventional start and stop bits as required.

The poll and selection capability is shown in Fig. 7.5. Note that the polling frame contains a specific *POL* character along with the tributary address; similarly there is a specific *SEL* character for select frames. The appropriate responses in these cases are EOT to indicate nothing to send, and ACK or NAK for acknowledgments. In addition to SEL, the protocol contains provision for *Fast Select* (FSL) where the select frame is followed immediately by the message without waiting for a ready indication from the tributary station, and *Broadcast Select* (BSL) for selecting all stations at once. There is also a variety of timeouts and disconnect procedures but such details are not necessary for our introductory understanding of the protocol.

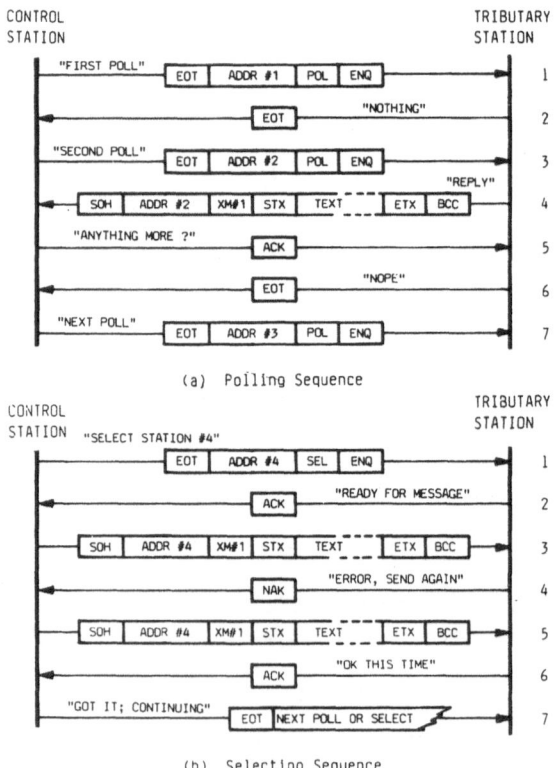

Figure 7.5. Poll/Select multipoint operation.

BSC and Poll/Select are both important protocols in their own right, and they contain many concepts fundamental to other data link protocols as well. They are widely used in existing data systems but the inherent HDX operation is a severe restriction in many modern distributed computing situations where there is no clear central primary station, and where high throughput in both directions is essential. For example, the long round-trip delay of over half a second for satellite links makes the average throughput and response time for an HDX protocol totally unacceptable. In the next two sections we shall consider the two main types of FDX protocol, again through the use of practical examples.

7.3. Byte-Oriented Protocol

The two character-oriented protocols of the preceding section were both HDX. They resolved the transparency problem by resorting to the insertion of the DLE character to distinguish control characters from data patterns. There are, of course, other ways to handle data transparency, and there are also various ways to make protocols operate FDX. In this section we will study one important example of protocols that provide transparency by simply *counting* the number of data bytes and including this count in a specific field of the frame header. Then, so long as this count is received correctly, the receiver need only count the indicated number of data bytes with no regard whatsoever for the inherent data patterns.

To provide FDX operation this protocol basically allows for acknowledgments of multiple frames to ride "piggyback" in a specific field of returning traffic frames. This scheme is based on each outstanding frame having a unique sequence number that is actually used to indicate which frame is being acknowledged. Clearly this adds additional overhead to each frame; in fact, for short transaction-type traffic with very short data fields, the overhead may be significantly larger than the actual data. However, for long blocks of data the procedure is increasingly efficient, and for links where bandwidth or response time is critical an FDX protocol may be the only viable choice.

The particular protocol to be considered here is the DEC *Digital Data Communications Message Protocol* (DDCMP), since it is the most popular and pervasive of the commercial byte-count data link protocols. Figure 7.6a shows the basic frame format for data transfer. Note first that, including the *two* CRC error check codes, there are ten bytes (80 bits) of overhead for each frame, not counting any initial SYN characters required when operated synchronously. The SOH byte, coded 10000001, starts the header of all Data Message frames. The next header field is 14 bits long and contains the actual *COUNT* of the number of data bytes in the frame. This, then,

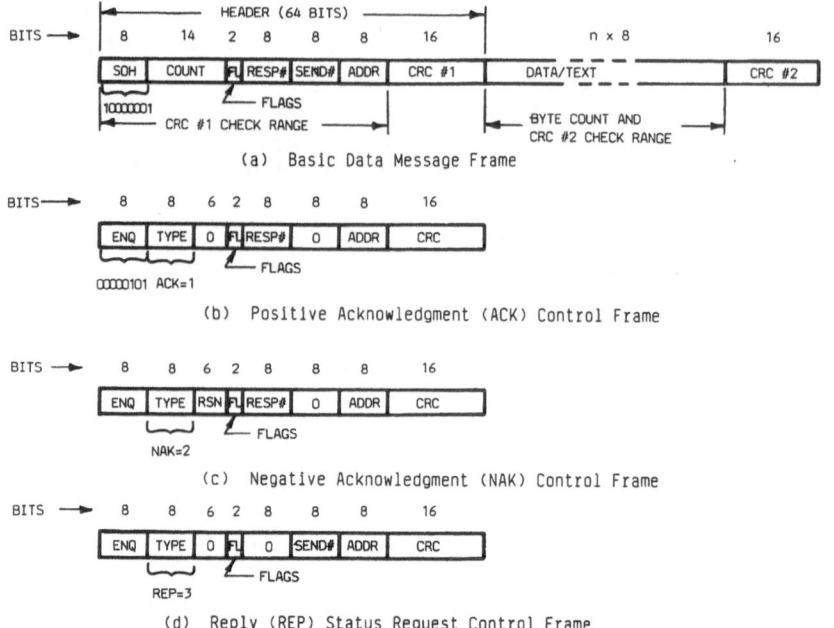

Figure 7.6. DDCMP message frame formats.

allows for a maximum of $2^{14} - 1 = 16383$ bytes of data. Longer messages must be divided into blocks and sent in multiple frames. For the maximum length data field, the overhead is only $10/16393 = 0.06\%$, but it is 50% for a ten-byte message. The remaining two bits following the count field are two *Flags* (FL). The first is called *Quick Sync* (QSYNC) and is used to facilitate receiver synchronization when there is a short gap before the next frame is to be sent. The second bit, the *SELECT* flag, is used primarily for polling and selecting by allowing the direction of transmission to be reversed for line turnaround.

The next two eight-bit fields provide the mechanism for FDX operation through sequential numbering of all transmitted and received packets. Each transmitted frame is numbered successively from 0 to 255 (modulo $2^8 = 256$), and this *send sequence number* (SEND#) is placed in the header. This ensures proper sequencing of the frames at the destination, and provides a unique (modulo 256)† identification number for the acknowledgment. If the frame is received without error, then the destination station may

---

† The sequence numbers will start over after reaching 255, i.e., ...254, 255, 0, 1, 2, ..., but will be unique provided no more than 256 frames are outstanding in the network at any given time.

acknowledge it by placing the frame sequence number into the *response number* (RESP#) field of the next frame of return traffic. The same procedure is used in the opposite direction. Since acknowledgment of any frame implies acknowledgment of all preceding frames, a continuous flow of frames may occur in both directions simultaneously, i.e., operation is FDX. If there is no immediate return data traffic on which to piggyback the acknowledging sequence numbers, then a separate *Positive Acknowledgment* (ACK) control frame is returned. This frame format is shown in Fig. 7.6b where the first byte of a control frame is always *ENQ* (00000101), and the following *Type* field is coded 00000001 to indicate the control packet is a positive acknowledgment. After the two flag bits, the response number goes into the RESP# field indicating the sequence number of the last frame correctly received. In general, up to 256 frames may be received before an acknowledgment is sent, provided the delay does not exceed a transmitter time-out period. This type of operation is descriptively called *pipelining*.

The eight-bit *Address* (ADDR) field allows the control station to address up to 255 tributary stations† in a multidrop configuration. For point-to-point links, the address is immaterial and arbitrarily set to 1. Finally, there are two check code fields of two bytes each, with the actual codes being computed by the same CRC algorithm. The first check is on the header, since it is relatively long and contains the byte count for the data field. The second check is on just the data field itself and is not present in the various control frames. Finally, the data field is totally transparent and is only required to contain the integral number of bytes specified by the count field.

If an error is detected in the receiver CRC check procedures, then it will return a *Negative Acknowledgment* (NAK) control frame having the format of Fig. 7.6c. It contains a six-bit code in the *Reason* (RSN) field to indicate the nature of the error, e.g., which check code failed to check. In addition, the RESP# field indicates the last correctly received frame prior to the error detection. Thus it provides the same information as the ACK frame. After sending a frame, the transmitter will set a timer and only wait this predetermined time for an acknowledgment before taking action. For example, if frames are received out of sequence or in error, the receiver may simply do nothing and let the transmitter timeout. The transmitter then requests the status at the destination using a *Reply* (REP) control frame of the form shown in Fig. 7.6d. The SEND# field contains the sequence number of the last unacknowledged frame sent, and upon receipt, the receiving station will send an ACK frame if it has correctly received the indicated frame (and those preceding it). Otherwise something is clearly wrong, and the proper response is an NAK containing the last sequence

---

† Address 0 is reserved, and not used for tributary addressing.

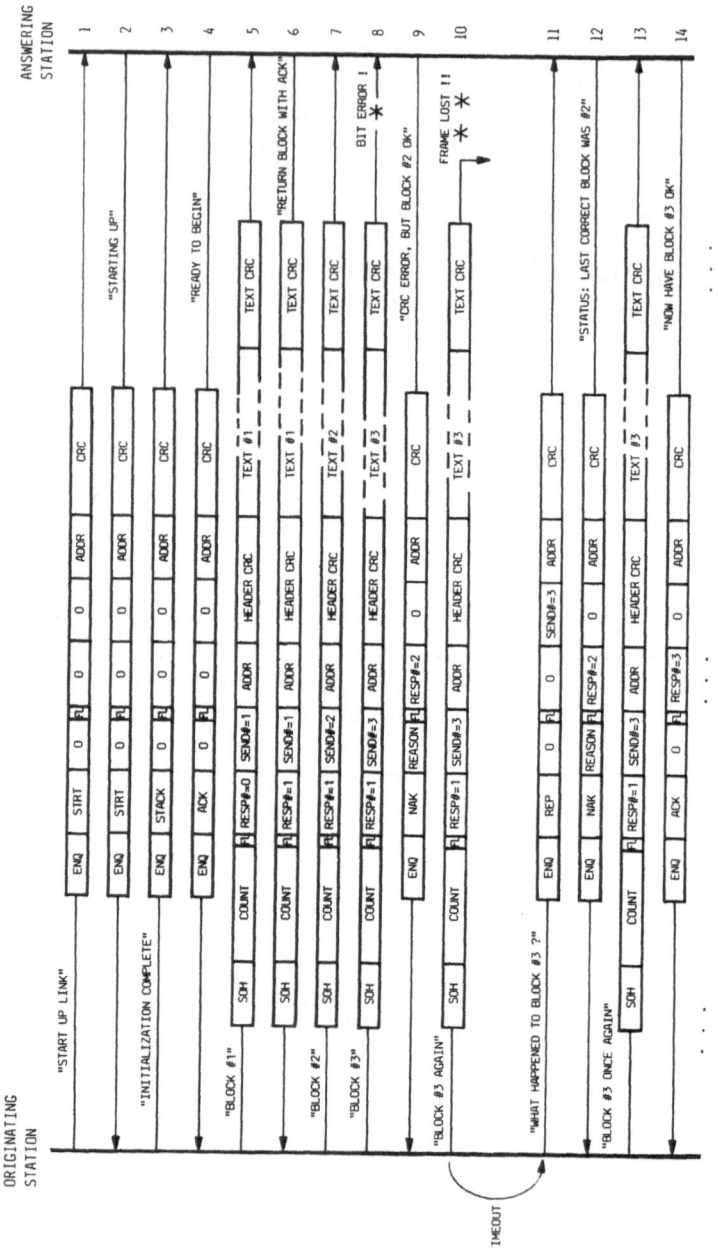

Figure 7.7. DDCMP point-to-point operation.

number that was correctly received. The transmitting station can then resend all the remaining unacknowledged frames.

To initialize a link, a *Start* (STRT) control frame is sent and returned, thus initializing all sequence counters and timers. The STRT frame is acknowledged with a *Start Acknowledge* (STACK) frame, after which the link is ready for nomal operation. Figure 7.7 shows an entire communications sequence for a representative data link using DDCMP, including some error procedures. Note (line 10) that when frame #3 is lost (or garbled beyond recognition) in transmission, an REP frame is used to obtain the status of the receiving end and then the necessary frames are retransmitted.

Polling is performed with DDCMP using the SELECT flag bit. On a multipoint network the control station sends to a tributary by simply sending a frame addressed to the tributary. This corresponds to what was previously called selecting. To poll a tributary, the control station will set the SELECT bit to 1 in the last frame of whatever traffic is being sent. Since every DDCMP frame has this flag and the control station will always initiate a polling sequence, then this type of polling is always possible.† When the tributary is through sending its traffic, it will set the SELECT bit in its last frame, thereby reversing the line again and relinquishing it to the control station. The same procedure is used on point-to-point HDX links to turn the line around.

In addition to the data and control frames, there is a group of *maintenance messages* which all start with a DLE character instead of SOH or ENQ. They are used for various "housekeeping" functions such as testing and downline loading, but they will not be considered here. There are also a few other details of DDCMP for restarting failed links and handling timeouts. However, the discussion and example of this section offer a fairly comprehensive view of how the protocol actually operates. DDCMP is relatively complex compared to BSC or Poll/Select, but it offers much higher performance with FDX capability, inherent transparency, and the ability to have many unacknowledged frames outstanding in the network.

While BSC and Poll/Select were developed some years ago to connect relatively simple (by today's standards) configurations of terminals to a central computer, DDCMP is part of a sophisticated hierarchy of protocol layers designed to handle networks of many different types of terminal devices and multiple minicomputers—all interconnected with each other. This basic philosophy accounts for many of the differences between the two types of protocol. We shall use our knowledge of data link protocol in Chapter 10 as the foundation for studying such complex protocol hierarchies, or network architectures. In this chapter, however, we next turn to the third category of data link protocol.

---

† Actually, the SELECT bit is used for polling, selecting, or a combination of both at the same time.

7.4. Bit-Oriented Protocol

In addition to character stuffing and byte counting, there is a third major approach to data link protocol transparency known as *bit stuffing*. This is the approach used by the majority of modern, high-performance, FDX protocols because of its simplicity and efficiency. The basic idea is to define a single *flag byte* that is then used to indicate uniquely the beginning and ending of each frame. Obviously, for transparency, the flag byte pattern must not be allowed to occur in the bit patterns of the frame contents between these opening and closing flag bytes. This is assured by continuously monitoring the outgoing data stream, and modifying it whenever the flag pattern might occur. The process involves inserting an extra bit at each such occurrence, hence the name bit stuffing. The receiver, in turn, must be able to reverse the process and remove any stuffed bits in an unambiguous manner. The overall transparency mechanism is quite simple and easily implemented compared to, for example, character stuffing as with BSC.

We shall study bit-oriented protocols through the specific example of the IBM *Synchronous Data Link Control* (SDLC) protocol [3]. First introduced in 1974, SDLC is not only widely used in the ubiquitous IBM data systems, but it has provided the pattern upon which many other important data link protocols and standards have been based. In particular, it is nearly a subset of the ISO *High-Level Data Link Control* (HDLC), which is fast becoming an internationally accepted standard. However, being older, SDLC is somewhat less complex and more easily understood than the related protocols, and thus provides a very appropriate vehicle to introduce all bit-oriented protocols. In a later chapter we will consider related protocols such as ANSI ADCCP, ISO HDLC, and CCITT X.25 LAP/LAP B, for which SDLC will provide an excellent foundation for study.

SDLC operates FDX by employing the same piggyback acknowledgment approach as DDCMP and can be used on point-to-point or multipoint, switched or nonswitched lines. It also has a mode of operation for local loop configurations. As the name implies, SDLC is entirely synchronous. Polling and selecting are performed with a special bit called the *Poll/Final* (P/F) bit, in a manner similar to the SELECT flag bit in DDCMP. The key to data transparency is the unique *flag byte*, which is the particular pattern 01111110. This flag is always the first and the last byte of every frame, and is not allowed to occur anywhere in the remaining bits of the frame. To ensure this condition, the transmitter monitors the outgoing data stream between flag bytes, searching for any pattern of a zero followed by five consecutive ones, i.e., ... 011111 .... Whenever this situation occurs, the transmitter inserts an additional zero after the fifth one, thus nullifying the possibility that a flag byte pattern could inadvertently occur. This *bit stuffing* is done as the last step before transmission; so it applies to header and trailer fields as well as the actual data field, making everything transparent. At the receiver the same monitoring procedure is done in reverse. After receiving five consecutive ones the next bit is examined. If it is zero, it is deleted since it must have been stuffed by the transmitter. However, if it is a one, then one more bit is examined to confirm that it is a zero; hence the flag bytes are detected at the beginning and end of each frame. Should there be more than six consecutive ones, an abort condition is indicated. For example, a string of from 7 to 14 consecutive ones indicates that the current frame should be aborted immediately, and more than 15 consecutive ones puts the line in an idle state so long as the ones continue to be transmitted. This forced idle condition is different from the normal idle when there is no current traffic, with the latter indicated by continuous flag bytes rather than continuous ones (the mark hold condition).

There are three basic types of frame in SDLC, all of which are similar in format. They all open and close with the flag byte, and all contain an address field, a control field, and a CRC error check field. Figure 7.8a illustrates the *Information Transfer* (I) frame, which is used for data transfers. After the opening flag byte there is an eight-bit *Address Field*. In SDLC, as in the preceding protocols, there is the fundamental notion of a control, or *primary*, station and a tributary, or *secondary*, station. With SDLC this is true even for point-to-point links. Only one primary is allowed, and it has the responsibility of controlling the link, but there may be many secondary stations such as on a multipoint configuration. The address field *always* refers to the secondary station, since communication is always between the primary and a secondary.† When the primary sends, it puts the secondary address in the address field; and when the secondary replies,

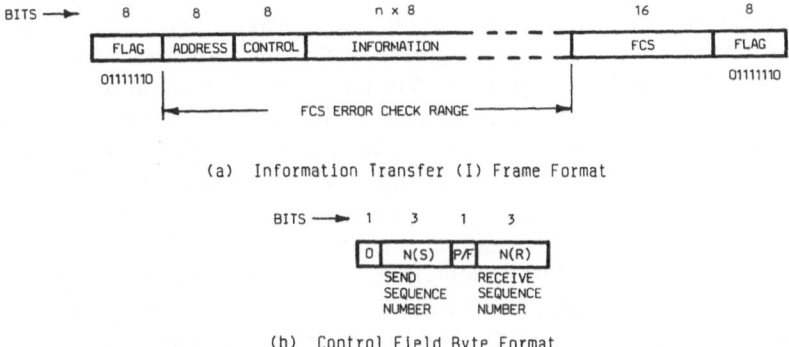

(a) Information Transfer (I) Frame Format

(b) Control Field Byte Format

Figure 7.8. SDLC information transfer (I) frame format.

† This is not strictly true in the SDLC *loop mode*, where a frame is passed from adjacent station to station, but the same addressing scheme is still used.

it also puts its own address in the field. There is also a provision for *group addressing* of selected groups of secondary stations and for *broadcast addressing* to all stations listening. For point-to-point operation the address is not critical, but it is often selected according to the X.25 convention of assigning the primary address 00000001 and the secondary, 00000011.

Following the address field is the eight-bit *Control Field*, which is always present but varies in its subfields according to the frame type. For the I frame the control field is shown in Fig. 7.8b, and is identified by a zero in the first (leftmost) bit position. Next is a three-bit field for the *Send Sequence Number* N(S) which allows transmitted frames to be numbered from 0 to 7, i.e., modulo 8. Since only I frames carry data packets, they are the only type of frame having an N(S) field. The last three bits of control field contain the *Receive Sequence Number* N(R) used to piggyback acknowledgments in a manner similar to DDCMP. There are some differences, however. Besides allowing only seven frames to be acknowledged at a time, the number in the N(R) field of SDLC indicates the *next frame expected* by the receiver, rather than the last one actually received as in DDCMP. The receipt of a particular N(R) value implies acknowledgment of all frames up to, but not including, the transmitted frame having a send sequence number equal to N(R). Finally, in between the N(S) and N(R) fields is a one-bit flag called the *Poll/Final* (P/F) bit. When set to 1 in the last frame of a transmission sent from primary to secondary, it acts like a *poll bit* (P) in that it requires the secondary to transmit. In effect, it reverses the direction of transmission. The same bit may be set in some frames from the secondary in response to polls from the primary. In this latter direction it is called the *final bit* (F), since it is not set until the last frame of the response solicited by the primary station's poll. Thus the P/F bit operates in a manner quite similar to the SELECT flag of DDCMP. For every poll (P = 1) frame sent by the primary, there should be a corresponding response (F = 1) frame from the secondary. This pairing of P and F bits provides an additional means of error detection and recovery known as checkpointing.

Following the control field there is an *Information Field*, containing user data or a packet from higher protocol levels, which is present only in the I frames. It may be of any length, restricted only by the physical limitations of the particular equipment, the decreasing effectiveness of the error check code with the length of the block being checked, and an SDLC restriction that the field be an integral number of bytes. This last restriction is not present in the more advanced link protocols such as HDLC, but it is usually easy to pad out the end of a transmitted data block to ensure that the total number of bits is an exact multiple of eight. Notice that, because of bit stuffing, there is absolutely no restriction on the bit patterns of the information field. The next field contains a two-byte error check code known as the *Frame Check Sequence* (FCS), which is a variation of the conventional CRC check code described in the Appendix. By modifying the dividend and then the resulting remainder and using the complement of the result, some additional protection is afforded against certain unusual data patterns such as long strings of zeros. The FCS is computed for the combined address, control, and information fields and is applied before any bit stuffing is performed. It thus protects the entire frame from errors that are detectable. Finally, the FCS is followed by the final flag byte to end the frame. If another follows immediately, it is permissible for this ending flag to serve also as the beginning flag of the next frame, but there must always be at least one flag byte between frames.

The second basic type of frame in SDLC is the *Supervisory* (S) frame. S frames are used to acknowledge I frames when there are no returning I frames on which to piggyback the acknowledgments, and also to control the flow and retransmission of I frames. S frames do not carry data themselves but otherwise have a format similar to the I frames, as shown in Fig. 7.9a. The address and FCS fields are exactly the same as for the I frame, while the control field is as shown in Fig. 7.9b. The first two bits are always 10 to uniquely identify an S frame, and the P/F and N(R) fields are the same as in the I frame. Thus when an S frame is used for an acknowledgment in lieu of a returning I frame, the sequence number of the next frame expected goes in the N(R) field, thereby acknowledging all transmitted frames having sequence numbers through $N(R) - 1$ modulo 8. The remaining two bits of the control field allow the coding of four different types of supervisory messages, but only three are used in the conventional SDLC protocol.† *Receive Ready* (RR) S frames, coded 00, are used for acknowledging I frames as indicated above, and may be sent by either the

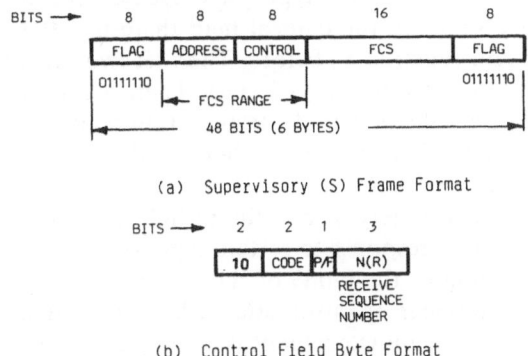

(a) Supervisory (S) Frame Format

(b) Control Field Byte Format

Figure 7.9. SDLC supervisory (S) frame format.

---

† The fourth code, 11, is called Selective Reject and is supported in most of the more advanced protocols that are based on SDLC. It is intended primarily for satellite links which have long transmission delays, as discussed in Chapter 10.

primary or secondary station. RR frames are also used by the primary with P = 1 for polling and by the secondary with F = 1 for responding to a poll when there is no traffic to send. Finally, RR may be used to indicate that a station is now ready to receive after a temporary interruption of transmission such as to clear a busy condition.

The remaining two codes are 01 for *Receive Not Ready* (RNR) and 10 for *Reject* (REJ). RNR is sent by either primary or secondary to indicate a temporary busy condition which precludes the acceptance of any more frames. It carries an acknowledgment in the form of an N(R) count, along with the P/F bit capability for the respective stations. RNR is cleared with either an RR, an REJ, or an appropriate I frame. REJ is used to request the retransmission of previously transmitted frames beginning with sequence number N(R), while acknowledging all preceding frames. An REJ is normally cleared when the requested frame N(R) is correctly received. The operation of SDLC with the various S frames is illustrated in Fig. 7.10, where the control field content is indicated symbolically by the form "I, N(S), P/F, N(R)" for I frames, and "RR, P/F, N(R)" for S frames. Note particularly the use of the P/F bit (lines 4 and 6) and the checkpoint recovery after the FCS error (line 5) is detected, as well as the use of the RNR S frame (lines 11 and 12). It will also be instructive to verify that all of the N(S) and N(R) numbers are correct. Although the example is for point-to-point connections, polling on a multidrop line is done in an identical manner with, of course, the appropriate secondary station address used for each station.

The third and final type of SDLC frame is the *Unnumbered* (U) frame, for which the basic format and control field are shown in Figs. 7.11a and b, respectively. As indicated, some types of U frame have an information field, while others do not. In the control field there is a P/F bit that has the same position and function as before. The first two bits are 11, which distinguishes the U frame from S and I frames. There are no send or receive sequence numbers associated with U frames (hence the name); so the remaining five control field bits are available for coding up to 32 command, response, or user-defined operations.

U frames provide a variety of additional link management functions such as link initialization, response to certain types of error conditions, and the setting and changing of the mode of the secondary station. They may also be used for the transfer of information when no sequence number is desired; however, if an acknowledgment is required, it must be provided by the user or by a higher protocol level. For example, an *Unnumbered Information* (UI) frame may carry status, link parameter, timing, or initialization information in its information field. A link disconnect may be requested by the secondary with a *Request Disconnect* (RD) response frame and can be effected by the primary station with a *Disconnect* (DISC) command.

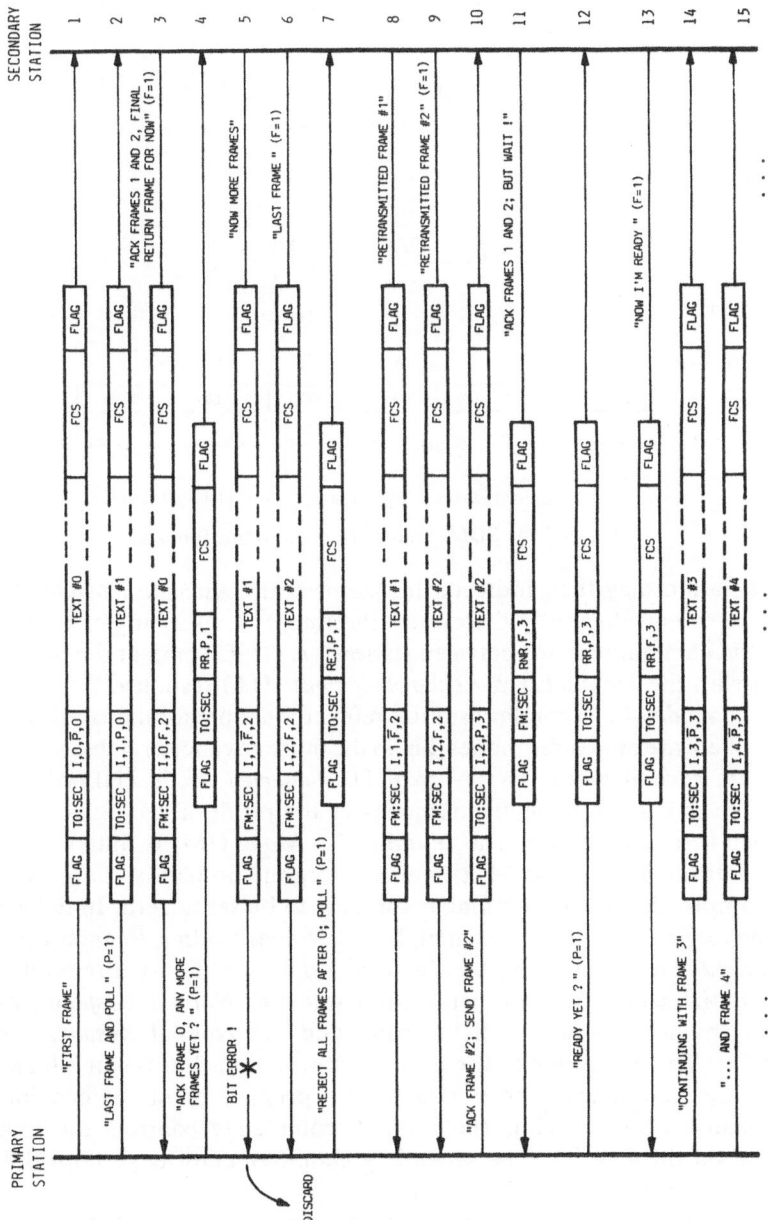

Figure 7.10. SDLC point-to-point operation.

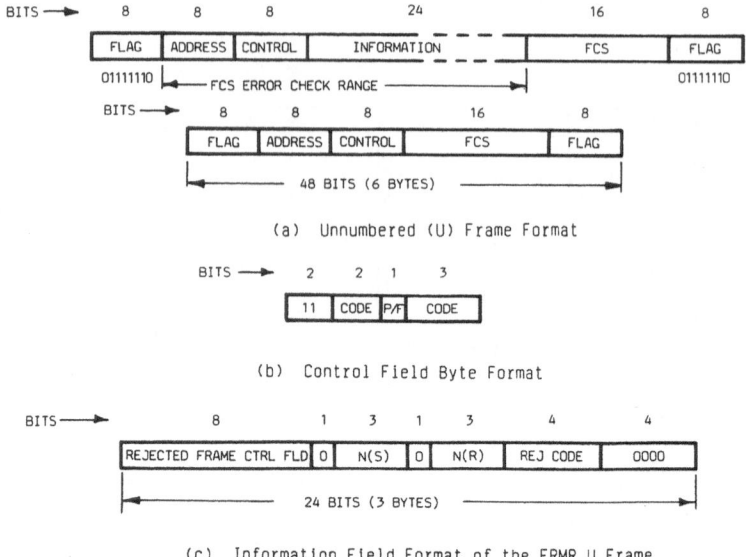

(a)   Unnumbered (U) Frame Format

(b)   Control Field Byte Format

(c)   Information Field Format of the FRMR U Frame

Figure 7.11. SDLC unnumbered (U) frame formats.

Either station may then indicate this status with *Disconnect Mode* (DM). DISC does not physically disconnect the line, but rather logically disconnects the communications between stations. A DISC command is acknowledged with the *Unnumbered Acknowledgment* (UA) response.

The *mode* of a secondary station refers to its operational condition. In SDLC there are only three modes, but in the more advanced related protocols there are several more. The *Normal Disconnected Mode* (NDM) occurs when the link is disconnected, typically as the result of a DISC command or when first powered on. The *Initialization Mode* (IM) is initiated by the *Set Initialization Mode* (SIM) command from the primary station, causing, for example, the sequence number counters to be set to zero. Initialization may be requested, but not initiated, by a secondary with a *Request Initialization Mode* (RIM) U frame, to which the primary will normally send SIM. After initialization the link can be put into the *Normal Response Mode* (NRM) by the primary with the command *Set Normal Response Mode* (SNRM). This is the mode for normal information transfer in which any secondary station must be polled by the primary before it can initiate information transfer. Thus the primary completely controls the link by polling via the P/F bit. The secondary should acknowledge both SNRM and SIM frames with the UA frame.

U frames are also used for certain error conditions that cannot be corrected by simply retransmitting the erroneous frame. Such conditions include the receipt by a secondary station of a frame having an invalid control field, an information field that is too long for the receiver buffers, or an N(R) number that is completely out of sequence. In such cases the secondary will send a *Frame Reject* (FRMR) response, with the information field of FRMR indicating the reason for the reject and containing data from the rejected frame. As shown in Fig. 7.11c, the three-byte field contains the control field of the rejected frame, the secondary's present values for N(S) and N(R), and finally four bits indicating possible error causes. Once a secondary station has sent an FRMR, it will continue to respond with the same frame until the primary resets the mode; e.g., SNRM, since the presumption is that retransmissions will not resolve the problem. There are a few other U frames in SDLC for testing, identifying new stations, and operating in the loop configuration, but they will not be elaborated upon here. The example of Fig. 7.12 should help clarify usage of the U frames and illustrate the operation of SDLC on a multipoint link. Note particularly the link set up with RIM/SIM and SNRM (lines 1-6), the pairing of P and F bits, and the use of FRMR (line 11) for the erroneous frame in line 8.

In the *loop mode* SDLC always operates HDX with a single primary and multiple secondaries. Transmission always goes in the same direction (clockwise or counterclockwise) and the loop is totally controlled by the primary. To send, it addresses the destination station with a frame that is passed around the loop until it reaches the addressee, where it is copied and then passed on around until the primary removes it. A "go-ahead" pattern is then introduced which allows the proper secondary to send a response frame. The loop mode is important for hub-go-ahead polling. It is frequently used for local equipment interconnection around a single building or small complex of buildings, in which case the bit clock information is conveyed with the signal through a baseband coding technique known as Non-Return to Zero Inverted, or NRZI. For local applications, however, there is now a much more sophisticated approach known as the Local Area Network which will be studied in Chapter 11. The various SDLC extensions such as HDLC do not include the loop mode; therefore it will not be considered further here.

Bit-oriented protocols like ADCCP and HDLC include SDLC as a subset,† and have some additional modes of operation not included in SDLC. In particular they contain the notion of a *balanced* configuration of two peer stations, each of which has the *combined* properties of both primary and a secondary. Each *combined station* can transmit both command and response frames, resulting in a *balanced* link control capability. In the *Asynchronous Balanced Mode* (ABM) either station may initiate transmission and exercise control of the link. This mode is highly advantageous

---

† There are a few features of SDLC such as the loop mode not included in these advanced protocols, so SDLC is not strictly a subset. For most practical applications, however, it can be considered a proper subset.

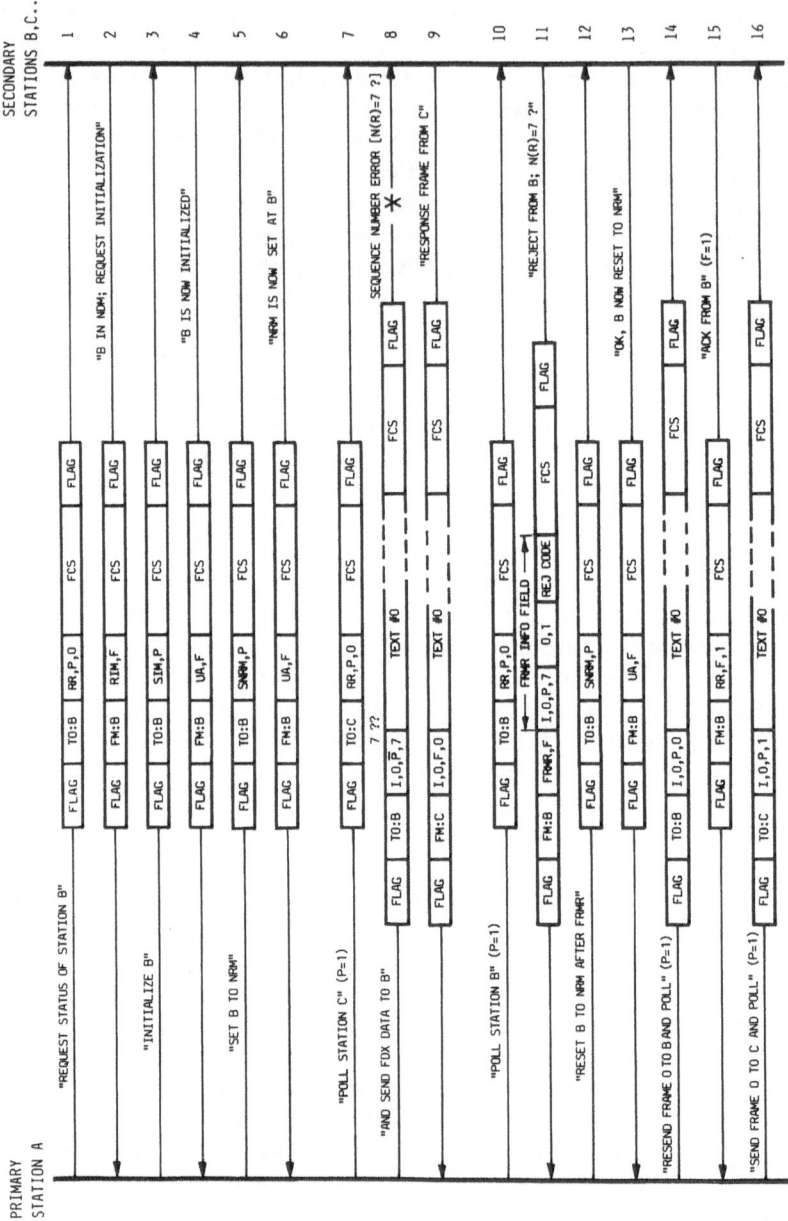

Figure 7.12. SDLC multipoint operation.

in multiply connected network configurations, since stations at nodes interior to the network may have to handle FDX traffic on several links simultaneously and thus act as primary and secondary at once. It is also possible with these protocols to extend the control field to accommodate seven-bit sequence numbers, allowing up to 127 unacknowledged frames, a procedure which is widely used on satellite links.

While SDLC is now a rather old protocol, it is basic to a great deal of installed IBM equipment. This, of course, has resulted in its use by many other vendors of IBM-compatible equipment. Thus, SDLC is an extremely important data link protocol in its own right and will remain so for many years. Even more important from a pedagogic standpoint is the influence on public data network protocol standards such as X.25. We shall return to this subject later in Chapter 10 on network architecture.

7.5. Comparisons and Conclusions

The four protocols that we have studied in this chapter all differ significantly. Not only were they developed by different vendors at different times, but they were intended for different applications. For example, BSC works well for batch computer communications with the older HDX modems like the 201 and 202. Synchronization is easily maintained, and line turnarounds are not a critical factor with relatively long messages such as an application program output. For high-speed applications requiring much back-and-fourth transaction traffic, an FDX protocol is more appropriate. For such cases SDLC provides efficient communications between a large central mainframe (the primary station) and multiple remote terminal devices (the secondary stations). With this type of application all communication is with the primary, just as SDLC requires.

On the other hand, Poll/Select is oriented towards interconnecting business machines with special purpose computers, such as the banking systems, which are a traditional Burroughs market. Here there is typically a lot of polling and transaction traffic, so overall response time is more important than the line data rates. Features like asynchronous or synchronous operation, addressing, sequence numbering, and error checking are highly desirable in business applications where accuracy is important but operators are often not computer oriented. Finally, the DDCMP protocol is oriented towards networks containing multiple DEC minicomputers and sophisticated terminals. Thus, features like synchronous FDX operation and extensive data link control capabilities are important. The extensive capability to remotely load and execute system programs at unmanned

Signetics

Microprocessor Products

SCN2652/SCN68652
Multi-Protocol Communications Controller (MPCC)

Product Specification

DESCRIPTION
The SCN2652/68652 Multi-Protocol Communications Controller (MPCC) is a monolithic n-channel MOS LSI circuit that formats, transmits and receives synchronous serial data while supporting bit-oriented or byte control protocols. The chip is TTL compatible, operates from a single +5V supply, and can interface to a processor with an 8 or 16-bit bidirectional data bus.

FEATURES
- DC to 1Mbps or 2Mbps data rate
- Bit-oriented protocols (BOP): SDLC, ADCCP, HDLC
- Byte-control protocols (BCP): DDCMP, BISYNC (external CRC)
- Programmable operation
  - 8 or 16-bit tri-state data bus
  - Error control – CRC or VRC or none
  - Character length – 1 to 8 bits for BOP or 5 to 8 bits for BCP
  - SYNC or secondary station address comparison for BCP-BOP
  - Idle transmission of SYNC/FLAG or MARK for BCP-BOP
- Automatic detection and generation of special BOP control sequences, i.e., FLAG, ABORT, GA

- Zero insertion and deletion for BOP
- Short character detection for last BOP data character
- SYNC generation, detection, and stripping for BCP
- Maintenance mode for self-testing
- TTL compatible
- Single +5V supply

APPLICATIONS
- Intelligent terminals
- Line controllers
- Network processors
- Front end communications
- Remote data concentrators
- Communication test equipment
- Computer to computer links

BLOCK DIAGRAM

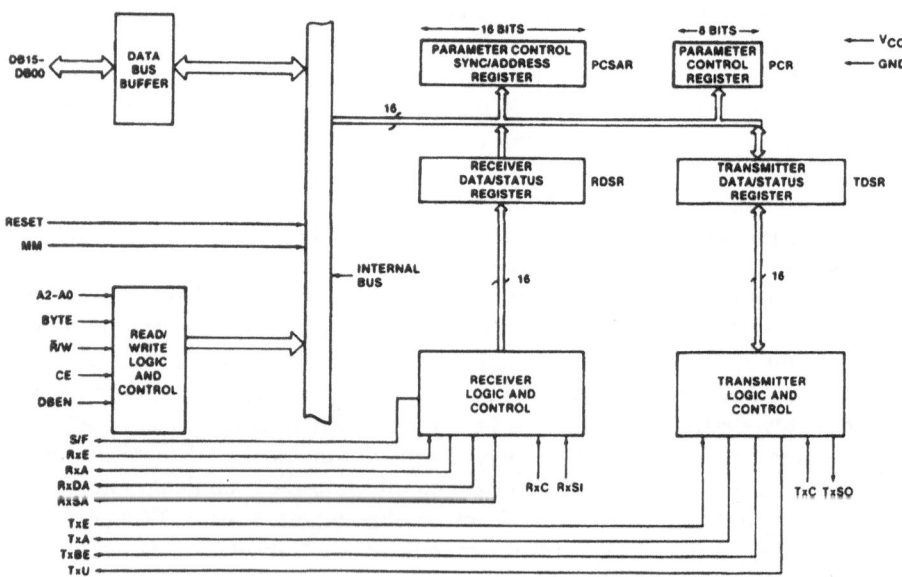

Figure 7.13. BSC/SDLC/DDCMP/data link protocol controller.(Courtesy of Signetics, Inc.)

computer sites is also appropriate to this DEC environment. Thus, the selection of a particular data link protocol is dependent on many factors, and there is certainly no universally optimal choice. To put it in the vernacular—different strokes for different folks!

In our description of data link protocol, there has been minimal emphasis on the actual hardware and software implementations. As already noted, there are many commercial LSI *communications protocol controller* chips that perform data link protocol functions. These sophisticated devices can handle a wide range of essential functions, often for several different protocols with a single chip. Figure 7.13 shows a high-performance protocol controller that can handle essentially all functions for BSC, DDCMP, and SDLC.

This chapter has considered data link protocol mainly in the context of simple point-to-point and multidrop configurations, rather than as a component of a much larger network with many data links. The approach has been through representative examples, since this is an easy way to obtain a practical understanding of the subject. All of these protocols are, of course, closely related to the physical interface protocols of Chapter 6 as well as to the modems and the telephone system of the preceding chapters. In effect the data link protocol operates on top of the physical bit stream provided by digital lines or modems operating on analog lines. It will be instructive at this point for the reader to look back over the three chapters of Part II with the objective of integrating the material into a comprehensive understanding of data link protocols.

As we have emphasized, a large amount of existing data communications is point-to-point over simple data links. This, however, is slowly changing as extensive digital networks become more and more prevalent. Even so, the data link is still the fundamental segment in such networks, and it then becomes even more important for computer and communications professionals to understand the operation of data link components, interfaces, and protocols. In Part III we shall take the same general approach to large data networks that we have taken in Part II to data links. That is, we will develop the basic building blocks and design considerations, and then add the necessary protocol for the entire network to transfer data effectively between users.

Appendix: Error Checking with Parity and CRC Techniques

Coding may be performed on binary data for such purposes as security, bandwidth reduction, spectral shaping, incorporating synchronization information, and error control. Our concern here is with this last application, for which a great many coding techniques have been proposed. They are commonly classified as either *block* or *convolutional* codes. In general such coding may be done only to *detect* errors, or it may also *correct* certain errors. The latter function is obviously more complex but is widely used in applications such as simplex links where requesting the retransmission of erroneous data is not feasible, or at least less attractive than the additional correction coding complexity. For example, convolutional codes for *forward error correction coding* have been historically used on satellite links and are now beginning to be used for very high-speed telephone line modems.

For the types of data link protocols considered in this chapter, retransmission of well-defined blocks of data is almost always possible. Hence, our concern will be with block codes that only *detect* errors. In particular, we shall look briefly at *parity check codes* and at *cyclic redundancy check codes*.

The simplest type of parity check is the addition of one more bit to each transmitted character so as to make the total number of binary ones in the character odd for *odd parity* (or even for *even parity*) as previously discussed in Chapter 6. This character parity is sometimes called a vertical redundancy check (VRC). It works well when there are occasional single-bit errors, but it will fail to detect two-bit errors in the same character (or any even number in fact). Since much of the noise on telephone lines (particularly dial-up) is impulse noise, errors often occur in *bursts*. Assuming a burst has equal probability of causing an even or odd number of bit errors in a character, then the rate of detection will be only 50%. This can be improved in a number of ways. A common approach is to add a *longitudinal redundancy check* (LRC) character to the end of the normal data with character parity. If this data is visualized as being organized into a block with one character per row, then the LRC bits are simply the parity bits for each *column* of the block including the last column consisting of character parity bits.

Table 7.1, with seven-bit ASCII characters, illustrates the VRC and LRC using odd parity for both. The resulting parity-coded message would be transmitted in the text field of a data link frame as a continous stream of eight-bit characters. The receiver would then check both vertical and longitudinal parity, requesting a retransmission if any part of the check fails. This technique is effective for isolated single-bit errors due to the VRC, and for burst errors due to the LRC.† It is interesting to note that limited error *correction* is also possible. For example, a single erroneous bit in the entire block is pinpointed by the row in which the VRC fails and the column where the LRC fails. There are many more powerful parity-type codes, such as the well-known Hamming code [4]. However, there is also

---

† It is apparent that any single burst no longer than one character (including the parity bit) will always be detected. Some longer bursts may not be detected, however.

Table 7.1. Vertical and Longitudinal Redundancy Error Checking

Message	ASCII Code							VRC Parity	Total Ones
STX	0	0	0	0	0	1	0	0	(1)
P	1	0	1	0	0	0	0	1	(3)
A	1	0	0	0	0	0	1	1	(3)
R	1	0	1	0	0	1	0	0	(3)
I	1	0	0	1	0	0	1	0	(3)
T	1	0	1	0	1	0	0	0	(3)
Y	1	0	1	1	0	0	1	1	(5)
ETX	0	0	0	0	0	1	1	1	(3)
LRC Parity →	1	1	1	1	0	0	1	1	
Total Ones →	(7,	1,	5,	3,	1,	3,	5)	(5)	

an accompanying increase in implementation complexity and bandwidth. This has historically been a major constraint, although it is less so today with inexpensive VLSI processors and controllers.

An alternative coding approach that is both highly effective and easily implemented in hardware or software is the *cyclic redundancy check* (CRC) code. These block codes are theoretically based on the notion of digits being coefficients. For example, the string 1100101 would become

$$G(x) = 1x^6 + 1x^5 + 0x^4 + 0x^3 + 1x^2 + 0x^1 + 1x^0 = x^6 + x^5 + x^2 + 1$$

In fact, we can represent an entire message by such a polynomial $G(x)$. Computation of the CRC checksum $R(x)$ is based on binary division by a carefully selected *generating polynomial* $P(x)$ of degree r, i.e., $P(x) = x^r + \cdots + 1$. The dividend is $x^r G(x)$, which is just the actual message with r zeros appended. All arithmetic is done modulo 2 (exclusive-or logic), so there are no carries and subtraction is the same as addition.† Finally, $R(x)$ is taken as the remainder of this division process; thus

$$\frac{x^r G(x)}{P(x)} = Q(x) + \frac{R(x)}{P(x)}$$

where the quotient $Q(x)$ is never used.

To illustrate the actual computation using four-bit characters, consider the two-character message $G(x) = 1100\,0110$ with the generating polynomial $P(x) = x^4 + x + 1$. We first form $x^4 G(x) = 110001100000$, then divide using

---

† Mathematically, these polynomials are defined over the binary field GF(2).

modulo 2 arithmetic:

$$
\begin{array}{r}
11010001 = Q(x) \\
P(x) = 10011\overline{\smash{\big)}\,110001100000} \\
\underline{10011} \\
10111 \\
\underline{10011} \\
01001 \\
\underline{00000} \\
10010 \\
\underline{10011} \\
00010 \\
\underline{00000} \\
00100 \\
\underline{00000} \\
01000 \\
\underline{00000} \\
10000 \\
\underline{10011} \\
0011 = R(x)
\end{array}
$$

Hence the checksum is 0011. The transmitted block of the message with the checksum appended is therefore $x^4 G(x) + R(x) = 1100\ 0110\ 0011$. Note that the checksum length is equal to the order of $P(x)$. To check for errors, the receiver uses the same generator polynomial $P(x)$ to divide the received block including the checksum. If there are no errors, this remainder will be zero, as can be verified with the previous example. Conversely, errors are indicated by any nonzero remainder. However, it is possible for there to be undetected errors even when the remainder is zero. If the generating polynomial is properly selected and the data block length is not excessive, then this probability will be quite small but not zero. Selection of $P(x)$ is based on rather involved algebraic theory, and some standard generating polynomials for data protocols are

DDCMP, BSC (EBCDIC)	$P(x) = x^{16} + x^{15} + x^2 + 1$
SDLC, HDLC, CCITT†	$P(x) = x^{16} + x^{12} + x^5 + 1$
ATT Digital Superframe†	$P(x) = x^6 + x + 1$
BSC (Transcode)	$P(x) = x^{12} + x^{11} + x^3 + x^2 + x + 1$
Ethernet/IEEE 802	$P(x) = x^{32} + x^{26} + x^{23} + x^{22} + x^{16} + x^{12}$
	$\quad + x^{11} + x^{10} + x^8 + x^7 + x^5 + x^4$
	$\quad + x^2 + x + 1$

† These protocols actually use a Frame Check Sequence (FCS), which is computed like the CRC except that $x^r G(x)$ is augmented with ones rather than zeros, and the resulting remainder is complemented bit-by-bit before transmission. The receiver algorithm is similar, with the error-free remainder being a non-zero constant instead of all zeros.

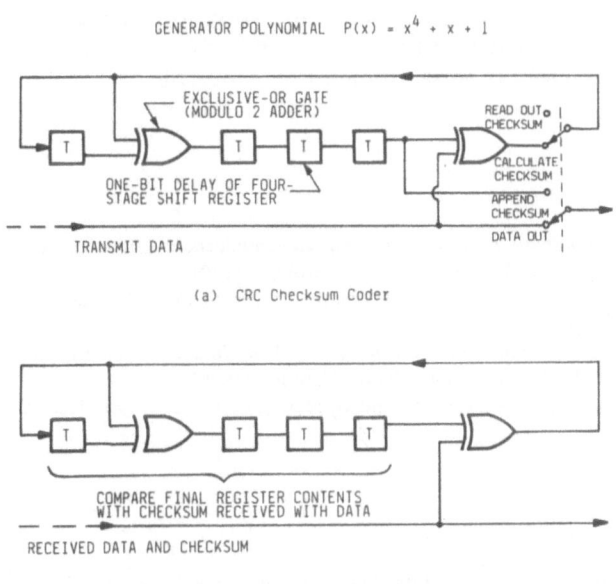

GENERATOR POLYNOMIAL $P(x) = x^4 + x + 1$

(a) CRC Checksum Coder

(b) CRC Checksum Decoder

Figure 7.14. CRC coder/decoder circuits.

These CRC checks are highly efficient, typically detecting around 99% of all burst errors, all single- and double-bit errors, and many other combinations as well. Unlike parity checks, the overhead is always a fixed number of bits regardless of the message length.

From an implementation standpoint the CRC algorithm is executed by very simple binary shift registers with a few exclusive or logic gates, or the equivalent software operations. Figure 7.14 shows both the transmitter and receiver circuit diagram for our previous example with $P(x) = x^4 + x + 1$. The register in (a) is initialized to an all-zero state, then shifted as each message bit is received for transmission. When the last message bit has been shifted in, the resulting CRC checksum $R(x)$ contained in the register is appended to the message by shifting both switched up. This process can be shown to perform exactly the same mathematical operation as the long division of $x^4 G(x)$ by $P(x)$.

At the receiver, shown in Fig. 7.14b, the incoming message is divided by the same polynomial $P(x)$ by passing it through an identical register. This produces another checksum which will match the transmitted one if there are no detected errors. Normally the transmitted checksum is simply shifted on through the receiver register which, on completion, will contain non-zero values if any errors are detected. Otherwise it is presumed that there are none. From this example it is clear that the CRC implementation is very simple, particularly in view of its power to detect errors. All of these attractive features account for the widespread use of CRC error checking in the vast majority of modern data link protocols.

References

1. *General Information—Binary Synchronous Communications* (3rd ed.), Manual No. GA27-3004-2, IBM, Research Triangle, North Carolina (1970).
2. *Basic Data Communications—Line Control Procedure* (Vol. 2), Burroughs Corporation, Detroit, Michigan (1976).
3. *IBM Synchronous Data Link Control—General Information* (3rd ed.), Manual No. GA27-3093-2, IBM, Research Triangle, North Carolina (1979).
4. R. W. Hamming, Error detecting and correcting codes, *Bell Syst. Tech. J.* 29(2), 147–160 (1950).

Suggested Readings

D. E. Carlson, Bit-oriented data link control procedures, *IEEE Trans. Commun.* **28**(4), 455–467 (1980).

A nice overview of the major features and operation of SDLC and related protocols.

E. R. Berlekamp (ed.), *The Development of Coding Theory*, IEEE Press, New York (1974).

Reprints of the key papers in coding theory, including Hamming and CRC codes.

W. W. Peterson and D. T. Brown, Cyclic codes for error detection, *Proc. IRE* **49** (1), 228–235 (1961).

The classic paper on CRC codes, and still one of the best and most easily understood treatments of the subject.

A. Goldberger, A designer's review of data communications, *Computer Design* **20**, 103–112 (1981).

This is an excellent tutorial summary of modems and data link protocols in a single, concise trade journal article at a simple, nontechnical level.

Check Your Understanding of Chapter 7—True or False?

1. The DDCMP frame header has its own checksum because of the COUNT field.
2. Acknowledging a frame in SDLC also acknowledges all previous frames.
3. Control characters of transparent data link protocols are not passed by modems.
4. BSC is poorly suited for satellite links because it is an HDX protocol.
5. SDLC U frames never have sequence numbers.
6. Poll/Select provides fields for addressing, sequence numbers, and polling or selecting.

7. Three-bit sequence numbers are nearly always inadequate for satellite links.
8. Parity checks are always applied to blocks rather than streams of data.
9. The DDCMP SELECT flag can be used for polling rather than selecting.
10. The main drawback of the FCS is the large amount of number crunching required.
11. BSC is a synchronous protocol requiring at least two SYN characters for each frame.
12. Synchronous data link protocols will not run on links with asynchronous FSK modems.
13. SDLC allows any data pattern to be sent in the information field or the header.
14. In Poll/Select, polling is done by setting the Poll (P) bit to $P = 1$.
15. The basic data link data unit is called a fracket (frame/packet) in DDCMP.
16. Transparency is obtained in BSC by counting the number of bytes in the data field.
17. Piggyback acknowledgments allow DDCMP and SDLC to operate FDX.
18. Bit stuffing will always insure data transparency on error-free SDLC links.
19. For FDX SDLC, the REJ and FRMR frames result in the same action being taken.
20. Only an SDLC primary station may initiate a transmission, even when operating HDX.
21. The effectiveness of a CRC check depends heavily on the generating polynomial used.
22. For point-to-point SDLC operation, the P and F bits always occur in pairs.
23. Error checks are usually contained in the headers of most data link protocol frames.
24. The BSC WACK command aborts the link connection since something has gone wacky.
25. The SDLC P/F bit indicates whether a frame has Passed or Failed the CRC error check.

III

Wide Area Networks

III

8

Data Communications
Network Components

8.1. Basic Data Network Concepts

Dating back to their origins in the 1960s, computer data communications networks have experienced unrelenting growth, not only in the United States, but throughout the modern world. This growth has been driven by many factors including electronic and computer technology, the proliferation of both mainframe and small computer systems, and a favorable economic and business environment.

Early data networks were relatively simple, as typified by computer time-sharing services. Here one or more centrally located mainframe computers were shared at different times by many remote users, with communications lines typically asynchronous over point-to-point low-speed modem links. By the late 1960s multidrop lines with HDX synchronous protocols were being used. As the number of users and traffic volume grew, it became desirable to replace groups of long-haul low-speed links from terminals in the same geographical region with single high-speed links that could be accessed locally and shared by the entire group of regional users. The technique employed for such line sharing was, of course, multiplexing. With increased reliance on time-shared computers came the commensurate need for data network reliability, and this need was normally filled by providing multiple routes through the centralized network. Then not only did the user have an emergency backup route, but the network itself could assign user traffic so as to spread out, or *level*, the overall traffic load. This type of alternate network traffic routing quickly led to the use of *circuit switches* at internal network nodes, and the use of *network control centers* to manage and monitor the increasingly complex networks.

By the early 1970s computer technology developed to the point that most large users had their own *host computers*, often in various sites throughout the country or the world. The need to communicate among many distributed sites, rather than always with one central site, motivated the evolution of centralized networks into *distributed networks*. The *distributed network topology* tended to consist of many multiply connected nodes, and *message switches* were increasingly employed to improve bandwidth utilization and response time. As the cost-to-performance ratio of memory and digital electronics declined and switching theory advanced, it soon became feasible to use *packet switches* at the nodes, thus further improving reliability and bandwidth utilization. As the amount of communications processing at each host computer increased, vendors began to provide colocated front-end processors to perform just the communications functions thereby freeing the host for the user applications for which it was intended. By the mid 1970s the basic form of these large general purpose *wide area networks* (WAN) tended to be a high-quality multiply connected, high-speed, store-and-forward, switched *backbone network* interconnecting many smaller and simpler *local access networks*, which usually had inexpensive tree topologies and very limited switching capabilities. These local networks gained access to the backbone network through special nodes known as *interface processors*.

Although we have used the evolution of time-sharing networks as an example, analogous developments have occurred in other fields using large data networks. The physical composition of such networks has many commonalities. They typically consist of point-to-point leased analog or digital telephone links connected to switching nodes, with each link employing a *data link protocol* with its inherent error correction. However, additional high-level *end-to-end protocols* are also necessary to ensure that user traffic gets through the network efficiently, is routed to the proper host, is accessible to the intended application program exactly when needed along with any required files or parameters, and is finally displayed in the proper format for the ultimate (human) user. The complete set of these protocol layers for a network is called a *network architecture*.

In this chapter we shall introduce the key data network nodal components and their interconnection via the point-to-point links that were established in the preceding chapters. The general nature of both large and small networks is introduced and then illustrated with a careful selection of operational network examples. Included are major networks from the public, private, and government sectors. The material in this chapter on components, along with that of the next two on topology and protocol, is designed to provide a comprehensive understanding of the key factors involved in the design and operation of computer data communications networks of all types.

8.2. Conventional Multiplexers

Multiplexing is the conventional method of sharing scarce or expensive communications circuits. It allows the circuit bandwidth to be divided into multiple channels, each of which is then available to a separate user. This allocation may be done in many ways, but for telecommunications we shall distinguish three basic categories of multiplexers (MPX): namely, *frequency division*, (FDMPX), *time division* (TDMPX), and *statistical* (STMPX).† Examples of the first two have already been encountered in our study of the domestic telephone system and will be considered in more general terms in this section. Statistical multiplexers are the subject of the next section.

Although widely used for many years, multiplexers are not standardized to any significant degree among vendors. However, there are some key characteristics common to all multiplexers. First, at the destination the multiplexed traffic must always be separated, or *demultiplexed*, and sent to the proper user. And the same multiplex/demultiplex operation must be performed in the reverse direction for HDX or FDX communications. Thus multiplexers always occur in pairs, one at each end, and the term "multiplexer" invariably implies both a multiplexer and a demultiplexer in a single unit.‡ Note, however, that the units on each end do not have to be physically identical, but only compatible. For example, one unit could be a stand-alone hardware device with the other bring realized in software within a computer or FEP. Multiplexers are also located between a number of DTEs on one side and the line or a modem on the other. Thus the multiplexer must look like a modem at each DTE interface, and on the line

---

† We shall use the notations MPX, FDMPX, TDMPX, STMPX for the actual multiplexers, and denote the multiplexing processes by FDM and TDM.

‡ Sometimes *muldem* is used for such a unit, but not extensively in practice.

side must either drive an analog line directly or, if digital, must look like a DTE to a modem.

Figure 8.1 indicates the use of multiplexers. Without multiplexing the seven HDX terminals require seven separate cross-country leased lines, along with 14 low-speed modems. Although modems can be purchased and depreciated (capitalized), the many leased lines represent a considerable and recurring cost over the lifetime of the connections. With a pair of standard eight-port multiplexers, all the terminal traffic can be combined over a single 9600-bps leased line with sufficient unused capacity for one spare channel. Only one line need be leased, and the pair of modems and multiplexers is a one-time expense. For many practical situations the line savings is enough to justify the multiplexer cost in a matter of months. And in the future there is every indication that the cost of the hardware will continue to decrease much faster than telephone line costs. On the other hand a line, modem, or multiplexer failure will prevent all terminals from communicating unless some type of backup link is provided. Note that the multiplexed line must operate FDX, since different HDX terminals may be sending or receiving traffic at any given time.

Frequency-division multiplexers divide the available line bandwidth into multiple frequency bands by using conventional bandpass filters. Each input channel is then assigned by the FDMPX to a unique band and the corresponding signals modulated to fit into this band for transmission. At the

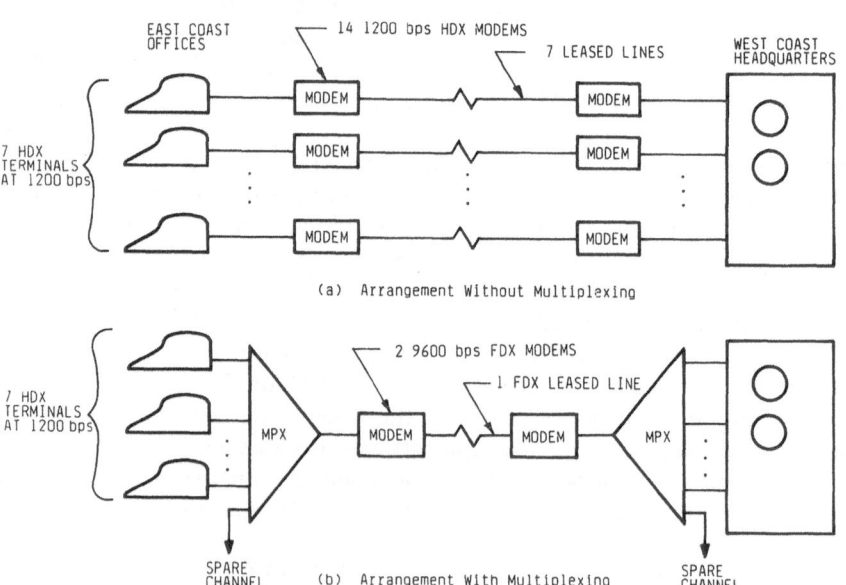

Figure 8.1. Line savings with multiplexing.

destination end the FDMPX demodulates each channel and delivers the resulting baseband signal to the appropriate DTE. FDM is an old and well-established technique upon which the world's telephone carrier systems were built. It is also used in conventional two-wire FDX modems such as the 103 and 212, although this is a rather specialized variation with only one HDX channel in each direction.

Perhaps the most important advantage of FDM is its inherent simplicity. As Fig. 8.2 shows, it is possible to design the filtering so that a number of different channel bandwidths are provided. However, between adjacent bands it is necessary to provide a *guard band* to allow for sufficient filter roll-off; otherwise there may be crosstalk interference between poorly separated adjacent channels. The traffic in each FDM channel is completely autonomous. It can be analog or digital, synchronous or asynchronous, and at any data rate up to the limit imposed by bandwidth and SNR constraints. Each channel can also be independently dropped or inserted at intermediate locations without affecting any other channel.

Table 8.1 gives the frequency allocation for a standard telephone company multiplexer providing eight 150-Bd channels suitable for low-speed Teletype transmission. Although channel spacing is 340 Hz, the channel bandwidths are approximately 150 Hz each, which makes the total available bandwidth only 1200 Hz. The remaining bandwidth is used for guard bands; so this FDM system actually uses only about half of the total bandwidth for transmission. Frequency-division multiplexers also suffer from the classical analog filter ailments of component tolerance and aging and require considerable modifications to change a particular channel bandwidth. Finally, the multiplexer itself provides no inherent error control or performance monitoring capabilities.

Although FDM has a number of disadvantages, it is relatively simple and robust, allows practically any type of input, and is rather inexpensive. Consequently, it is still used today for many low-speed low-density applications as well as for the analog telephone carrier system. However, it is being

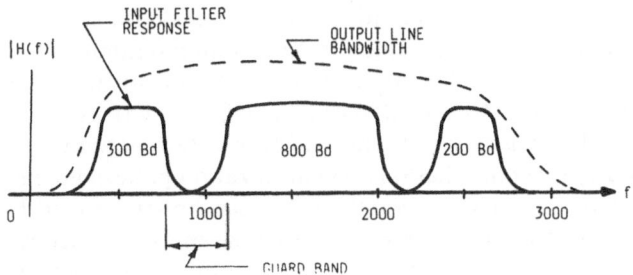

Figure 8.2. FDMPX output spectrum.

**Table 8.1. AT&T 150-Bd FDM Channel
Bandwidth Allocations**

Channel	Center Frequency (Hz)
1	680
2	1020
3	1360
4	1700
5	2040
6	2380
7	2720
8	3060

Channel spacing = 340 Hz
Channel bandwidth = 150 Hz
Total usable bandwidth = 1200 Hz

steadily superseded by time-division multiplexing techniques for practically all aspects of data communications.

Time-division multiplexers operate by briefly switching all available bandwidth among the input channels in some prescribed manner. Each sample from a given input is placed in a unique *time slot* in the outgoing composite signal, and at the destination TDMPX, the contents of the time slot are removed and routed to the appropriate destination. These samples may be amplitude-modulated pulses or even short analog signal bursts, but we shall restrict our consideration to the more common case where the inputs are digital. Then the samples are usually taken to be either one bit or one character in length. This is illustrated by Fig. 8.3, where in part (a) the three input channels are *bit interleaved* and in (b) they are *character interleaved*, assuming four bits per character. The resulting time-division multiplexed (TDM) signal is then transmitted by a conventional modem. In order for the receiver to determine where each channel is located in the received bit stream, it is necessary to allocate some of the bandwidth to a *frame synchronization pattern*. This should uniquely define the beginning of each repetitive cycle through all inputs; i.e., each *frame*. For example, in Fig. 8.3b there is a two-bit frame synchronization pattern followed by three four-bit characters making up each 14-bit frame. In the original 193-bit T-carrier system TDM frame described in Chapter 4 there are 24 character-interleaved channels with a frame synchronization pattern of only one bit. It is obviously essential that this synchronization pattern be distinguishable from the data. In some cases uniqueness can be guaranteed by restricting the allowed data patterns, as was the case when PCM voice was carried in all channels of the original T1 frame. In most cases, however, the data patterns cannot be sufficiently restricted, and it is necessary to search

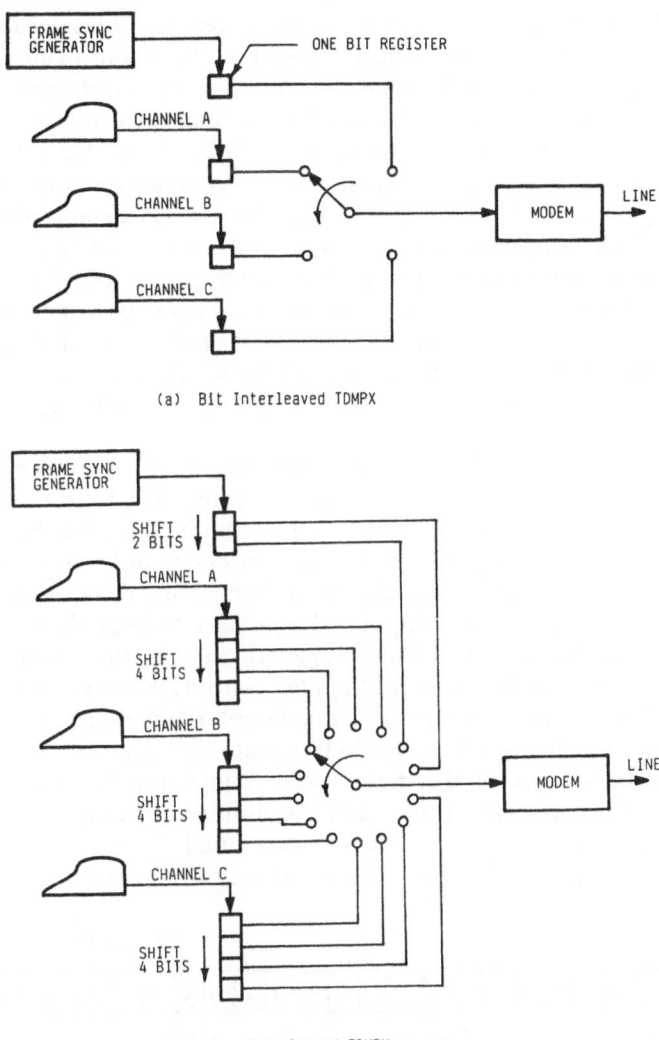

(a) Bit Interleaved TDMPX

(b) Character Interleaved TDMPX

Figure 8.3. Bit and character interleaving.

successive frames to confirm the recurring pattern. Frame synchronization should ideally be acquired quickly and maintained throughout the transmission [1]. This depends strongly on how the frame synchronization patterns are chosen and positioned in the frame, with the particular technique used in practice being vendor dependent. Loss of framing, of course, implies that the receiver data is in error and should initiate some type of alarm condition.

The frame formats for the multiplexing schemes of Figs. 8.3a and 8.3b are shown in Figs. 8.4a and 8.4b, respectively. We again assume four-bit characters and denote, for example, the second character of input B by the bits $(b_{21}, b_{22}, b_{23}, b_{24})$. In the bit-interleaved case each frame is four bits long, including one frame synchronization (S) bit. Note that it takes four of these frames to make one character. In comparison, the character-interleaved frame is 14 bits long, including a two-bit synchronization pattern (S_1, S_2). These frames are transmitted continuously, and simply contain idle characters or bits when there is no traffic from an input. The resulting wasted bandwidth is, in fact, one of the main disadvantages of conventional TDM. Many interactive terminals are actually sending characters only a small portion of the time, maybe 5%, so bandwidth efficiency can be low. However, for steady traffic such as telemetry or large file transfers TDM is very efficient.

Conventional TDMPXs are inherently synchronous, since in the composite data stream the sample in each channel slot has a fixed time relation to all other samples. The *bit clock* is typically obtained via the interface clock lines from the modem, where it may originate or be recovered from the received data stream. The multiplexer then subdivides this clock down to the proper subrate for each input and supplies it through the corresponding interface such as RS-232C. Thus the general rule that the modem supplies the transmit and receive clock still applies with TDM links. If this is not the case, then it is necessary to use more complex techniques such as pulse stuffing to reconcile any differences in frequency and phasing between clock signals on either side of the multiplexer. Asynchronous terminal inputs must also be synchronized to match the multiplexer outgoing rate. The TDMPX may do this, for example, by occasionally inserting an extra stop bit if the input rate is too slow or deleting one if the input is too fast. Of

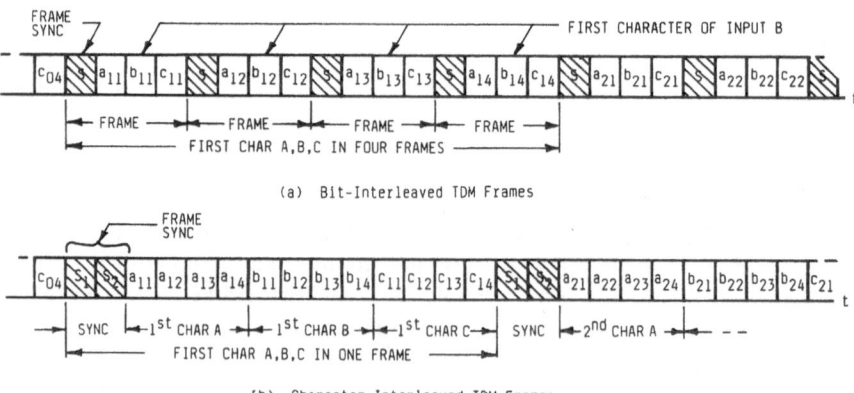

(a) Bit-Interleaved TDM Frames

(b) Character-Interleaved TDM Frames

Figure 8.4. Time-division multiplexed data streams.

course, the asynchronous characters must also be momentarily buffered for some fraction of a bit interval in order to align it with the assigned time slot in the outgoing data stream. At the destination TDMPX the stop bits must be put back in their original form before the character is delivered to the DTE.

If we now assume the usual case of synchronous TDMPX inputs, it is clearly necessary that the input data rate bear a rational relationship to the outgoing aggregate data rate. For example, if the latter is 9600 bps, then viable input rates might be 4800, 2400, or 1200 bps but not 3100 or 7000 bps. Thus there is some restriction on input data rates, and any inputs not meeting those rate restrictions must be retimed before multiplexing. However, all inputs do not have to be at the same rate, as Fig. 8.5 shows. Note that every other time slot is assigned to input A, every fourth to B, and only one slot of the eight-slot frame to C. For clarity we have also allocated one full slot to the frame synchronization pattern, although using 12.5% of the bandwidth in practice would be inefficient.

A TDMPX usually obtains *bit* synchronization from the associated line modem and uses that rate to derive the various input clocks. These may then be supplied directly to a colocated DTE or to a remote DTE via a modem link with the modem using the TDMPX clock. This conventional clocking arrangement is shown by the dotted lines in Fig. 8.6. Besides the bit clock a TDMPX must have *frame* and perhaps *character* synchronization, and this is invariably recovered and used internally. There is also a need for internal buffering since the lower-speed inputs must always be positioned to fit into the high-speed output, and vice versa.

Conventional general-purpose TDMPXs use RS-232C and RS-449 interfaces on their CCITT equivalents, although V.35 is still used in many instances when rates are above 19.2 kbs. To the line modem the TDMPX must look like a DTE, but to the input DTEs it has to look like a DCE (modem); hence they are "two-faced."† Each input port must be individually configured for the usual parameters; e.g., data rate, word length,

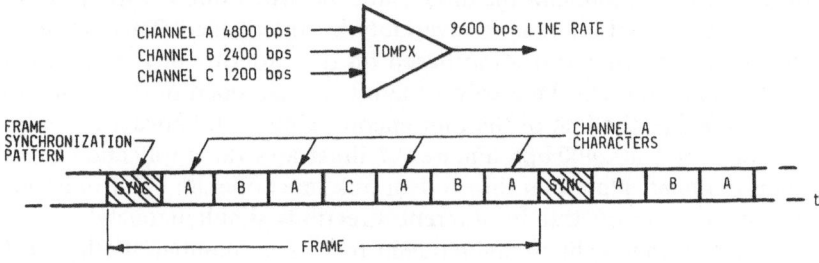

Figure 8.5. TDMPX synchronous input rate relationships.

† This comment is intended to apply to the device itself, and not to the TDMPX vendor!

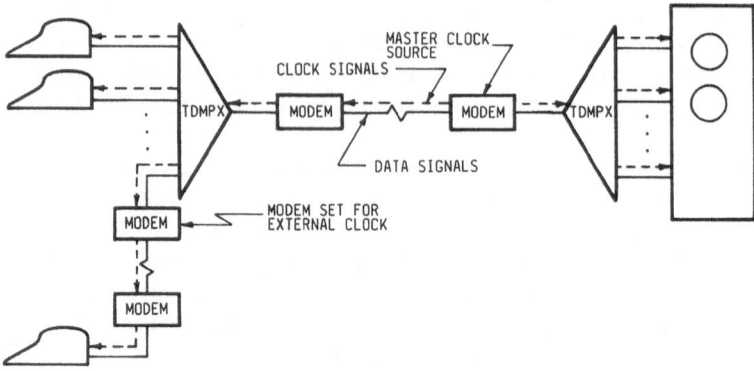

Figure 8.6. TDMPX clocking arrangements.

parity, etc. The same is true of the line modem port, although this may not be apparent when the modem is supplied internally with the multiplexer.

Multiplexers may have other useful features including loopback testing, error detection, alarm capabilities, and the ability to automatically recognize the data rates and formats of asynchronous inputs. TDMPXs can also operate on polled multidrop lines by assigning time slots dynamically to those stations being polled. Finally, the more sophisticated microprocessor-based TDMPXs can be designed with the capability to receive new operating programs and data remotely over the data link, a procedure known as downline loading.

There are a number of other useful data communications techniques related to multiplexing that can be very useful in certain situations [2]. Many high-speed modems, primarily 9600 bps and above, contain a built-in multiplexing capability. The particular input rates possible with these *split-stream* or *multiport* modems depend on the vendor, but for the most prevalent 9600 bps case common rates are 4800, 2400, 1200, and sometimes 7200 bps. In these modems the inputs must be synchronous with the modem clock, and the multiplexing is invariably bit interleaved. This provides an inherent frame synchronization based on the ordering of the bits used to encode each baud. At the receiver the bits are decoded in the same order; e.g., at 9600 bps the first of the four encoded bits could always be assigned to input port 1 at 2400 bps. Figure 8.7 illustrates the application of split-stream modems. Note that the modem link must operate FDX in order for different ports to operate in different directions simultaneously.

In addition to split-stream modems there are a number of other devices that provide line savings. *Modem-sharing units*, or modem multipliers, are addressable remote units that allow multiple remote terminals to be addressed individually over a single modem link. In operation the device

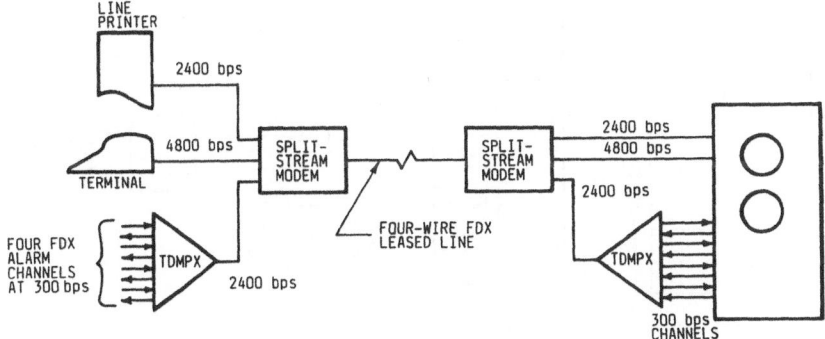

Figure 8.7. Use of split-stream modems.

recognizes the address of traffic intended for each terminal, then provides all necessary interface functions to the addressed terminal. Thus the single line and host computer port are sequentially shared among all the terminals much as in a multidrop configuration rather than simultaneously shared as with multiplexing, and only one unit is required rather than two. Similar devices are available for local use without intervening modem links. Known as *cluster controllers* and *line sharing units*, they provide sequential access for groups of local terminal devices by supplying the necessary intelligence. Operation is typically with a computer of some sort.

Patch panels are essentially digital switching devices that can result in line savings by connecting only active terminal devices to available lines. The name dates back to the old manual telephone plug boards, but our concern here is patching digital rather than analog signals. Basically, the device allows connections among terminals, lines, and computers to be easily set up, monitored, changed, and taken down without having to physically change to cabling. *Intelligent patch panels*, commonly called *port selectors*, are programmable, thus allowing users (or user programs) to request connection to any desired port. This may be a particular terminal or computer port, or it may just be any port in a *port class* capable of handling the user's requirements. Typically the number of terminals will exceed the number of computer ports; so a terminal may be unable to get immediate connection. Port selectors may then provide *port contention* by putting the request on a waiting list (queue) and making the connection as soon as a suitable port is free. Various other control and monitoring features may also be available with port contention devices. A related contention device often used with port selectors is the *telephone rotary*, which allows users to dial a single number and be connected to the next available modem in a *modem pool.* These various digital switching devices are illustrated in Fig. 8.8.

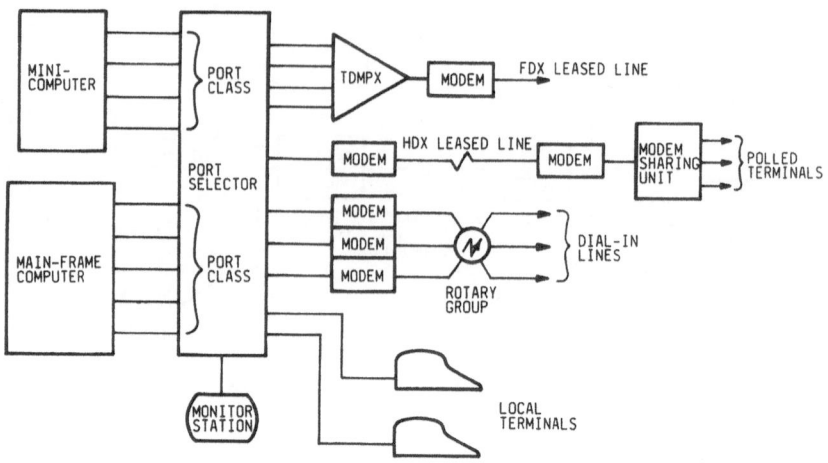

Figure 8.8. Digital switching devices.

Time-division multiplexing and port selection are clearly important capabilities in many potential data communications applications, and it is logical to consider combining such functions in a single "switching multiplexer." Such a device is well within the capabilities of microprocessor-based design. In fact, in addition to adding switching and other nice features, it is possible to improve the bandwidth utilization of TDMPXs significantly by using statistical input selection techniques implemented by sophisticated internal processors. These statistical multiplexers are currently replacing a variety of older equipment, and are the subject of the next section.

8.3. Statistical Multiplexers

As data communications usage grows, more and more applications involve short back-and-forth *transactions* such as point-of-sale terminals and electronic messages. Since the average overall rate of such *terminal-oriented* data input and output is quite slow relative to the peak rates, the fixed time slot assigned by a TDMPX carries actual user data only a fraction of the time. In general, the bandwidth of a conventional TDM must be assigned to handle the *peak* rates and is used very inefficiently whenever the *average* rate is much less than the peak. This is an increasingly common characteristic of today's computer data communications traffic.

An obvious solution to this problem is to have the multiplexer assign bandwidth, e.g., time slots, *dynamically* only to those inputs that actually have traffic waiting to be sent. One method of doing this is to allow the TDM slot lengths to vary dynamically: if an input has no traffic for a particular frame, then only a minimal *terminator code* is sent. Otherwise, longer slots up to some maximum length are provided for inputs with more traffic waiting to be sent. The other approach is to add an input channel identifier, or *tag*, to any outgoing frame slots and assign no slots to those inputs with no current traffic. This latter case is shown in Fig. 8.9 along with the corresponding TDMPX frame for comparison, again assuming four-bit characters. Note that despite the two-bit channel identifier tags there is a net bandwidth savings with the STMPX. If there is a similar average savings over all the frames, then an additional terminal might be added without changing the 4800-bps line data rate.

In either slot assignment approach there must be sufficient *intelligence* in the multiplexer to select the active inputs and make the proper bandwidth assignment for each frame. It is then possible to increase significantly the number of inputs to the point where the total, or *aggregate*, rate of all inputs is much greater than the *composite* multiplexer line output rate. The ratio of aggregate to composite rates is often called the "compression ratio." The multiplexer is thus gambling that the *statistical average* rate of all input rates will remain less than the output capacity. Consequently, this type of multiplexer is commonly called a *statistical multiplexer*† (STMPX). Conceptually it can be thought of as a switch that intelligently jumps around to

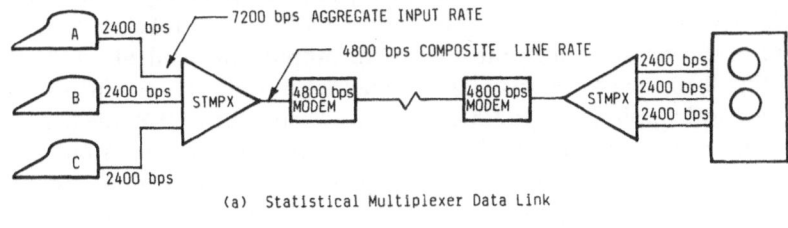

(a) Statistical Multiplexer Data Link

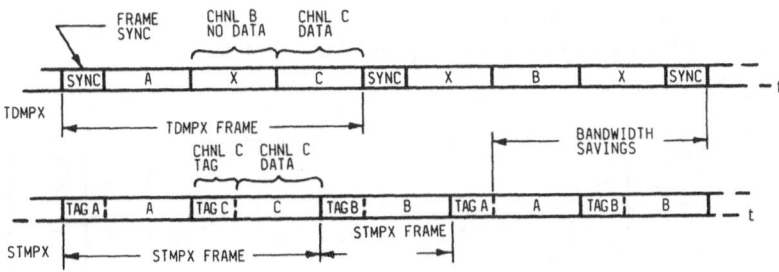

(b) Conventional and Statistical TDM Framing

Figure 8.9. Statistical multiplexing concept.

† The terms "intelligent," "demand," "concentrating," and "asynchronous" multiplexer are also sometimes used.

contact only those inputs having data to send, rather than rotating to all inputs sequentially as in Fig. 8.3.

Since the inputs are independent, there is no way to guarantee that the STMPX line output capacity will never be momentarily exceeded. Therefore, buffering must be available for each input so that brief overloads can be smoothed out until sufficient output slots become available. The price paid for this bandwidth efficiency is, of course, an unpredictable *buffer delay*, and this delay can be a serious problem in some real-time applications where 50–100 msec can be critical. However, for most terminal-oriented traffic it is not a limiting factor.

The other major consideration with buffering is *overflow*. If the input traffic remains above the statistical average for too long the buffers will become full. The STMPX must recognize when such a situation is developing and stop further input before there is any loss of data. This *input flow control* can be done either with interface control lines such as turning *CTS off* in RS-232C, or by sending the special control character *XOFF* to the corresponding terminal. When the buffer size is sufficiently reduced, input is resumed by turning *CTS back on* or by sending *XON*† to the terminal. STMPXs also generally provide *output flow control*, by which a terminal device such as a printer can tell the local STMPX to signal the *remote* STMPX to flow control the data source. This is commonly done by sending XON/XOFF or characters representing RS-232C control signals across the link to the remote end. Both input and output flow control are illustrated in Fig. 8.10, where the printer on channel C indicates a full print buffer either by dropping RTS in RS-232C or by raising the BUSY line in a parallel interface. Even though the local STMPX receives this signal immediately, there is an inherent delay in sending it to the remote end and actually effecting the flow control, during which additional traffic may be received. This is especially true on satellite links. Since the printer cannot accept this traffic on arrival, the local STMPX should buffer it up until the printer is ready, then deliver it before removing the flow control.

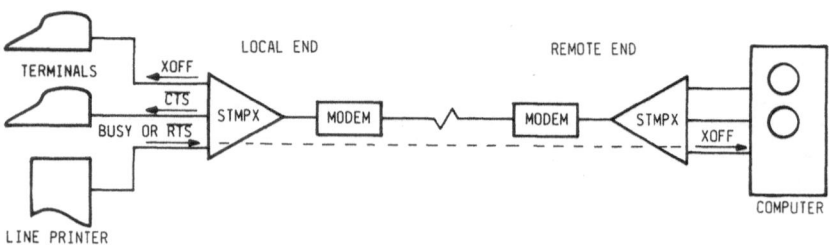

Figure 8.10. Input and output flow control.

† In ASCII XON and XOFF are actually the two control characters DC1 and DC3, respectively.

In actual practice the vast majority of STMPXs use a *data link protocol frame* such as SDLC or HDLC to convey each frame of multiplexed data. This offers a number of advantages, not the least of which is the wide availability of inexpensive protocol controller chips. As Fig. 8.11 shows, the data characters and tag from each input to be multiplexed are put into the information field, which is, of course, variable in length. Then the usual flag, address, control, and error check fields are added. The error check range includes all of the multiplexed channels, and frames received with bit errors are retransmitted. This provides an essentially error-free channel to the users. The control field provides the usual sequence numbers for piggyback acknowledgments. Control frames (S and U for SDLC) may also be used for various status and management functions including output flow control and reconfiguration. The maximum size of these data link frames is largely determined by the physical characteristics of each STMPX, but error probability and retransmission times are also factors. Frames are actually transmitted either when they are full or after a predetermined time-out interval, whichever comes first.

As microprocessor power has increased in recent years, there has been a trend towards increased intelligence in data communications equipment in general, and in statistical multiplexers in particular. This intelligence allows many sophisticated features to be incorporated at a small incremental cost. In fact, in addition to taking over a large share of the traditional FDMPX and TDMPX roles, they are also capable of performing an increasing array of functions heretofore performed by minicomputer-based communications processors. Here we shall discuss a number of such standard and advanced features and capabilities.

Among the standard features of the STMPX is the inherent ability to *combine asynchronous and synchronous inputs,* since all inputs are buffered and treated the same inside the data link frames. For asynchronous characters it is common practice to remove the start and stop bits during transmission to save bandwidth. It is also commonplace to use simple *run length coding,* where long strings of the same character, such as blanks, are sent as a single character and a repetition code. More sophisticated and computationally intensive data compression techniques, such as variable-length Huffman coding, have also been used to conserve bandwidth, and a number of very efficient data compression devices are now available. *Data rate*

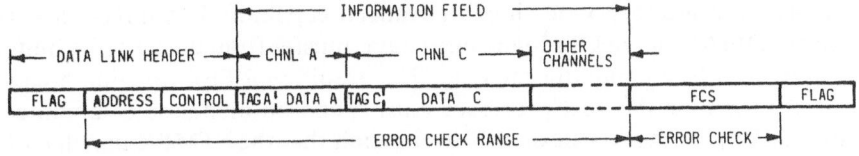

Figure 8.11. Typical STMPX frame format.

conversion is easily implemented on STMPXs and allows a terminal to communicate with a variety of other devices without concern for their individual rates. Any resulting throughput discrepancies are easily resolved with the STMPX flow control mechanism. Many STMPXs will also automatically determine the data rate of a new input and adjust accordingly, which is called *autobaud*. *Code conversion*, e.g., from ASCII to EBCDIC, is also possible, and the code may even be recognized automatically by some multiplexers.

Another important STMPX feature is *echoplex*. When a local computer terminal operator types a character, the character normally appears immediately on his display. In actuality, the character is sent to the computer and then *echoed* back to the terminal, and this echo is the displayed character. Problems may arise with this character echoing when the terminal operates with a remote computer where there is significant transmission delay, particularly over a satellite link. Then there is an annoying pause in the display of each typed character, known as the "sticky keyboard" effect. This annoyance can be alleviated by an STMPX since it is feasible for the local multiplexer to echo the characters back immediately and then assume full responsibility for their delivery to the remote computer. This works even with transmission errors since they are detected by the data link protocol and the frame is then retransmitted.

STMPXs also usually now have a control or *supervisory port* through which a local or even a remote operator can configure each port individually using software commands rather than mechanical straps or switches. Performance monitoring and statistics collection may be done via the supervisory port for characteristics like buffer loading, delays, error rates, alarms, and modem performance. Tests and diagnostics may also be run through the supervisory port. In addition to the conventional modem loopback tests discussed in Chapter 6, local and remote analog loopbacks are possible at each line output port, and digital loopbacks can be effected at *each* terminal input port individually. Thus a step-by-step loopback test sequence can be used on statistically multiplexed links to isolate any degraded or failed components.

In addition to the preceding standard features found on most STMPXs, there is an increasing variety of more advanced features that may also be available. In fact, it is becoming difficult to distinguish the more advanced STMPXs from such major network components as front-end processors or remote concentrators. One simple advanced capability is a *direct memory access* (DMA) channel for high-speed data transfer from the STMPX buffer directly to a host computer memory. This traditional FEP function greatly reduces the time the host processor must spend on mundane communications tasks. Another traditional FEP function that the STMPX may handle is *polling*. If remote STMPXs are used at each drop location, then each poll will access all terminals connected to the polled multiplexer. Such a polling configuration is shown in Fig. 8.12, in which each remote multiplexer acts as a controller for the cluster of terminals connected to it.

Port selection and *port contention* are also available with statistical multiplexing. The more sophisticated of these data switching capabilities allows any pair of local and/or remote input ports to be selected upon request by the user. If the desired port or port class is not available, the request may be put in a waiting queue and the requestor notified. Port selection, for example, would allow a user to execute a large program involving data base accesses, routines on different host computers, and multiple output devices without having to reestablish his channel for each operation. Advanced STMPXs may also have *dual data links,* which can provide twice the point-to-point data rates of a single voice-grade line without the cost of a wideband line. Delay can be minimized by balancing the load across both links, and in the event of a line failure the remaining one provides an automatic backup. If the two data links are connected to different locations and port selection is available, then more network-like connections are possible. An example is shown in Fig. 8.13. Terminal B is multiplexed with the host in the conventional way, but the channel of terminal C *bypasses* the host node and connects over a second link to terminal D. The ports of terminals E and F are switched by the local STMPX. The figure also shows the *tandem connection* of a lower-speed multiplexer A to the host over a single STMPX channel. Many STMPXs can also have their internal programs and parameters *downline loaded* over their data links from a central site. This capability allows changes and reconfigurations to be made easily throughout the network by a single operator.

Figure 8.14 gives the specifications for a typical commercial statistical multiplexer. It should be noted that, although the basic approach of most

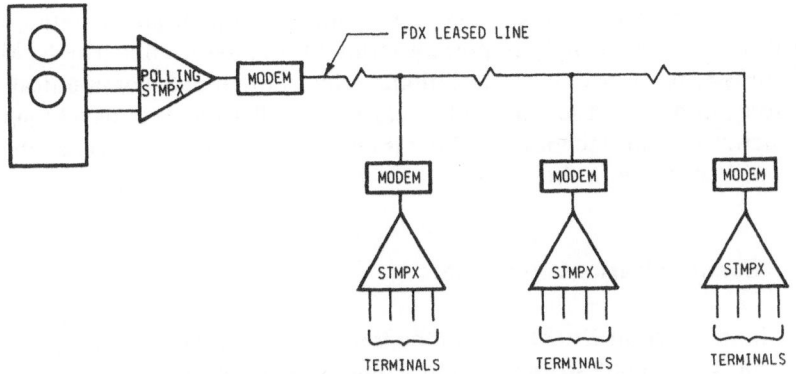

Figure 8.12. Multidrop multiplexing configuration.

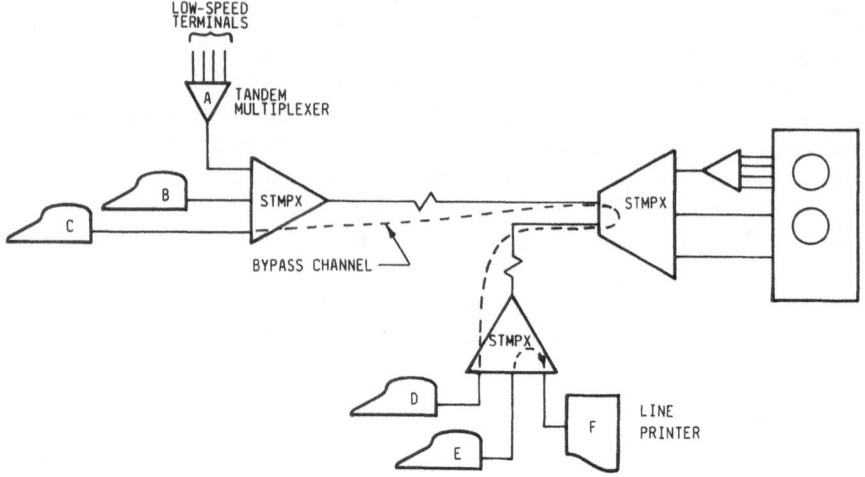

Figure 8.13. Advanced statistical multiplexer functions.

vendors is similar, there are no accepted standards for STMPXs. Thus compatibility should not be expected between different vendors and should be carefully verified among different product models from the same vendor.

Other advanced statistical multiplexer features include protocol conversion, the ability to act as a gateway between two separate networks (particularly with one a local access network), and even the assembly and disassembly of data packets from synchronous terminal inputs. However, the discussion of these functions will be deferred until the next section where they are presented in their traditional context of communications processors.

Statistical multiplexing is a rapidly growing and highly versatile technique for modern computer data communications. Although it may be less efficient than conventional TDM for continuous file-oriented traffic, it is ideal for the increasingly important terminal-oriented traffic. The STMPX has already taken over many traditional roles of other data communications devices. And since the capabilities closely parallel the new developments in microprocessor technology, there is every indication that this expanding role will continue in the future.

8.4. Communications Processors

In this section we shall begin the transition from the basically point-to-point links covered so far to more complex and versatile networks to be studied in the remainder of this book. However, these are not two distinct

SWITCHING MICROPLEXER®
Distributed Data Switching System
8, 24, and 48 Port Models

Features:

■ Integral switching and contention with 4 user definable port types:

- Selecting
- Contending
- Contention
- Dedicated

■ Automatic alternate routing

■ Two data links for flexible network configurations

■ Data link rates up to 19,200 BPS

■ User definable resource identifications

■ Interactive terminal interface providing a menu of additional commands usable during an established session

■ Successive call requests permitted for dial-in users without reconnection to network

■ Network security

- Dual level Supervisory Port protection
- Restricted access port groups

■ Single point network control through System Supervisory Port

■ Intelligent connection procedure

■ Global message of the day

■ Comprehensive hardware and software diagnostics

■ Statistical multiplexing of both asynchronous and synchronous data including SDLC

Statistical Multiplexing

The SWITCHING MICROPLEXER Systems use statistical multiplexing techniques to concentrate any combination of up to 48 ports onto 1 or 2 data links. Both asynchronous and synchronous data can be multiplexed together. Any 5, 6, 7, or 8 bit asynchronous code plus a parity bit can be accommodated, including ASCII, Baudot, BCD, and codes for 9-bit graphics terminals. Synchronous protocols include SDLC and BISYNC.

Only asynchronous ASCII ports can be switched.

SWITCHING MICROPLEXER Products

The SWITCHING MICROPLEXER unit communicates with the full SWITCHING MICROPLEXER product line including the more powerful QUAD SWITCHING MICROPLEXER unit and the NETWORKING MICROPLEXER System.

SPECIFICATIONS

Ports:
Up to 8, 24, or 48 ports per unit with speed, data code, and traffic flow control independently programmable for each port.

Data Speed:
Standard asynchronous speeds are 50, 75, 110, 134.5, 150, 300, 600, 1200, 1800, 2000, 2400, 3600, 4800, 7200, and 9600 BPS.

Asynchronous Data Codes:
5, 6, 7, or 8 data bits plus parity; 1, 1 5, or 2 rest bits.

Synchronous Data Codes:
IBM BISYNC, IBM SDLC, ICL Synchronous, Honeywell VIP, CDC UT-200.

Data Base Memory Protect:
Rechargeable batteries maintain the user-programmed data base in memory for approximately 90 days without AC power. Batteries will automatically recharge during 16 hours of normal operation.

Interfaces:
EIA RS-232C; CCITT V.24/V.28; RS-423; RS-422 balanced; MIL-STD-188C; MIL-STD-188-114 balanced and unbalanced.

Maximum Aggregate Input:
76.8K BPS for 8 port SM8 models (8 x 9600 BPS). 230.4K BPS for 24 port SM24 models (24 x 9600 BPS). 460.8K BPS for 48 port SM48 models (48 x 9600 BPS). All ports can operate simultaneously at 9600 BPS.

Memory Capacity:
RAM: up to 208K bytes.
Data Base: 2K bytes.
Program: up to 200K bytes.

Data Links:
Two synchronous links with maximum data rate of 19,200 BPS on 1 data link or 9,600 BPS on each simultaneously.

Data Link Control:
Synchronous X.25, level 2 based, with dynamic block size adjustment.

Data Link Error Control:
16 bit cyclic redundancy check (CRC) character with selective go-back-N ARQ protection.

Figure 8.14. Representative statistical multiplexer specifications. (Courtesy of Timeplex Inc.)

concepts. As we have seen, point-to-point links can be multidropped, multiplexed, and connected in tandem to produce embryonic networks adequate for many practical data communications applications. Conversely, even the largest *wide area network* (WAN) is made up of point-to-point links connecting various kinds of nodes. Thus our rather extensive treatment of point-to-point data links will provide an essential background for understanding more sophisticated network concepts.

There are several key dimensions to a complete grasp of networks. Perhaps the topological layout is the most basic, while the operational protocol is the most advanced. Both of these topics are treated in detail in the next two chapters. Here we shall focus our attention on the major components that constitute network nodes, i.e., the devices that are linked together and to various terminals to make the actual network. In general, such nodes are specialized programmable minicomputers known collectively as *communications processors* (CP).

There are a number of different types of CPs that, in the past, have been fairly distinct. However, this distinction is rapidly blurring as increased processing power at lower costs allows the combination and extension of many traditionally separate functions. CP technology, like the rest of data communications, is also being driven by the old tradeoff between line cost and processing cost. With the notable exception of national governments, few owners of WANs are able to purchase all of their transmission lines outright. The more common situation is for them to lease analog or digital lines from a common carrier such as the telephone company, and add their own nodal processing to create a *value-added network* (VAN). Economically, the leased line costs are a recurring expense that tends to increase over time, while the DP equipment at the nodes is a one-time capital cost that has decreased dramatically in the past. Thus CPs are fundamental to the operation and cost effectiveness of the modern data communications network.

Despite the lack of distinct demarcation, we will group our study of CPs into the three conventional categories: front-end processors, concentrators, and switches. It is certainly not uncommon for these functions to be combined in a single unit, but the basic functions can usually still be distinguished conceptually at least.

A *front-end processor* (FEP) is a specialized computer collocated with one or more host computers, and designed specifically to perform communications functions that the host would otherwise have to do. First introduced in the early 1960s, this arrangement can be highly efficient since data communications tends to be rather sporadic and interrupt driven in nature. Without an FEP, such real-time traffic from a large number of input lines can consume huge amounts of the host processor time, leaving proportionally less for it to perform the user application-oriented operations for which it was intended. Alternately, a front-end offloads all such chores, taking from and delivering to the host only the actual data with just the format, code, and speed that is most efficient for it. High-speed transfer between host and FEP is usually done over a *DMA channel*, thus further lessening the communications load on the host. From the host's point of view, the FEP looks like any other peripheral device.

FEPs have been traditionally offered by the major mainframe computer vendors for their own particular product lines, so there is little standardization from one vendor to another. Furthermore, since modern FEPs are invariably programmable,† even different units of the same model may be set up to operate quite differently. However, there are many functions that characterize front-ends in general. The *conversion* of character codes, data rates, and formats allows many different types of terminals and nodes to communicate easily with the host computer without any changes by the host itself. A similar function is asynchronous-to-synchronous conversion. FEPs also do various *control functions* throughout the entire network such as performance monitoring and statistics collection, error control and recovery, network reconfigurations, network security and access control, and down line loading of new programs and parameters for remote devices. *Polling* is also commonly done by FEPs for both locally and remotely located terminals. *Buffering* is usually provided both for load leveling on the outgoing links, and for error recovery in the event of a serious failure where there is loss of data. FEPs can handle *autodialing* and *autoanswering* for dial-up links.

A major function of the FEP is the correct execution of data link and end-to-end *network protocols.* This includes all of the conventional data link functions like CRC checking, retransmission timeouts, frame sequencing, acknowledgments, and link setup and disconnection. End-to-end functions may include routing table updating and dissemination to all nodes, orderly flow control throughout the network, packet sequencing and reassembly, and various error control functions to maintain the required degree of end-to-end integrity of the data.

Of course, not all FEPs perform all of these functions, but all perform some of them. The classical example of an FEP is the IBM 3705.‡ This FEP was introduced in the mid-1970s to run with the IBM System/370 mainframe. The operating system, known as *Network Control Program* (NCP), executes polling, buffering, timeouts, code conversion (to/from EBCDIC), and error detection. It also collects performance statistics, runs diagnostics, and initiates error recovery procedures. The 3705 with NCP works with a broad range of IBM terminals and other systems oriented to many diverse applications including banking, POS, manufacturing, and data acquisition. Common protocols supported are asynchronous (start-stop), BSC, and SDLC.

Intuitively, *concentrators* are devices that take the traffic from a number of input lines and somehow "squeeze" it into a smaller number of output lines having less total capacity. STMPXs are concentrators by our definition, whereas TDMPXs are not because the aggregate input and output capacities are equal—there is no concentration. For network applications it is often

---

† Early devices were hardwired and called *communications controllers* rather than FEPs.
‡ The 3705 FEP has now been superseded by the more powerful 3725, but the basic functions are quite similar.

very cost effective to put a large concentrator in a distant location, perhaps another city or country, where there is a large number of low-speed terminals. Such a *remote concentrator* can control the terminals through polling, addressing, protocol and code conversion, error control, and logging. Then it can combine the aggregate traffic on a few high-speed long-haul links back to the host FEP. This is the same general idea as statistical multiplexing, but typically on a larger scale. Remote concentrators usually have some buffering capability to smooth out the peak traffic rates and for error recovery, and may also employ data compression coding to further improve their bandwidth utilization.

From even this brief functional description it is apparent that there is much similarity between concentrators and FEPs. This is also true for multiplexers, and there is now a significant overlap between smaller concentrators and advanced statistical multiplexers. Figure 8.15 indicates some representative interconnections between a remote concentrator, and an assortment of terminal devices.

The third application of CPs is *switching*. This could be either packet, message, or circuit switching, or even some hybrid form. However, we shall restrict our discussion here to the former two, i.e., to *store-and-forward* (S/F) switching, and note that circuit switching was introduced in Chapter 4 in connection with telephone switches. S/F switching is a fundamental

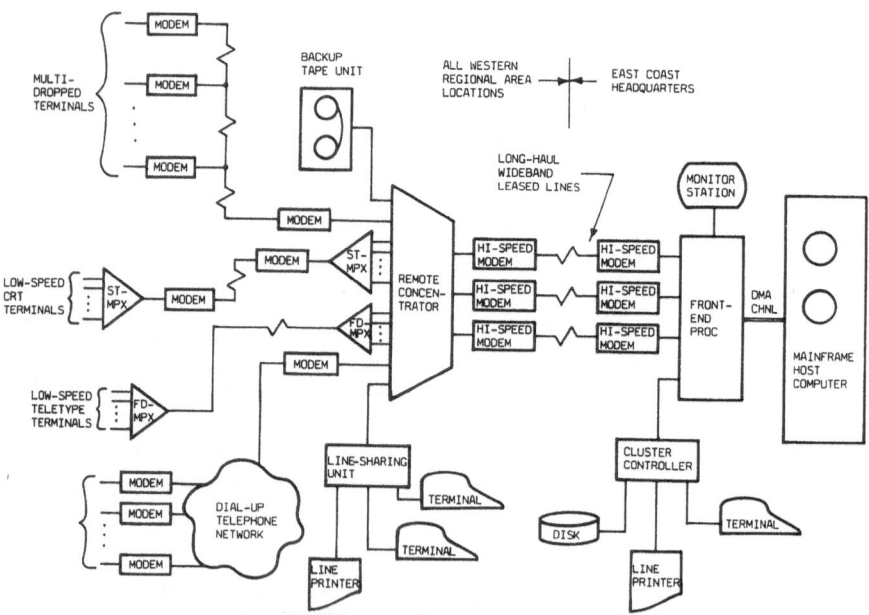

Figure 8.15. FEP and remote concentrator connections.

function of most data communications network nodes. Message switches require extensive disk space to buffer complete messages (of unknown lengths) before switching them to an available outgoing line. The resulting delay can be considerable, even a matter of days when there is a large amount of higher priority traffic. It may also be desirable in some instances to intentionally store messages at nodes for later delivery, perhaps to another time zone. In packet switching, on the other hand, long messages are subdivided for transmission into small packets of typically a few thousand bits. Each packet traverses the network independently and is only briefly buffered in the nodal processor's *main memory* before being switched to an available outgoing line. At the source and destination nodes each message must be *disassembled* into packets for transmission, and the arriving packets *assembled* back into the original message. This *packet assembly/disassembly* (PAD) function is generally done by another CP performing very specialized concentration or front-end processing.

In order to do S/F switching, nodal processors must know in which direction to switch a received packet or message. This is usually determined from a nodal *routing table*, which contains the best route out of the node to reach any given destination. These tables are determined from routing algorithms, which may be executed locally by each node or may be computed and distributed from one central site. Routing is covered in detail in the next chapter.

In addition to switching and routing, nodal processors may do many other useful tasks. These include all aspects of data link protocol handling for the attached links, error control and recovery, logging, performance statistics collection for network monitoring, diagnostic tests, and sometimes concentration of low-speed traffic.

There are two other types of nodal processors that are important in many networks. First is the *interface processor* (IP) which constitutes the node at which the user interfaces to the (backbone) network. This interface may be as simple as a dial-up modem link or as complex as an extensive local network supporting hundreds of terminals. In addition to the usual nodal switching functions, an IP may also act much like a concentrator for low-speed user terminals. Protocol and code conversion are possible, and security-oriented functions like authentication and encryption are now becoming common. The other specialized nodal processor of interest here is the *gateway processor*. Gateways are communications processors that connect two different networks together by providing the proper nodal characteristics to each. This function involves a great deal of translation, e.g., protocols, addresses, speeds, code sets, and routing tables. There may also be extensive administrative requirements such as billing and statistical collections. Seemingly routine actions like acknowledgments and error recovery can become quite involved when they must be effected across several large networks. Synchronization at different levels can also be a major consideration of a gateway node. To help clarify these various roles, Fig. 8.16 gives an example of the relative interconnection of different types of communications processors in a simple packet-switched wide area network. We have assumed that all the long-haul lines are digital so that no modems are required. Note also that the STMPX attached to the remote concentrator is capable of packetizing the data from its inputs.

Having now introduced the main types of multiplexers and communications processors, we shall next take a brief look at some fundamental ways of physically interconnecting them to form networks. There is, of course, an infinite variety of different node and link combinations. The intent here is not to present an exhaustive enumeration of all such cases but rather to look briefly at a few instructive combinations of component interconnections based on the more common topologies. The more basic issue of just how the network should be designed for a given situation is covered in the next chapter.

Simple computer networks are often connected with a *tree topology*. As the name suggests, the network is made up of point-to-point and multi-drop links connected together like the branches of a tree but containing no loops. The network spreads outward from a central node which is often a host computer or a switch. It is also frequently a multiplexer that feeds into another network. Since there are no closed loops, there is only one path between any two nodes; and if any node or link fails, then all outward nodes are isolated from the central node. Even worse, if the central node

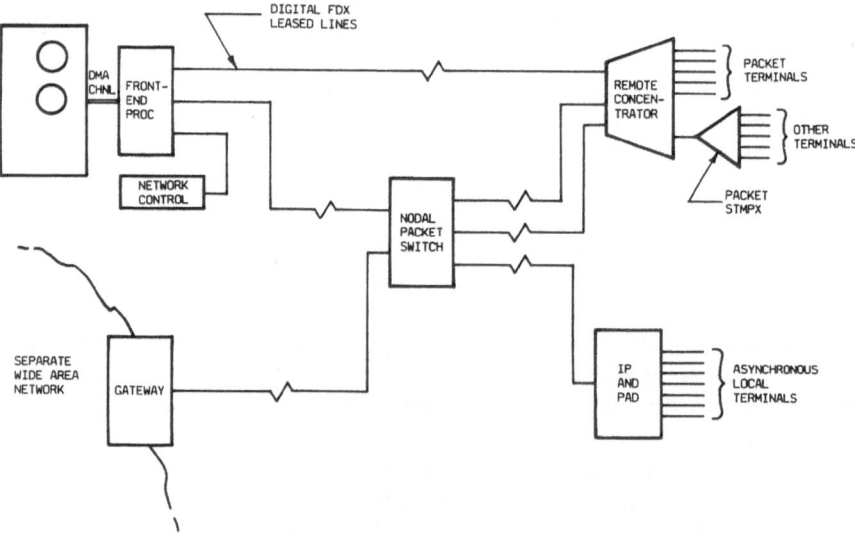

Figure 8.16. Network functions of communications processors.

fails the network will be unable to function at all. Thus tree networks are most appropriate when either the overall reliability is very good or else reliability is not very critical and occasional failures can be easily tolerated.

Figure 8.17 is an example of a tree network for accessing a host computer from a number of different remote terminal locations. The factory and regional sales locations are connected to the host computer at headquarters by leased analog and digital lines operating at the rates indicated. All remote traffic is concentrated through the 56 kbps STMPX. Several other multiplexed and multidropped branches extend on out from it to the firm's factory, engineering, administration, and sales offices. The FEP here must perform statistical multiplexing as well as the more conventional buffering, polling, timing, code and format conversion, and protocol handling. Note that there is no redundancy, so any node or link failure will disable at least part of the network. The severity increases as the central node is approached, and in the worst case of an FEP failure the entire network will be inoperative. Dial backup may be indicated if, for example, the warehouse alarm system is deemed vital.

The classical computer data communications network is centralized about a single large mainframe computer. Each of a limited number of local and remote terminals accesses this host computer via a point-to-point data link, with the resulting arrangement known as the *star topology*. Figure 8.18 illustrates such a star network with leased analog telephone lines. All the

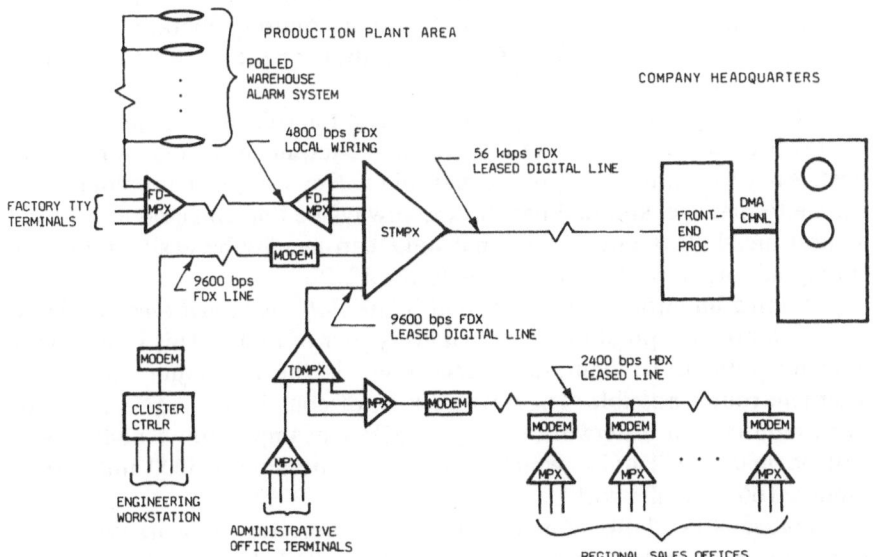

Figure 8.17. Centralized tree network with FEP.

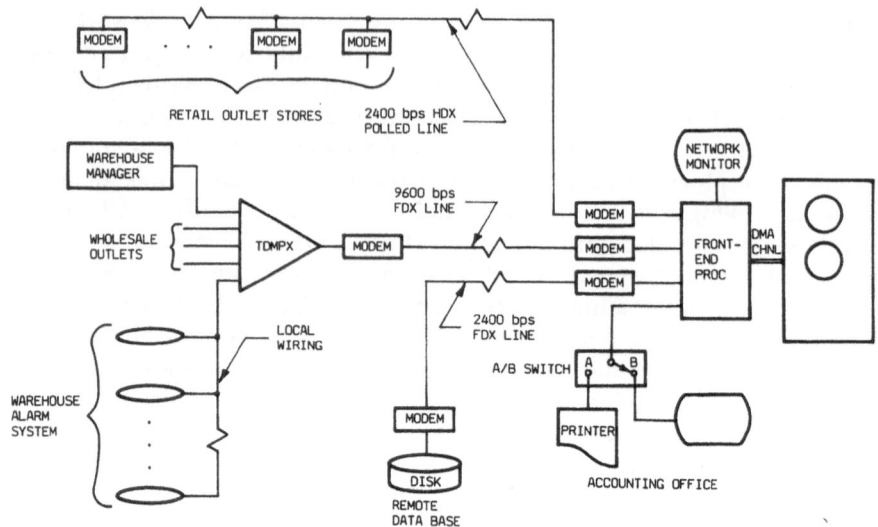

Figure 8.18. Centralized star network with single host computer.

retail outlets are multidropped in one 2400-bps line, while the warehouse
and wholesale outlets are on a separate 9600-bps line. There is also a local
link to the accounting office. Since this latter traffic is always light, with
only one hardcopy made per day, the printer and CRT terminal share the
same RS-232C line through a simple two-position (25 pole!) switching
arrangement known as an A/B switch. The accounting operator manually
turns the switch from B to A for the duration of the printing, and then
returns it to B.

Note that with the star topology, as with the tree, there is little redundancy. If a link fails, then all the associated equipment is isolated from the
host. An FEP failure is again catastrophic. However, star topologies are
simple to manage, and dial backup can always be implemented as a hedge
against any link failure. Additional links can usually be easily added by
simply adding more I/O port cards to the FEP.

A third important network layout is the *fully connected topology*. Here
there is a point-to-point link between every pair of nodes. This is obviously
a highly reliable configuration, since there is the maximum number of
alternate routes possible. The drawback is, of course, the huge number of
links required for a network of any size. For example, for n nodes there
will be $n(n-1)/2$ links, which turns out to be almost 5000 links for a
modest 100-node network!

Figure 8.19 shows a five-node fully connected network with two
minicomputers and a remote data base. Note that there are $5 \times 4/2 = 10$
links required. Communications can be maintained between any pair of

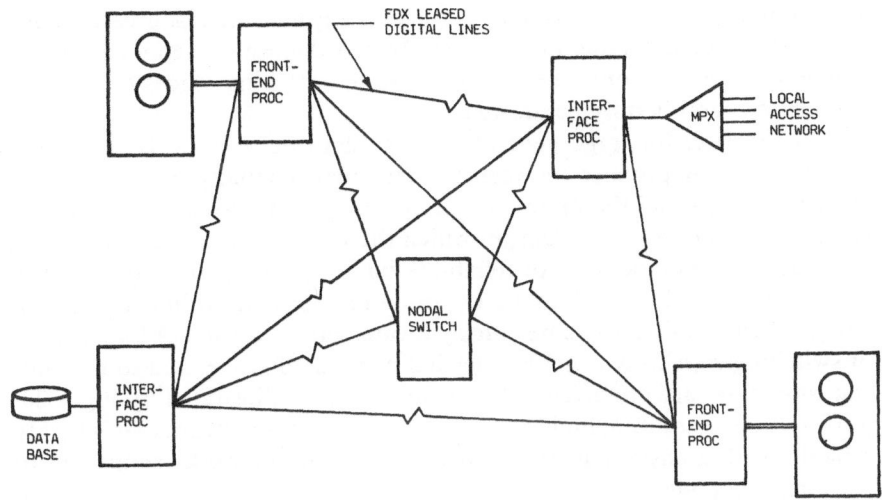

Figure 8.19. Fully connected minicomputer network.

nodes so long as at least one workable link remains in every possible partitioning of the network such that one of the communicating nodes is in each partitioned half. Furthermore, under normal operation there is a direct link to every destination node, so no intermediate switching, routing,

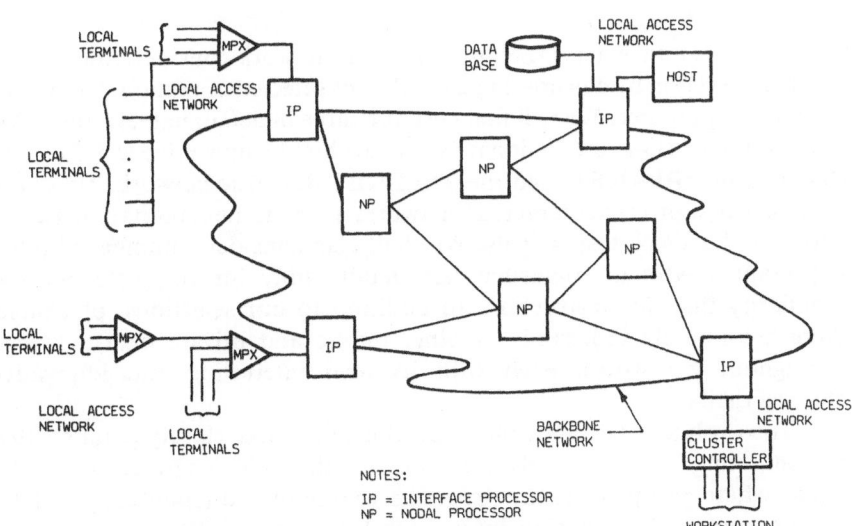

Figure 8.20. Backbone and local access topologies.

or buffering is necessary. Any node can access any network resource directly, and all have equal status in doing so. In the domestic telephone system switching hierarchy, only the highest class of switching offices (class 1 or regional centers) is fully connected.

In practice, for large general purpose data networks it is inevitably necessary to compromise between the maximal reliability of a fully connected topology and the minimal cost of a tree or star network. The result is a *multiply connected topology*, in which there are enough alternate routes to ensure the required level of reliability but at an acceptable cost. On the other hand, local access networks are often located within a building complex where failures can be quickly located and corrected. A high degree of switching capability is neither needed nor cost effective, and so tree and star topologies are preferred. The overall result, as illustrated by Fig. 8.20, is a large, decentralized, multiply connected *backbone* network that is highly reliable, with many small tree-shaped *local access* networks connected to the interface nodes.

We shall consider these various topological network design issues in more detail in the next chapter, along with such protocol concerns as routing and flow control. Before doing so, however, it will be instructive to conclude this chapter by surveying a selection of major operational computer data communications networks in the final section.

8.5. Data Network Examples

We shall now extend the rather general network considerations of the preceding section by considering a number of selected operational networks of major importance. We will first describe three in detail; namely, the SITA international airline reservation network, the highly innovative U.S. Defense Department ARPANET, and the TELENET domestic network. These are examples of a private commercial network, a government military network, and a public VAN, respectively. We shall also consider a number of other important WANs in somewhat less detail, more for completeness and familiarity than for instruction. In addition to our selections, of course, there are literally thousands of other public and private data networks throughout the world, each with its own interesting and innovative characteristics.

The evolution of data communications networks closely parallels that of computing. The very early computers of the 1950s operated in a *batch mode*, where groups of entire programs were buffered (spooled) and later all run without interruption by a central mainframe. The corresponding network had a *star* topology with the few remote terminals and workstations normally supplied by the computer vendor. In the 1960s computers began operating in the *time-sharing mode*, with multiple programs running concurrently. These time-shared systems often used extensive *tree* networks to reach a large number of occasional users in many remote locations. As the cost of computing decreased enormously in the 1970s, *distributed computing* allowed users to share not only multiple host computers, but also a wide variety of other resources such as data bases, graphical output devices, and message services. The corresponding networks tended to be *multiply connected* with intelligent switching and data processing capabilities.

Switching has also evolved in a parallel way. The classical voice telephone network was, as we have seen, circuit switched. And this approach persevered as telephone channels began to carry data. For telegraph (Telex, TWX, etc.) systems the historical technique is message switching, going all the way back to the manual operator with the characteristic green eyeshade. However, the economic imperative of modern value-added carriers to minimize leased line costs, combined with technological advances, has led to a distinct preference for packet switching for large data networks.

Today the vast majority of wide area data communications networks are multiply connected and there is a strong trend towards packet switching. This is particularly true of those public and private VANs that must carry a general mix of traffic with little control over the inputs. As the examples will illustrate, the standard form is a large multiply connected backbone network fed by numerous smaller centralized local access networks.

The Societé Internationale de Télécommunications Aéromatiques, or *SITA*, network [3] is a worldwide data network to provide data communications services for its member airlines in some 150 countries. Begun in 1949 as a message switched telegraph and then teletype network, it now claims to be the world's first and largest private packet switched network, although it also still supports message switching. SITA services include the exchange of operational and administrative information and airline seat reservations. Since the larger airlines all have their own reservation systems, this latter service is utilized by smaller airlines. Finally, SITA participates (frequently, based on personal experience) in a lost baggage tracing service.

The current SITA network is built around a multiply connected backbone known as the *High-Level Network* connecting 16 switching nodes located in a major city of each of 10 nations. Data links were originally 4800-bps leased lines, but have now been mostly upgraded to 9600 bps over terrestrial, satellite, and undersea cable lines. Where traffic volume is sufficiently large, multiple 9600-bps lines or 56-kbps digital circuits are used, and there are plans for some T1 links between switches in the same city.

The nodal switches are actually mainframe computers with their own FEPs, and in addition to switching they interface to extensive local access networks and reservation services at various airline facilities and airports.

The backbone network runs on a proprietary SITA protocol known as P1000, which defines node-to-node packet transfer, routing, and error control procedures. For example, routing is determined by fixed tables in each node that are only changed a few times per year. Congestion is handled by simply discarding excess packets and relying on the eventual retransmission from the source. Flow control can be effected by reducing the rate at which agent terminals are polled by the network.

The local access networks are primarily tree networks of time-division multiplexers, remote concentrators, and minicomputer CPs (satellite processors) that connect all regional agent terminals and other airline departments to the corresponding backbone node. For example, the two New York nodes concentrate traffic from both North and South America over terrestrial and satellite links. The various local access nodal devices do such functions as terminal polling, data concentration, reformatting, protocol conversion, and diagnostic testing.

Traffic on the SITA network is divided into two categories. *Type A* (conversational) traffic is real-time, high priority data from office and reservation terminals. It consists predominantly of short, inquiry-response messages, and network response time is below 3 sec for such transactions. Long messages are subdivided into 128-character blocks, each of which is transmitted as an independent packet with its own address, sequence number, priority code, and error check. At the destination such blocks are reassembled into the original message before delivery.

Like the P1000 backbone protocol, SITA also uses proprietary synchronous access network protocols. All airline reservation computers use the same P1024 protocol, which ensures compatibility among them. Agent terminals are invariably polled, using P1024B for IBM and P1024C for Univac systems. Other vendor systems must be made compatible with either protocol.

SITA also has a second category of traffic known as *Type B* (nonconversational). This is basically lower-priority, low-speed teletype traffic which is message switched. It is used for communications about flight status, flight operations, safety, lost luggage tracing, and other operational and administrative requirements. Although response time is much slower than that of Type A traffic, there is extensive error control to guarantee correct message delivery.

The overall SITA network requires a high degree of reliability, since commercial air travelers require accurate, prompt reservations worldwide. One can imagine the "Catch 22" situation where a failure of the reservation system prevents the repairman from purchasing a ticket to fly to the location of the failure. In actuality, there are extensive procedures for monitoring and testing suspect links, and also plans for responding to various types of disasters.

Over the past 25 years the SITA has evolved into an increasingly more sophisticated and extensive network. The latest stage in this evolution is the Advanced Network, which includes a major revision of the network protocols for improved performance, simplified user access, and easier future expansion. Also included is a network control system in Paris where all network monitoring, control, testing, and reconfiguration can be done from one central location. SITA is a clear example of a dynamic private data communications network that is highly functional and continually evolving to incorporate new technologies and provide new services.

The U.S. Department of Defense *ARPANET* [4] is, in a real sense, the granddaddy of today's packet-switched networks. Back in the mid-1960s the Defense Department's Advanced Research Projects Agency (ARPA) funded some initial experiments on communication between pairs of computers which, although not particularly significant in themselves, led to the realization that multiple computers might be interconnected with telephone lines. Funding to investigate this distributed computing concept was provided, and in 1969 the highly experimental ARPANET began initial operations with four nodes. It employed the entirely new concept of *packet switching*, which was originally proposed as a method of increasing the survivability of military communications in wartime.

Many of the early ARPANET nodes were located at university and commercial research installations, and the network initially served as a testbed for developing techniques for effective communications among many different computer types. The initial nodes used a minicomputer CP known as the *Interface Message Processor* (IMP). Each IMP was able to handle up to four hosts and their attached terminals. For nodes without any host, a different CP called the *Terminal Interface Processor* (TIP) was developed to handle up to 64 terminals of various types. The nodes were then multiply connected with 50-kbps leased wideband lines so as to provide at least two paths between every pair of nodes.

Functionally IMPs and TIPs executed the conventional interface processor functions—switching concentration routing, data link error control, high-speed communication with the hosts, and PAD interfacing functions for terminals connecting directly to a TIP. Host-to-host packets were limited to 126 bytes, five of which comprised the packet header, and messages were limited to eight packets. Interface processors then put these packets into a data link frame with fields for addresses, sequence numbers, piggyback acknowledgments, receive buffer size, and error checks.

Initial experiments with the network exposed a number of lockout problems, such as when the network packet flow became frozen because of various full buffer conditions. These findings led to the use of *buffer allocation requests* for flow control of multiple packet messages. A station wishing to send such a message was first required to request sufficient buffers at the destination node. Upon correct receipt of an entire message, the destination automatically reallocated the buffers until there were no more messages, at which time the buffer allocation was refused and transmission ended. Routing was handled by having each node (IMP) estimate its own delay based on buffer occupancy, then exchange this information with adjacent nodes at less than one-second intervals. Routing tables were then updated dynamically every few hundred milliseconds based on this local information. Packets were routed for estimated minimal transient time according to these tables.

Higher-level protocols were developed for host computer and user communications. The Host-to-Host protocol, originally implemented in the ARPANET Network Control Program (NCP) host software, provided for the establishment and management of a *logical channel*, i.e., an apparent path to the user through the network. Over this path several application protocols were provided to perform complex functions frequently needed by users. These include file transfer (FTP), remote job entry (RJE), electronic mail (MTP), graphics (NGP), and terminal protocol conversion (Telnet). The latter used the concept of a *virtual network terminal protocol*. Every user's terminal traffic was converted into this virtual terminal form for transmission and then converted into the appropriate format for the particular destination terminal. This resolved the problem of communication's between a large number of otherwise incompatible terminals using the same network.

In 1975 the network was fully operational with some 50 nodes and 100 hosts, including satellite links to Europe and Hawaii. In that year operational control of ARPANET was turned over to the Defense Communications Agency (DCA) by the prime contractor Bolt, Beranek and Newman, Inc. Although the network began carrying operational military traffic, research and improvements continued. The next year marked the introduction of a new Pluribus IMP capable of handling dozens of hosts and hundreds of terminals, and the packet format was modified for 16-bit word lengths. In 1979 the routing algorithm was completely changed to provide table updates every ten minutes using more accurate global rather than local transit time information [5]. By 1982 there were 100 nodes.

Closely related to ARPANET is the new *Defense Data Network* (DDN) [6]. As ARPANET developed in the 1970s, a number of other military networks began using the same packet-switched technology. After an aborted attempt to upgrade the existing military AUTODIN data network, it was decided in 1981 to interconnect a number of these networks and ARPANET, with the result called DDN.

The DDN includes a number of new characteristics relative to ARPANET. Leased line rates have been increased to the more prevalent 56-kbps rate offered by DDS. A sophisticated end-to-end encryption device known as the Internetwork Private Line Interface (IPLI) was developed to allow classified traffic to be sent over nonsecure networks. Additional encryption is used on all backbone links and some host access lines as well. Tight security, along with priority levels, is of course a primary requirement of military networks. The most significant of the new developments for DDN were made with respect to the protocols. Around 1978 a new Host-to-Host protocol known as *Transmission Control Protocol* (TCP) was formally introduced to replace NCP where a reliable communications path was required between users in different networks. Closely related to TCP is the *Internet Protocol* (IP), which essentially provides packet routing between network gateways. TCP/IP are distinct protocol layers, and are now used by many networks besides the DDN.

ARPANET has now clearly developed into an important operational military network, but of greater significance here is the role it has played in the development of wide area packet-switched network technology. It served for many years as an ideal research vehicle, and the results were widely published. The consequent effect on international standards has been substantial. Not only has ARPANET spawned numerous other military networks, it has also been a model for many public VANs including TELENET, which we shall consider next.

The *TELENET* packet network is a direct result of the ARPANET development work in the early 1970s. In fact, it is purported that in 1972 some of the ARPANET developers attempted unsuccessfully to interest AT&T in building a public packet network before starting an independent venture. In any event, the TELENET VAN began operation in 1975 in seven U.S. cities with a network control center at its Virginia headquarters. The initial network consisted of distinct backbone and local access networks, and used the new CCITT X.25 packet protocol instead of the more specialized ARPANET protocol. Dial-up access was supported to reach occasional and remote users. At first traffic was generally from information retrieval services, but business and government soon became significant users.

The network was designed in three hierarchial levels. Level I was the backbone network with nodes in major U.S. cities. Each node was multiply connected to at least two other backbone nodes with 50-kbps and later 56-kbps trunks. The level I nodes also served as connection points for level II nodes, each of which concentrated traffic from roughly one or more telephone area codes. Access to these nodes was from level III nodes over a tree network of concentrators and time-division multiplexers, which are now predominantly statistical multiplexers.

The initial packet switches were based on commercial minicomputers, but the network soon shifted over to the TELENET Processor (TP) family of nodal processors based on a custom designed computer. A current version of this communication processor is the TP 4000, which can be configured as a packet switch, concentrator, multiplexer, or a terminal-handling device, and the newest TP 4/II "superswitch" can switch up to 224 64-kbps ports. The level I and II nodes are actually a highly redundant cluster of packet switches that are themselves multiply connected for maximum reliability. Nodes are fully programmable, and can be downline loaded from the network control center for start-up or reconfiguration.

TELENET may be accessed by the customer in a number of ways. The most common way is to dial an access number and follow the conventional modem protocol for asynchronous terminals. Private network ports may be obtained for asynchronous or synchronous inputs. For host computers and heavily used terminals there are dedicated access ports that handle traffic up to 56 kbps. Finally, very large customers may obtain the Private Packet Exchange access with concentration equipment located on the customer's premises. Although TELENET supports the X.25 packet interface, most terminals are not packet-mode devices. Consequently, the network access nodes provide a *packet assembly and disassembly* (PAD) function for non-packet-mode users. Rather than use the virtual terminal approach as ARPANET did with Telnet, TELENET defines each terminal by a long list of some 60 parameters for such mundane things as terminal speed, character echoing, control characters, page length, and data forwarding conditions. The node is given a list of these parameters for each connected terminal, and uses it to convert all incoming traffic from any other terminal into the format specified for the addressed terminal. Although such a PAD function is now specified by CCITT in connection with X.25, the TELENET *Interactive Terminal Interface* (ITI) PAD actually predated the corresponding CCITT standard and, although similar, is somewhat more extensive. For a number of common host computer types, TELENET provides software to enable packet-mode operation.

Routing is done via routing tables that allow each node to dynamically select suitable outgoing links with the greatest unused capacity, thus providing a relatively even distribution of traffic throughout the network at the expense of suboptimal delay. The network control center also occasionally reruns the network design algorithms to ensure a near-optimal topology as the network terminals and traffic patterns change. The X.25 protocol provides for end-to-end error procedures as well as for the conventional HDLC data link error control. Flow control is accomplished by using the same request-for-next-message concept as with ARPANET.

User data is normally grouped into blocks of up to 128 characters and each block is combined with a conventional X.25 packet header to form a data packet. The packets are then individually routed through the network within a series of data link frames. At the destination PAD the packets are reassembled into the original message, checked for errors, and then delivered to the destination terminal. From the user's point of view there appears to be a dedicated *virtual circuit* through the network, although internally packets are continuously multiplexed (statistically) on the links. There are many other operational details of TELENET, but we have touched on the salient features. The X.25 protocol will be studied further in Chapter 10, as will be the PAD function.

In 1978 TELENET was purchased by GTE, who proceeded to invest heavily to upgrade and expand the network. In 1981 satellite links were added and T1 links began to be introduced. Today it is among the largest public VANs in the world and provides a wide spectrum of services to users with all types of terminals. For example, many personal computer users send electronic mail and access various data bases through TELENET. At the other extreme, huge government agencies use TELENET to interconnect hundreds of locations throughout the country. TELENET can be accessed directly or by a modem link from dozens of foreign countries, and it has gateways to many other domestic and foreign packet networks.

The present *TYMNET* common carrier data network [7] originated in 1970 to provide remote access to a computer time-sharing service owned by Tymshare, Inc. User-owned computers were gradually added, and in 1977 Tymnet, Inc. was tariffed as a common carrier VAN to offer data communications services to the public. The network has grown to over 1300 nodes serving some 300 metropolitan areas in the U.S. and 40 foreign countries. It also has gateways to many other major data networks. The minicomputer nodal processors, called TYMNET engines, are interconnected by leased backbone links at 9.6 and 56 kbps. Users may access the network with a wide selection of asynchronous and synchronous protocols and rates from 110 to 9600 bps.

Internally the network uses a proprietary packet protocol that segments long messages and combines short ones to get fairly consistent packet sizes. Packets may be restructured at intermediate nodes. User access, routing, flow control, and general network management are done by a program named SAM (Supervisor in Active Mode) resident in one node. Dormant versions of SAM are located in various other nodes, capable of taking over control should the active one fail.

TYMNET is a major U.S. provider of public and private network services to many large firms and agencies, as well as to individuals and on-line data base services. There is also an electronic mail capability. After several false starts the network was acquired by McDonnell-Douglas in 1984.

The Telecom Canada (formerly TCTS) *DATAPAC* packet-switched network began commercial operation in 1977. At that time it was highly innovative, being based on the new and incomplete CCITT X.25 packet protocol. It provided switched digital service, in contrast to the earlier private-line DATAROUTE network also run by TCTS. DATAPAC uses multiprocessor minicomputer-based nodal switches connected primarily by 56-kbps backbone links. The network control center is located in Ottawa.

At present there are well over 2000 user connections at speeds of 110 to 9600 bps. Terminals can be either X.25 packet-mode or character-mode, with access via a PAD in the latter case. Users are offered both normal and priority grades of service. Initially there was considerable debate over the use of datagrams or virtual circuit, with the result being datagrams between nodes supporting virtual circuits between end users. Gateways are provided to many major national and international public data networks including TELENET and TYMNET, and most are based on the CCITT X.25 standard which is completely compatible with the X.25 internal protocol.

TRANSPAC is the French national public packet switched network, and also uses the CCITT X.25 protocol with virtual circuits. It began operation in 1978, and now has around 30 nodes throughout France, plus international gateways to most other major public packet networks. The network control center is in Rennes. User access is either direct for packet-mode terminals or via a PAD otherwise. Access speeds range from 50 bps for telex up to 48 kbps.

The backbone network uses 64- and 72-kbps links with at least two paths between every pair of minicomputer-based nodal switches. The original network design for 25,000 ports is now being expanded to eventually accommodate 100,000, with installation of a new and more powerful second generation switch now underway. TRANSPAC has proven to be a highly successful data network and is used by many large firms as well as smaller organizations and individuals.

The Japanese national *Digital Data Exchange* (DDX) network began operation in 1979 after extensive development based on two earlier prototype networks. DDX offers both circuit-switched (DDS-CS) and X.25 packet-switched (DDS-PS) services, with the latter instigated in 1980. The network now covers essentially all of the country, with some 13,000 terminals in over 100 cities.

The DDS-CS backbone network uses TDM concentrators and modified electronic telephone switches, all connected primarily with 1.544-Mbps links. User access is via CCITT X.20 and X.21 digital interfaces, or the analog equivalents. The DDX has steadily grown since 1979, and is now being upgraded with new generation concentration and switching equipment. Much of the underlying design philosophy is oriented toward future evolution to the ISDN with a complete range of voice, data, and video capabilities.

There are, of course, many other important and interesting data networks in existence which are not included here. In addition to large public data networks, there are literally thousands of private networks of all sizes, shapes, and colors. Our brief survey in this section should, however, provide a reasonable idea of how typical data networks are constructed and what services they might be expected to offer the user.

The material presented in this chapter is intended to provide a coherent transition from point-to-point links to more complex networks. Although we started out by considering only multiplexers, these basic concentrating nodes were then extended to include more general types of nodes capable of a wide range of networking functions such as switching, concentration, protocol handling, and interfacing to various terminal equipment. All conventional computer data communications networks are essentially comprised of such nodes interconnected by point-to-point data links, along with the appropriate protocols, or rules, for getting the user data through the network. To illustrate this point, overview examples of three major operational networks were presented along with brief surveys of several more.

With the transitional network introduction of this chapter behind us, we now turn to the two remaining essential considerations for networks; namely, the topological layout and sizing of the network, and the selection of a suitable set of protocols for its operation. The former consideration is the subject of the next chapter.

References

1. R. A. Scholtz, Frame synchronization techniques, *IEEE Trans. Commun.* **25**(8), 1204–1212 (1980).
2. E. G. Brohm, Sampling new technologies of network processors, *Data Commun.* **13**, 143–147 (1984).
3. Interconnect net navigates airline reservations, *Data Commun.* **11**, 99–107 (1982).
4. J. M. McQuillan and D. C. Walden, The ARPA network design decisions, *Computer Networks* **1**, 243–289 (1977).
5. J. M. McQuillan, I. Richer, and E. C. Rosen, The new routing algorithm for the ARPANET, *IEEE Trans. Commun.* **28**(5), 711–719 (1980).
6. L. C. Vandenberg, *DoD Record Networks: AUTODIN I, AUTODIN II, DDN*, SIT 662 Term Report, Southeastern Institute of Technology, Huntsville, Alabama, April (1983).
7. D. Bass, *Tymnet Network*, SIT 661 Term Report, Southeastern Institute of Technology, Huntsville, Alabama, February (1984).

Suggested Readings

R. Glasgal, *Advanced Techniques for Data Communications*, Artech House, Dedham, Massachusetts (1978).

A detailed, nontechnical description of many aspects of multiplexing equipment from an entirely practical viewpoint.

D. R. Doll, *Data Communications*, Wiley, New York (1978).

Although now a bit dated, there is a good discussion in Chapter 7 of the economic savings in line costs with multiplexing.

Check Your Understanding of Chapter 8—True or False?

1. For short, bursty traffic a TDMPX can waste a lot of bandwidth.
2. Most large backbone packet networks are fully connected.
3. Historically, the world's great telephone networks were built using FDM.
4. Split-stream modems provide two FDM channels for two-wire FDX operation.
5. Code and rate conversion are commonly done by statistical multiplexers.
6. A telephone rotary is actually a contention device.
7. A major disadvantage of FDM is that all multiplexed frequency bands are simplex.
8. Both ARPANET and TELENET are used by the U.S. government for operational traffic.
9. Telephone line TDMPXs typically have RS-232C interfaces for both input and output ports.
10. All input data rates need not be the same with either TDM or FDM.
11. Distributed networks were motivated by the trend to multiple hosts in organizations.
12. With FDM no bandwidth need be allocated for frame synchronization.
13. The PAD function can be done in a nodal switch of some packet networks.
14. Gateways are devices that interface secure classified terminals to public networks.
15. Remote concentrators and modern STMPXs now perform many of the same functions.
16. FEPs and modern STMPXs now perform many of the same functions.
17. The original TYMNET was a time-sharing service for centrally located main frames.
18. Patch panels, port selectors, and switching STMPXs all switch data circuits.
19. Multiplexers are always used point-to-point, and never for multidrop lines.
20. A fully connected network with 11 nodes will have exactly 55 links.
21. With STMPXs the aggregate input rate is greater than the line data rate.
22. Modem sharing units and line sharing units perform the same basic function.
23. STMPX flow control can be done either in-band or out-of-band.
24. Individual input data in a TDMPX frame is identified by a unique flag bit.
25. The airline network is so large that it is named Send Information to Anyone (SITA).

9

Network Design Techniques

9.1. Network Design Considerations

In this chapter we consider a number of fundamental network design issues and present some potential techniques for solving them. The major emphasis is on the topological layout of links and nodes and the traffic flow through them. However, before getting down to specific details we shall first use a simple example to establish some basic subproblems associated with data communications network design.

The great international retailing firm of Discount Dog has decided to build the first of a series of Mega-Stores, with each containing several acres of floor space. Within each store there will be dozens of departments, each with a central cashier's station and a single fire main connection. The first network problem is how to run the fire main through the ceiling so that all stations are connected with the least amount of pipe. The resulting tree network is called a *minimum spanning tree*. Next, consider the location of the aisles. For n stations there could be $n(n-1)/2$ direct two-way connections, but only a fraction of these will normally be used. The selection of this subset, and perhaps the station locations as well, is the *topological layout* problem.

Once the aisle locations are selected, the security and fire control personnel will need to determine the quickest way to reach any particular station from their own station. Assuming the time is proportional to distance, this is a problem of finding the *shortest path* through the network of aisles. Another basic network design problem is how wide to make the aisles. Wide aisles, of course, are more expensive since they reduce the display areas, but they also can carry more customers. The selection of aisle width, or customer capacity, is a problem of *capacity assignment*.

Given these aisle capacity limitations, two more network issues arise in a natural way. First is just how large a flow of customers is possible between any two stations, e.g., between the front door and the toy department on the first day of the great Discount Dog Dingbat Doll sale. This is the *maximal flow* problem. The other issue is that of directing customers along the aisles between pairs of stations and the control of customer flow along these routes to avoid congestion in the aisles. These are the *routing* and *flow control* problems.

Each of these problems in our example has an analogy in communications network design. Although the problems are presented individually, there is clearly a great deal of interrelation among them. For example routing is closely dependent on how capacity is assigned and how the flow control is done. Except for very simple networks, the *comprehensive optimal design* of entire networks is usually not feasible, and thus iterative and suboptimal design methods must invariably be used.

Since node locations are usually fixed by the required locations of terminals and network facilities, network design is predominately a question of first where to put links and concentrators and second what their data carrying capacities should be. For a given set of n nodes it is seldom necessary for them to be fully connected with all $n(n-1)/2$ possible links. Consequently, there is an inherent design tradeoff between equipment simplicity and low cost with fewer links, as opposed to high reliability and low delay with more links. Other common design criteria are throughput between source and destination nodes; protection against congestion at internal network nodes; survivability in the event of natural or man-made disaster; network control and repair capabilities; ease of expansion (or contraction) without excessive cost or down time; and finally a whole host of nontechnical constraints such as politics, competition, regulation, and managerial value judgments.

There are many viable ways to go about designing networks, but no completely general practical method exists. Over the years specific techniques have been developed for different network types, and they can be grouped into two major categories of *optimal* and *heuristic*. Intuitively, we would always prefer an optimal solution whenever practicable. However, this is not always as simple as it may seem. Completely optimal solutions to large networks with hundreds of nodes can be extremely difficult to formulate accurately. Even in those cases where the problem can be well formulated the resulting nonlinear optimization programs may be prohibitively expensive to compute. And even if a solution is viable, the question of just what "optimal" means is often ephemeral and difficult to quantify. Like beauty, it is often in the eye of the beholder. Should delay, for example, be actually minimized or should it just be constrained to stay below some maximum acceptable value? What parameters should be optimized, and what is the cost and relative importance of each? How will optimality change if in the future new nodes and links must be added or old ones modified? And, of course, how does one account for intangible factors such as personal biases, future user requirements, competitor actions, traffic changes, new technologies, and plain dumb luck? In fact, a great deal of experience and judgement is often involved in defining the optimal criterion, and the resulting network may depend heavily on the quality of such input. Even if all this can be well formulated at the beginning of the design process, there is a high probability that the formulation will change as the network is developed or, at least, after it begins operation. However, despite these potential imperfections, our optimal solution should be better than anything else we can obtain and at the very least will provide a good initial place to begin the actual implementation.

On the other hand, in many design problems it is not feasible to obtain a truly optimal solution because of either modeling uncertainties or excessive costs. Here we must be willing to settle for suboptimal, or Pareto† optimal, solutions provided they are not too far from optimal. There are two standard approaches to obtaining such solutions in practice: simulation and the use of heuristic algorithms. The first approach is *simulation*, where an analytical model of the network is written as a (computer) program and the model performance is observed under a wide range of anticipated inputs. Various model parameters are then systematically changed in a search for improved performance to the same sample of inputs. Over a large number of such iterations the designer should be able to develop a good feel for the range of possible network performance along with a suitable set of parameters. The latter might include link connections and capacities, buffer sizes,

---

† This is akin to the so-called "80/20 rule," that achieving the last 20% of optimality will account for 80% of the total cost and is often not worth the extra effort.

reliability under various component failure patterns, and comparisons with competing network services. Although simulation is widely used and can be highly effective for very complex networks, there are some caveats. Simulation depends directly on the effectiveness of the model and also depends critically on the choice of sample inputs. If the latter is not representative of those that will be encountered in practice, then we can well end up with a great solution—to the wrong problem! And even if the input samples are intelligently chosen there is always some statistical possibility that, just by bad luck, they will not indicate the representative network performance. Taking larger samples should tend to reduce the probability of such an eventuality but at a proportional cost in time and resources. As a rule of thumb, simulation is indicated when the complexity and level of detail of the network design are too great for strictly analytical methods to be used.

The second standard suboptimal approach is through the use of *heuristic algorithms*. These are usually based on common sense and clever insights, which hopefully result in designs that are acceptably close to optimal but with much less computational effort than for complete optimization. In view of our preceding discussion of optimality, it should seem plausible that in many practical applications a heuristic solution is entirely acceptable. It may even prove better than an optimal one that was poorly formulated or based on outmoded criteria. Furthermore, for large complex problems requiring numerous iterations over the course of the network design, heuristic approaches may be the only viable approach short of an extensive (and expensive) simulation. Consequently, in this chapter we shall present a number of heuristic algorithms as alternatives to optimal design procedures. As with simulation, however, heuristic algorithms are not an unmixed blessing. There is no absolute guarantee that the result will be "sufficiently" close to optimal; and since the optimal is obviously unknown, it may be difficult to estimate how close to optimal the result actually is. It is also, of course, always statistically possible that the outcome will be far from optimal and result in a terrible network design. The only way really to hedge against this is to understand the algorithm operation and limitations, and to always weigh the results against your own judgement and experience.

This chapter first looks at the *modeling problem* in some detail, since this is where any systematic design must begin. The primary techniques for network modeling come from the disciplines of *graph theory, mathematical programming*, and *queueing*. We first consider the representation of networks with *topological graphs* and develop a number of useful properties relating to path lengths, traffic flow, and reliability. The fundamentals of *linear programming* are also presented, since this most widely used optimization technique is directly applicable to many network problems and forms the basis for more sophisticated approaches to others. The intent is to only develop an intuitive understanding of linear programming, since the subject is covered extensively in the literature and complete computer routines are widely available.

Armed with these modeling tools, we then turn to actual network designs, starting with the local access network. Here we assume reliability is not the major consideration, but rather we want the most economical network that will do the job. This is invariably a *tree topology*, perhaps with constraints on the maximum link capacities and the number of drops on any one line. Such networks typically are centralized on one computer or interface nodal processor, but for large local access networks it may be desirable to use additional multiplexers or other concentrators to save on line costs and perform polling and protocol functions. Thus we shall also be interested in the *concentrator location* problem.

Next we move to the questions of routing, traffic flow, throughput, delay, and link capacity assignment. These are key design issues for both the local access and backbone networks. However, they are much more difficult to determine in multiply connected backbone topologies than for local access trees. Therefore, our focus will be on the former. We shall look at some common algorithms for generating global routing tables and also at some local and dynamic routing schemes. Then the limitations on network traffic flow are obtained and heuristic methods for increasing end-to-end flow are given. Some popular protocol flow control techniques are also described.

The issues of delay, link capacity, and nodal buffer size are closely interrelated and hence are treated together. The standard approach is through queueing theory, which is used primarily in its simplest $M/M/1$ form. The Appendix contains a brief review of the essential probability and random process theory.

Finally, in the last section we consider the comprehensive network design problem. Even for modest networks, the complete global optimization of a meaningful performance criterion subject to realistic constraints on resources is too large and complex to be viable. Consequently, the design problem is commonly divided into subproblems for which tractable methods exist.

Although the design techniques selected for this chapter are intended to be both instructive and practical, there are many other variations and alternative techniques. Much of the existing material on network design came directly from the extensive field of operations research, and much additional technique and insight can be found there. The mathematical level is also somewhat higher than the preceding (and following) chapters, but a concerted effort to master it will return considerable dividends if the reader is doing actual network designs. However, most of the chapter can be understood conceptually by the perservering reader without a full grasp of all the mathematics involved. In addition, the Appendix covers some specific mathematical techniques used in the chapter.

9.2. Graph Theory and Linear Programming

A *topological graph* $G(N, L)$ consists of two associated sets N of *nodes* (vertices, junctions, points) and L of *links* (branches, edges, lines, arcs) such that each end of a link terminates on a node. The correspondence to communications network nodes (terminals, switches, concentrators, PADs) and data links is obvious. A *subgraph* of $G(N, L)$ is a graph such that all nodes and links are subsets of N and L, respectively. The links may have associated directions, in which case the graph is *directed* (oriented); other-wise, it is *undirected*. These basic concepts are shown in Fig. 9.1a–9.1c, where n_i and l_j denote node i and link j, respectively. Undirected graphs will be our primary concern here, since they correspond to data networks

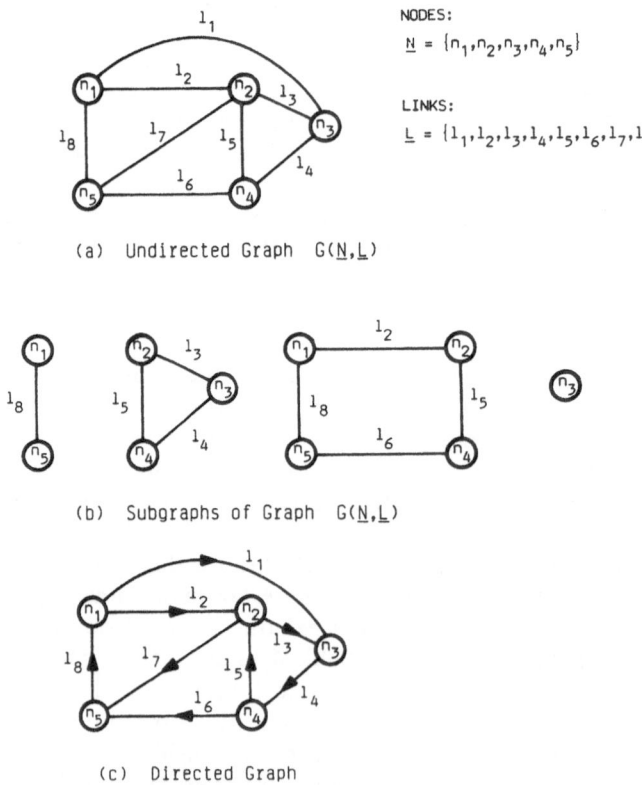

NODES:

$\underline{N} = \{n_1, n_2, n_3, n_4, n_5\}$

LINKS:

$\underline{L} = \{l_1, l_2, l_3, l_4, l_5, l_6, l_7, l_8\}$

(a)   Undirected Graph $G(\underline{N}, \underline{L})$

(b)   Subgraphs of Graph $G(\underline{N}, \underline{L})$

(c)   Directed Graph

Figure 9.1. Basic graphs and subgraphs.

with FDX or HDX data links, which are by far the most prevalent. There are, however, some simplex networks in operation, e.g., in telemetry, where directed graphs must be used.

A *path* (route, walk) through the network is a consecutive sequence of nodes and links that connects two specific nodes, which will often be the *source* (origin) and *destination* (sink) nodes. If a path starts and ends on the same node it forms a *loop* (circuit, cycle). A graph $G(N, L)$ is *connected* if there is some path between every pair of nodes, and a *component* (fragment) of $G(N, L)$ is simply a connected subgraph. A graph is *complete* if there is a link between every pair of nodes. Completeness thus corresponds to a fully connected data communications network. This, along with the notion of connectedness, is again illustrated with Fig. 9.2. Note that in the (a) part there may be many paths between the same two nodes but each can be uniquely described by its ordered set of nodes (or links).

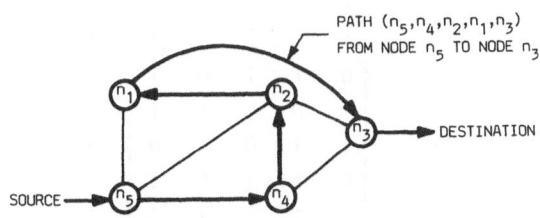

(a) One Possible Path Between Nodes n_5 and n_3

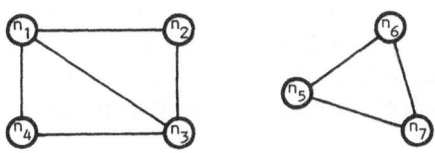

(b) A Non-Connected Graph with Two Components

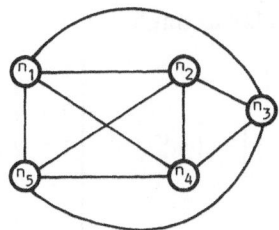

(c) A Graph that is Complete

Figure 9.2. Paths and connectedness of graphs.

With these fundamental definitions we can now develop some mathematical properties of graphs that will be helpful in network modeling. The *adjacency* (connectivity) *matrix* $\mathbf{M} = [m_{ij}]$ of an undirected graph $G(N, L)$ having n nodes is defined by its components

$$m_{ij} = \begin{cases} 1 & \text{if there is a branch} \\ & \text{between nodes } i \text{ and } j \\ 0 & \text{otherwise} \end{cases}$$

Thus $\mathbf{M}$ will be a square matrix with n rows and n columns. It will also be symmetric for undirected graphs, i.e., $m_{ij} = m_{ji}$, where the first subscript always indicates the row index and the second the column index. As an example the adjacency matrix for the graph of Fig. 9.1a is easily found to be

$$\mathbf{M} = \begin{array}{c} \text{nodes} \rightarrow \begin{array}{ccccc} 1 & 2 & 3 & 4 & 5 \end{array} \\ \begin{bmatrix} 0 & 1 & 1 & 0 & 1 \\ 1 & 0 & 1 & 1 & 1 \\ 1 & 1 & 0 & 1 & 0 \\ 0 & 1 & 1 & 0 & 1 \\ 1 & 1 & 0 & 1 & 0 \end{bmatrix} \begin{array}{c} \text{rows} \\ \downarrow \\ 1 \\ 2 \\ 3 \\ 4 \\ 5 \end{array} \end{array}$$

The adjacency matrix has some interesting properties. First, the number of ones in any row k equals that of column k, and this number indicates the number of links connected to, or incident on, node k. This number is the *degree* (valence) of node k, and is an important indicator of network reliability. For example, in $\mathbf{M}$ above the sum of column 2 is four so there must be four links incident on node 4. This is apparent from node n_4 of the corresponding figure.

A second, rather obvious, property is that any entry m_{ij} indicates the number of one-link paths between nodes i and j. What is not so trivial is that the entries of $\mathbf{M}^2 = [m_{ij}^{(2)}]$ indicate the number of two-link paths between i and j. Thus for our particular example

$$\mathbf{M}^2 = \left[\sum_{k=1}^{n} m_{ik} m_{kj} \right] = [m_{ij}^{(2)}] = \begin{bmatrix} 3 & 2 & 1 & 3 & 1 \\ 2 & 4 & 2 & 2 & 2 \\ 1 & 2 & 3 & 1 & 3 \\ 3 & 2 & 1 & 3 & 1 \\ 1 & 2 & 3 & 1 & 3 \end{bmatrix}$$

Since, for example, $m_{14}^{(2)} = 3$ there must be three distinct two-link paths between nodes 1 and 4. In fact, from Fig. 9.1a they are seen to be

(n_1, n_3, n_4), (n_1, n_2, n_4), and (n_1, n_5, n_4). Similarly, $m_{22}^{(2)} = 4$ implies four different loops of two links each starting and ending on node 2. Furthermore, the same idea extends to higher powers of $\mathbf{M}$; thus, for $\mathbf{M}^k = [m_{ij}^{(k)}]$ there are $m_{ij}^{(k)}$ paths of exactly k links between nodes i and j.

The next graph theoretic subject we shall consider is that of the tree and its dual, the cut-set. A topological *tree* of a graph $G(N, L)$ is a connected subgraph that contains no loops, as the name suggests. A network will in general have many trees of various sizes, but we will be interested mainly in *spanning trees*. These trees contain all nodes of a connected graph and thus must consist of precisely n nodes and $n - 1$ links, as illustrated in Fig. 9.3 for the example graph of Fig. 9.1. Note that for any spanning tree there is exactly one path between any pair of nodes, and removal (failure) of any link will disconnect the corresponding graph. Intuitively we would expect a tree network to be relatively inexpensive but at the price of low reliability.

A topological *cut-set* (cut, link cut-set) is a minimal set of links that will disconnect the graph if removed, but the replacement of any one link will reconnect it. It is rather obvious but very important to note that all network traffic from nodes on one side of a cut-set to those on the other side must flow through exactly the links of the cut-set. As with trees, a graph will generally have many cut-sets, even for a specific pair of nodes. Figure 9.4 shows a few of the cut-sets for our preceding example graph. There are many others that are not shown.

To characterize spanning trees mathematically we again use a matrix. The appropriate one here is the (link-to-node) *incidence matrix* $\mathbf{A}_a = [a_{ij}]$, which is defined for a directed graph of n nodes and l links by

$$a_{ij} = \begin{cases} +1 & \text{if link } j \text{ is incident on node } i \text{ and directed } out\ of \text{ the node} \\ -1 & \text{if link } j \text{ is incident on node } i \text{ and directed } into \text{ the node} \\ 0 & \text{if link } j \text{ does not connect to node } i \end{cases}$$

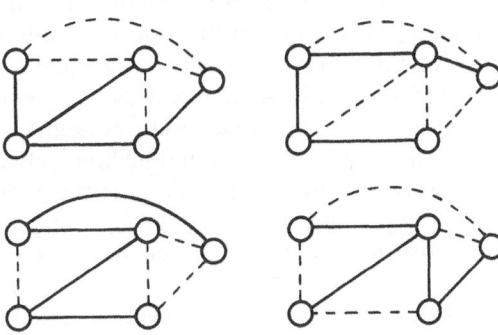

Figure 9.3. Some spanning trees of a graph.

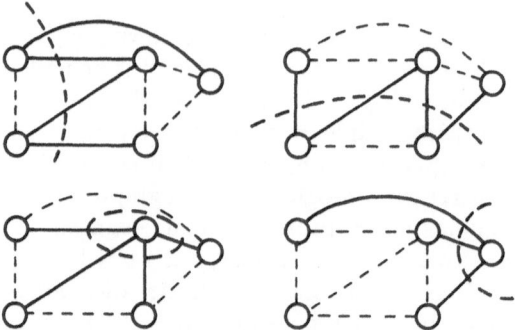

Figure 9.4. Some cut-sets of a graph.

For an undirected graph we define $a_{ij} = 1$ whenever link j is incident on node i with no regard for direction.

For the undirected graph of Fig. 9.1a again, we have $n = 5$ and $l = 8$, so

$$
\begin{array}{ccc}
& & \text{nodes} \\
\text{links} \rightarrow \quad 1 \quad 2 \quad 3 \quad 4 \quad 5 \quad 6 \quad 7 \quad 8 & & \downarrow \\
\mathbf{A}_a =
\begin{bmatrix}
1 & 1 & 0 & 0 & 0 & 0 & 0 & 1 \\
0 & 1 & 1 & 0 & 1 & 0 & 1 & 0 \\
1 & 0 & 1 & 1 & 0 & 0 & 0 & 0 \\
0 & 0 & 0 & 1 & 1 & 1 & 0 & 0 \\
0 & 0 & 0 & 0 & 0 & 1 & 1 & 1
\end{bmatrix}
&
\begin{array}{c}
1 \\ 2 \\ 3 \\ 4 \\ 5
\end{array}
\end{array}
$$

As with $\mathbf{M}$, the sum of the entries in each row indicates the number of branches incident on the corresponding node. Note that $\mathbf{A}_a$ has exactly two ones in each column so that any deleted row could always be reconstructed by simply entering a one for any column that does not already have two, and entering zeros elsewhere. Furthermore, $\mathbf{A}_a$ must have at least $n - 1$ links if the graph is to be connected, since this is what is required for a spanning tree. The implication of all this is that the rank of $\mathbf{A}_a$ is at most $n - 1$, and in fact can be shown to equal $n - 1$. Consequently we can define the *reduced incidence matrix* $\mathbf{A}$ as being equal to $\mathbf{A}_a$ with any row deleted.†
Hence, $\mathbf{A}$ has $n - 1$ rows and l columns.

There are two key relationships between $\mathbf{A}$ and the spanning trees of the corresponding graph. First, any nonsingular $(n - 1)$-square submatrix consisting of $n - 1$ columns of $\mathbf{A}$ corresponds to a spanning tree composed of those $n - 1$ links indicated by the $n - 1$ submatrix columns. Thus we

† This deleted row corresponds in electrical networks to the datum (ground, reference) node.

can, in theory, test for trees mathematically. Of more practical importance, however, is the second relationship: the total number of spanning trees for a directed graph with reduced incidence matrix A is given by the value of the determinant of the $(n-1)$-square product matrix AA^T; i.e., $\det(AA^T)$. If the graph is undirected, we can still obtain the number of spanning trees by temporarily assigning arbitrary orientations to the branches and then proceeding as before. Thus for our ongoing example of Fig. 9.1a, we arbitrarily assign link orientations and delete the last row of A_a to get

$$
\begin{array}{c}
\text{links} \rightarrow \\
\\
A =
\end{array}
\begin{array}{cccccccc}
1 & 2 & 3 & 4 & 5 & 6 & 7 & 8 \\
\end{array}
\begin{bmatrix}
1 & 1 & 0 & 0 & 0 & 0 & 0 & 1 \\
0 & -1 & 1 & 0 & 1 & 0 & 1 & 0 \\
-1 & 0 & -1 & 1 & 0 & 0 & 0 & 0 \\
0 & 0 & 0 & -1 & -1 & 1 & 0 & 0 \\
\end{bmatrix}
\begin{array}{c}
\text{nodes} \\
\downarrow \\
1 \\
2 \\
3 \\
4 \\
\end{array}
$$

After performing the matrix multiplication we obtain

$$
\det(AA^T) = \det \begin{bmatrix}
3 & -1 & -1 & 0 \\
-1 & 4 & -1 & -1 \\
-1 & -1 & 3 & -1 \\
0 & -1 & -1 & 3 \\
\end{bmatrix} = 45
$$

There are 45 different spanning trees for this example. To test for these 45 trees directly from A requires finding the determinants of $\binom{8}{4} = 70$ 4×4 submatrices, where the notation $\binom{n}{k} = n!/k!(n-k)!$. For example (l_5, l_6, l_7, l_8) is not a tree since

$$
\det \begin{array}{c}
\begin{array}{cccc}
5 & 6 & 7 & 8 \\
\end{array} \leftarrow \text{links} \\
\begin{bmatrix}
0 & 0 & 0 & 1 \\
1 & 0 & 1 & 0 \\
0 & 0 & 0 & 0 \\
1 & 1 & 0 & 0 \\
\end{bmatrix}
\end{array} = 0
$$

but (l_4, l_6, l_7, l_8) is because

$$
\det \begin{array}{c}
\begin{array}{cccc}
4 & 6 & 7 & 8 \\
\end{array} \leftarrow \text{links} \\
\begin{bmatrix}
0 & 0 & 0 & 1 \\
0 & 0 & 1 & 0 \\
1 & 0 & 0 & 0 \\
1 & 1 & 0 & 0 \\
\end{bmatrix}
\end{array} = 1
$$

Both cases are easily verified from the graph.

The number of different cut-sets can also be bounded mathematically by simply noting that of the n total nodes, a component created by the cut-set removal must contain either $1, 2, \ldots, n - 2,$ or $n - 1$ nodes, but for each cut-set with k nodes in one component there must be $n - k$ in the other. Hence the number of possible cut-sets for a graph with n nodes is bounded by

$$\text{total cut-sets} \leq \frac{1}{2} \sum_{k=1}^{n-1} \binom{n}{k}$$

with the bound achieved for complete graphs. For our continuing example with five nodes there are no more than $\frac{1}{2}(5 + 10 + 10 + 5) = 15$ cut-sets; in fact, it can be verified by inspection that there are actually 13.

In modeling data communications networks it is desirable to associate a numerical *weight* w_{ij} with the link between each node pair n_i and n_j. We then naturally call the result a *weighted graph*. These link weights may represent any number of quantities, such as the cost of installing or operating the link, the actual length of the link, the delay encountered in the link and link buffers, link reliability, or the actual or maximum traffic flow in the link. We shall usually refer to the link weights as *generalized* "costs" with the stipulation that they may not necessarily be monetary costs. We shall also use the terms *weighted graph* and *topological network* interchangeably, although the latter is often restricted to the case where the weights actually represent traffic flow. The particular choice of weighting factors will, of course, depend on the network we are modeling and on the selected criteria for optimality. The weights of a network can conveniently be displayed in a *weighting* or (generalized) *cost matrix* $\mathbf{W} = [w_{ij}]$. For the weighted graph of Fig. 9.5a it is easy to verify that the cost matrix is

$$
\begin{array}{c}
 & & & & & \text{nodes} \\
\text{nodes} \rightarrow & 1 & 2 & 3 & 4 & 5 & \downarrow \\
\mathbf{W} = & \begin{bmatrix}
- & 3 & 10 & \infty & 4 \\
3 & - & 6 & 5 & 1 \\
10 & 6 & - & 2 & \infty \\
\infty & 5 & 2 & - & 8 \\
4 & 1 & \infty & 8 & -
\end{bmatrix} & & & & & \begin{array}{c} 1 \\ 2 \\ 3 \\ 4 \\ 5 \end{array}
\end{array}
$$

Again note that the matrix is symmetric for an undirected graph, where the same link cost is assumed in both directions. Since we are not interested in a node sending traffic to itself, we indicate those entries with a dash, and for nodes with no connecting link the cost is taken to be infinite.

If we assume that the weights of a graph correspond to some generalized cost that we wish to minimize, then it is natural to ask what the least

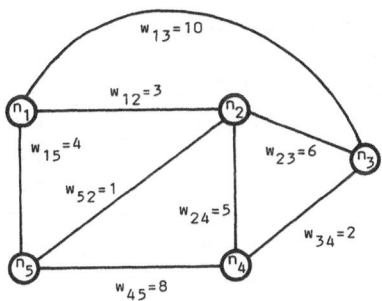

(a) Link Weighting for an Undirected Graph

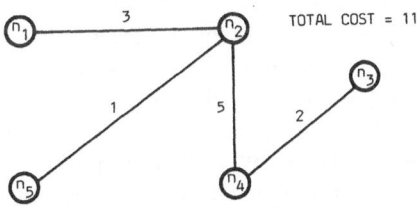

Figure 9.5. Weighted graph.   (b) The Minimum Spanning Tree

expensive subgraph is that connects all nodes. In fact, it will be a tree, for
if it were not we could always remove one branch of a loop to get a lower
cost. Any particular tree for which the cost is less than or equal to all other
trees is called a *minimal spanning tree* (MST). To find an MST, the procedure
is simple and often referred to as a "greedy" algorithm. It can be stated as
follows:

Algorithm 9.1 *(Prim Minimal Spanning Tree* [1])
1. *Start at any node, and connect it to an adjacent node having the
   least-cost link (if more than one such link, select any one arbitrarily).*
2. *Of all the yet unconnected nodes, connect one having the least-cost
   link to any one of the connected nodes (if more than one such link
   select any one arbitrarily).*
3. *Repeat step 2 until all links are connected; the result is an MST.*

For the example of Fig. 9.5a we find an MST starting at node 1 by connecting
nodes according to Algorithm 9.1. It is easily verified that the order of
selection is n_1, n_2, n_5, n_4, and n_3 with the resulting MST shown in Fig. 9.5b.
The resulting cost (total link weights) is 11 units. There is no other tree
with a lower cost. There are also some other algorithms for finding the
MST, but the one presented here is among the most popular. So long as

there are no constraints on the links, they will all give the same result. However, this equivalence is in general no longer true when constraints are present, as we shall see in the next section.

We now turn to the important topic of graph *connectedness*, which relates directly to the reliability and survivability of data communications networks. Recalling that the degree of a node is just the number of links connected to it, we can extend the idea by defining the *minimum degree of a graph* δ as the minimum degree of all nodes. Since each of l links must terminate on exactly two of n nodes, the *average degree* of the graph is just $2l/n$ links per node and obviously $\delta \leqslant 2l/n$.

Two indicators of the connectedness of a graph are cohesion and connectivity. For any given pair of nodes i and j, the *cohesion* λ_{ij} is the smallest number of links that must be disconnected (fail) before the graph becomes disconnected. Analogously, the *connectivity* ω_{ij} of nodes i and j is the smallest number of nodes that must be removed (fail) before disconnection. Again these definitions are extended to the entire graph by defining the cohesion λ and connectivity ω of the graph as the respective minima of λ_{ij} and ω_{ij} over all node pairs.

Intuitively, we would generally expect large values of δ, λ, and ω to indicate highly reliable corresponding networks, and vice versa. The relationship among them provides further insight. Since failure of a node always affects at least two links and often more, it it plausible that $\omega \leqslant \lambda$. Furthermore, we can always disconnect the graph by removing all links from a single node so $\lambda \leqslant \delta$. In fact, it has been shown [2] that†

$$\omega \leqslant \lambda \leqslant \delta \leqslant [2l/n]$$

From this result it appears that networks are more vulnerable to node than link failure and that this vulnerability increases as ω and λ decrease. This conclusion, however, is only true in a general sense [3].

It is now instructive to look at a few specific examples of connectedness. First, a tree network has $\omega = \lambda = \delta = 1$, so any node or link failure is fatal so far as disconnection is concerned. At the other extreme a complete graph has $n - 1$ links per node, and at least $n - 1$ of the $n(n - 1)/2$ links must then be removed to disconnect it. The graph remains connected so long as there are at least two nodes left; so we shall take the connectivity to be $n - 1$. Therefore $\omega = \lambda = \delta = n - 1$. In fact, it is straightforward to construct other graphs with $\omega = \lambda = [2l/n]$ for any integer value up to $n - 1$ [4]. Such graphs have *minimal vulnerability* in the sense that, assuming equal probabilities of any link or of any node failing, the chance of the graph becoming disconnected is minimized. For the example graph of Fig. 9.1a

† The notation $[x]$ denotes the integer part of x throughout the chapter.

it is easy to verify that $\omega = \lambda = \delta = 3$ and furthermore, $[2l/n] = [16/5] = 3$ also.

It is possible to calculate ω and λ for any graph simply by an exhaustive search. Fortunately, there are more efficient methods, the details of which we shall consider later. However, the basic procedure for λ is based on setting the capacity of all links to unity and then finding the cut-set with least capacity between all node pairs. This capacity then equals the cohesion. The same procedure will also find the connectivity provided a transformation is made on the original network. A less powerful but computationally much simpler approach is to test for a lower limit on connectedness. A graph is *k-connected* if $k \leqslant \omega$; so k bounds ω, λ, and δ. There is a simple test for k-connectivity. We first need to define, for any pair of nodes i and j, *node-disjoint paths* as those having no nodes in common other than, of course, i and j. Similarly, *link-disjoint paths* between i and j have no common links (but may have common nodes). These definitions are illustrated by Fig. 9.6. We can now state the test for k-connectivity of a graph:

Algorithm 9.2 *(Kleitman k-Connectivity Test [5])*.
1. *Start with any node, set $m = k$, and test to find m node-disjoint paths to each of the other nodes; if these paths cannot be found then $k > \omega$, so stop.*
2. *Remove the last node tested (and all incident links), decrement m by 1, select any node in the remaining graph, and find m node-disjoint paths to each remaining node.*
3. *Repeat step 2 until either $m = 0$, in which case $k \leqslant \omega$ and there is k-connectivity, or the test fails and hence $k > \omega$.*

As an example consider again the graph of Fig. 9.1a, which we shall test for 3-connectivity. Starting arbitrarily with n_5 as the test node, Fig. 9.7a shows the $m = 3$ required node-disjoint paths from n_5 to each remaining node. Upon deleting n_5 and selecting new test node n_3, we again are able to find $m = 2$ node-disjoint paths to each remaining node as in Fig. 9.7b. Finally, deleting n_3 and setting $m = 1$ results in the connected graph of Fig. 9.7c; therefore, we have verified that the graph is at least 3-connected. It is clearly not 4-connected since we cannot find four node-disjoint paths between any node pair; indeed, since $\omega \leqslant \delta = 3$, this is obvious. Consequently, a network corresponding to this example graph should continue to communicate among all remaining nodes even if any two of the nodes fail. This is not very surprising since the graph lacks only two links being complete (fully connected). As the redrawn version in Fig. 9.7c shows, adding the two dashed links makes the graph complete.

There are several other topics in graph theory that have application to network design, particularly with regard to traffic flow and flow constraints, but they will be more appropriately treated in the remaining sections of

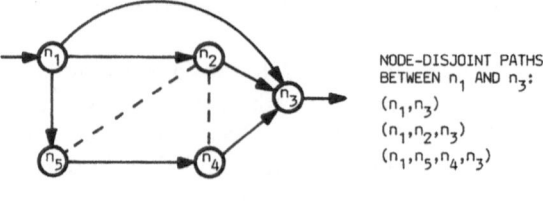

(a) Node-Disjoint Paths

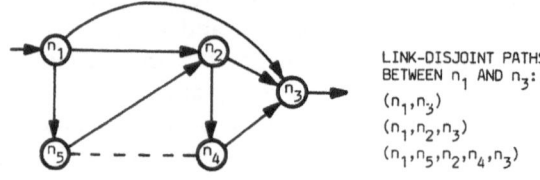

(b) Link-Disjoint Paths

Figure 9.6. Disjoint paths of a graph.

this chapter. The discussion so far should provide a good set of intuitive and mathematical design tools, and the advantages of using graphs to model computer data communications networks should be apparent. We now turn to the other major design technique of this section, linear programming, which has wide application to optimal network design.

The *linear program* (LP) is the most important optimization technique today, with application to problems in many disciplines. The term "program"† as used here refers to a mathematical algorithm and not a computer software program, although computer programs are invariably used to solve linear programs. Essentially, an LP is an optimization problem that has both a *linear* optimization criterion, i.e., *cost function*, and *linear* constraints. There are several major reasons for the popularity of the technique. First, many practical problems can be well formulated with linear models. Even clearly nonlinear problems can often be approximated by a series of linear subproblems, i.e., linearized, and solved linearly. Second, linear programs are far more efficient than general search methods, since they progress steadily towards the optimal solution and then stop. Either the optimal solution will be reached in a finite number of iterations, or the LP will discover that no optimal solution exists. Third, the standard simplex LP algorithm [7] is well suited for digital computer implementation, and problems with thousands of variables and constraints can be solved on conven-

† Actually, the technique grew out of efforts to improve military program planning; there is an interesting historical discussion [6] by G. B. Dantzig who developed the LP algorithm.

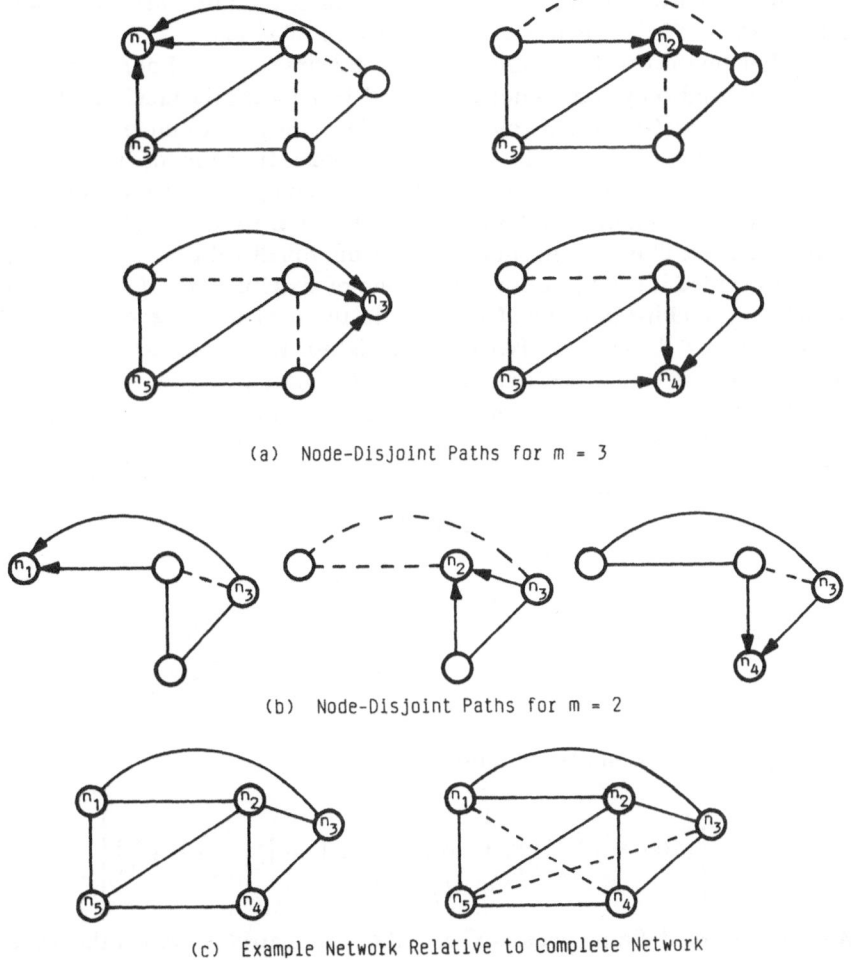

(a)   Node-Disjoint Paths for m = 3

(b)   Node-Disjoint Paths for m = 2

(c)   Example Network Relative to Complete Network

Figure 9.7. Network k-connectivity test.

tional mainframe computers. In fact, effective LP software is now becoming widely available for microcomputers.

Linear programming problems are most easily understood in the context of classical economics. Here we wish to maximize monetary profits (the "cost" function) subject to various constraints on available resources. Once the necessary intuition is developed, there should be no difficulty in transfering it to network problems. Thus consider the firm International Turkey that manufactures two different products P_1 and P_2, which are electronic modem boards that can be sold OEM† at a *unit profit* of $100 and $50,

† Original equipment manufacturer (OEM) means that the boards will be used as a component of another vendor's product.

respectively. If we make x_i units of P_i, $i = 1$ or 2, then the *total profit* will be $z = 100x_1 + 50x_2$, which is a linear function of x_1 and x_2.

Either board can be produced in a total time of one hour using the firm's new (and only) computer controlled Hyper-Hacker machine, but it can only hack out boards ten hours per day. Thus the first resource constraint is that $1x_1 + 1x_2 \leqslant 10$, which, again, is a linear function of x_1 and x_2. Although there are plenty of ordinary electronic components for the boards, P_1 and P_2 each contain a different Cruncher-Chip, CC_1 and CC_2, respectively, and these chips are in short supply from the distributors. In fact, the firm has decided it can use only a maximum of six Cruncher #1 and eight Cruncher #2 chips per day. This leads to the next resource constraints: $x_1 \leqslant 6$ and $x_2 \leqslant 8$. Since a Hyper-Hacker is not reversible, we can never produce a negative quantity, and the final constraints are $0 \leqslant x_1$, and $0 \leqslant x_2$.

This constrained, linear optimization problem can now be written mathematically as follows:

$$\text{maximize } 100x_1 + 50x_2 \text{ over } x_1 \text{ and } x_2$$
$$\text{subject to the constraints}$$
$$x_1 + x_2 \leqslant 10$$
$$x_1 \qquad \leqslant 6$$
$$x_2 \leqslant 8$$
$$x_1 \qquad \geqslant 0$$
$$x_2 \geqslant 0$$

or, using vector and matrix notation,

$$\max_{\binom{x_1}{x_2}} \left\{ (100, 50) \binom{x_1}{x_2} \, \middle| \, \begin{pmatrix} 1 & 1 \\ 1 & 0 \\ 0 & 1 \end{pmatrix} \binom{x_1}{x_2} \leqslant \begin{pmatrix} 10 \\ 6 \\ 8 \end{pmatrix}, \binom{x_1}{x_2} \geqslant \binom{0}{0} \right\}$$

More generally, defining vectors **x**, **c**, and **b**, and matrix **A**, we get the *linear program in canonical form*,

$$\max_{\mathbf{x}} \{ \mathbf{c}^T \mathbf{x} \, | \, \mathbf{A}\mathbf{x} \leqslant \mathbf{b}, \mathbf{x} \geqslant \mathbf{0} \}$$

Such a problem can in general be solved by any conventional LP computer routine, but here we shall solve it graphically to gain some intuition about the technique. Note, however, that graphical techniques are not practical except for the simplest case of two variables. As Fig. 9.8 shows, the optimal solution is $x_1^* = 6$ and $x_2^* = 4$, where the cost function z reaches its maximum possible value of \$800 within the constraint region. Any value of **x** that satisfies all the constraints is called a *feasible solution*. Notice in particular that *the optimal point must be at a vertex*. This is clearly true also in higher dimensions as well, so that the search for an optimal solution need only be

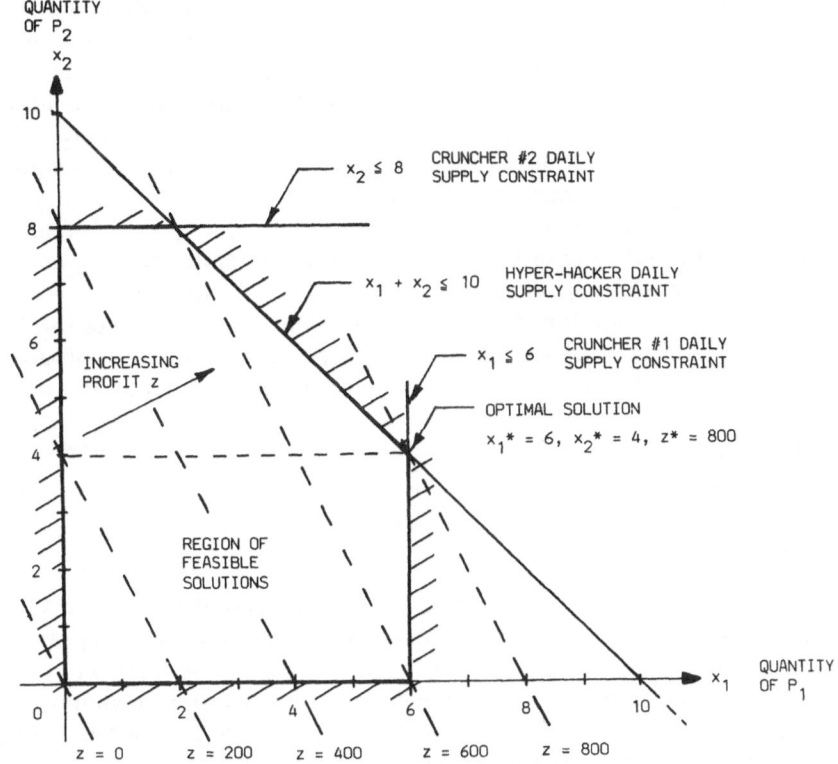

Figure 9.8. Graphical solution of linear program.

done over a finite set of vertices of the region of feasible solutions and never over a continuum of points interior to this region.

This simple example can now be generalized to explain how the *simplex* LP algorithm operates. First, only the vertices of the feasible region are considered, and such points are called *basic*. To get started, then, we must somehow find an *initial basic feasible solution*. Such a point is not obvious in general, but fortunately the simplex algorithm can be used to find its own initial basic feasible solution (usually called Phase I). Furthermore, if the constraints are such that no feasible region exists, i.e., the model is invalid or the problem insoluble, then the algorithm will determine this *infeasibility*. From such a point it then (Phase II) sequentially moves to that *adjacent* vertex which gives the largest increase in the cost function $z = c^T x$. This progression continues methodically until either z cannot be increased further and so is optimal, or it is determined that z can increase indefinitely and the problem is unbounded. This latter situation is, of course, due to a model with insufficient constraints. In any event, the algorithm will always

integer values, e.g., let $x_1^* = 6$ and $x_2^* = 3$. This is commonly done, but care must be taken that the result remains feasible. The solution should always be verified by substituting back into the constraints. There is also no guarantee that the result will be the optimal integer solution; it depends on which way the adjustment is made, as Fig. 9.9 shows. Fortunately, there are various optimal techniques such as "branch and bound" for computer solution of integer programs, both in the general case and for the common *0-1 case* where the variables must be binary, i.e., $x_i = 0$ or 1. A familiar example of the latter case is where the variables represent assignments. Then either the ith assignment is made ($x_i = 1$) or is not made ($x_i = 0$).

We have now developed the two key mathematical techniques for data communications network design, namely, graph theory and linear programming. The intent has not been to cover the fields in great depth, but rather to develop just those specifics essential to networks. The next section applies these techniques to the very common problem of designing networks with tree topologies.

9.3. Tree Network Design

The focus of this section is on the design of centralized networks that have tree topologies. Such networks find widespread application in practice, particularly where occasional link or node failures can be tolerated. Common telecommunications examples include access to a central computer or data base from keyboard terminals within a building or metropolitan area, local access links to a public packet network, and the actual telephone exchange plant network of distribution and feeder cables connecting the local central office and the customer premises. Other familiar examples are CATV distribution systems from the central headend location; metropolitan distribution networks of electrical power cables, water mains and sewer lines; and cross-country oil and gas pipelines. All of these networks are designed with tree topologies for two main reasons: cost and simplicity.

Clearly tree networks are very important and widely used. Even for multiply connected networks, tree subnetworks are used for routing and flow control. Fortunately, a good selection of both optimal and heuristic techniques has been developed for efficient tree design. In this section we shall consider some of the more useful ones. Before delving in, however, it is necessary to consider traffic flow and flow capacity constraints in networks. To do so we now return to the subject of graph theory.

In order that a topological graph $G(N, L)$ be a viable model for a computer data communications network, it is necessary to associate *flow* characteristics with the graph. Such a graph will be called a *network flow model*, or usually just a *network*. For each network link between two nodes i and j we shall in general associate a *flow* f_{ij} directed from i to j. We may also associate an upper limit on flow, i.e., a *link capacity* c_{ij}, with $f_{ij} \leq c_{ij}$. A lower limit could also be specified, but we shall take this to be 0. Finally there may be a cost per unit of flow h_{ij}, or a fixed link cost d_{ij}. Therefore, we could have link weights for actual flow, link capacity, cost, and perhaps length.

We also have flow relationships at the nodes. Flow input to, or output from, the network at node i is denoted f_i with $f_i > 0$ or $f_i < 0$, respectively, and called an *external flow*. We shall always assume *conservation of flow* at each node; i.e., counting external flows the instantaneous flow into the node always equals that out of it.† Thus no traffic is allowed to accumulate at a node. Although in an actual data network there is always some buffering, we assume that this is negligible in our model.

Now consider the problem of designing a tree network about one central site with known link capacity constraints. For example, this might be the case if modem types were specified a priori. If the traffic flow is so small that it is never constrained, then the solution is just the MST which can be found by Algorithm 9.1. Here we are interested in the *constrained minimal spanning tree* (CMST), i.e., the lowest-cost tree meeting all constraints.

We formulate the CMST problem first as an integer program. We assume the node costs and locations are fixed, and all possible total link costs d_{ij}, cost per unit flow h_{ij}, and capacities c_{ij} are known. Then we wish to minimize the total link cost subject to flow constraints $0 \leq f_{ij} \leq c_{ij}$. The central node is numbered 1 with terminals at nodes 2 through n, and respective external flows f_i. Finally, since our interest is in the presence or absence of a candidate tree link, we define an additional *binary* variable e_{ij} which is 1 if the link from i to j is present and 0 if not. Excluding branches that loop back to the same node (self-loops), the problem can now be stated as a *mixed integer program* in terms of the *continuous* f_{ij} and *discrete* e_{ij} variables:

$$\text{minimize } z = \sum_{\substack{i=1}}^{n} \sum_{\substack{j=1 \\ (j \neq i)}}^{n} d_{ij} e_{ij}$$

$$\text{subject to } \sum_{\substack{j=1 \\ (j \neq i)}}^{n} f_{ij} - \sum_{\substack{k=2 \\ (k \neq i)}}^{n} f_{ki} = f_i, \qquad i = 2, 3, \ldots, n$$

$$\sum_{\substack{j=1 \\ (j \neq i)}}^{n} e_{ij} = 1, \qquad i = 2, 3, \ldots, n$$

$$c_{ij} e_{ij} - f_{ij} \geq 0, \qquad i, j = 1, 2, \ldots, n \qquad (i \neq j)$$

$$f_{ij} \geq 0, \qquad e_{ij} = 0, 1, \qquad i, j = 1, 2, \ldots, n \qquad (i \neq j)$$

† In lumped electrical networks this is known as Kirchhoff's current law.

The restriction that $i \neq j$ ensures there are no self-loops. The first constraint is conservation of flow at node i. The second ensures that only one link is directed out of node i, hence forces the network to be a tree. The remaining constraints require the nonnegative link flows to be less than the corresponding link capacities, or zero if there is no link. The situation is illustrated for any node i by Fig. 9.10. The problem can now be converted into the usual form for computer solution by defining vectors $\mathbf{x}$, $\mathbf{c}$, $\mathbf{b}$, and matrix $\mathbf{A}$. An instructive example of the solution using the conventional branch and bound approach is found in the literature [8].

Optimal programs for even moderate spanning trees with a few dozen nodes can involve hundreds of constraints and variables when put into standard form. Thus it is highly desirable to have computationally efficient heuristic approaches. We shall therefore consider two common ones, the Kruskal and the Esau–Williams algorithms. These, along with the Prim MST algorithm, all have a similar basic concept and have been unified [9]. If there are no constraints, each algorithm will generate an actual MST, but the order of link selection will vary. When there are constraints, then the resulting *Heuristic* CMSTs may be different and, of course, there is no claim that they will be the optimal CMST. First, we introduce the algorithms with only flow constraints. We then add an additional limitation on the number of terminals per segment, e.g., drops, where a *segment* is defined as a subtree terminating at the central node.

The Kruskal algorithm basically adds the least cost remaining link that does not violate any constraints. Intuitively, it builds lots of little local trees that eventually merge into the solution tree. It is assumed that all node-to-node link costs are known in the form of the cost matrix $\mathbf{W}$, which is, of course, symmetric for an undirected graph. External flows f_i and link capacities c_{ij} are also known. Then the algorithm is as follows:

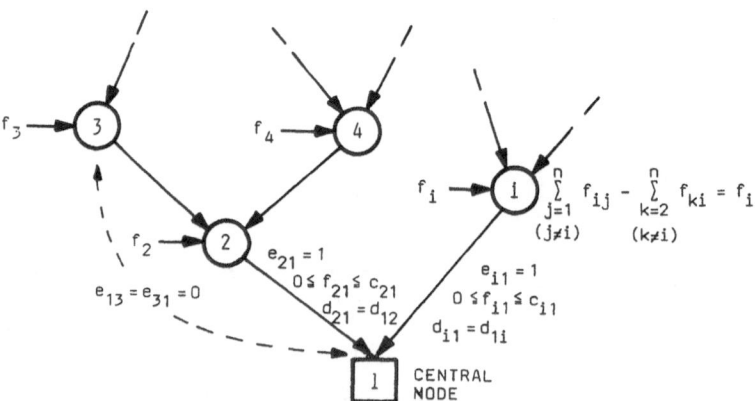

Figure 9.10. Tree network design constraints.

Algorithm 9.3 *(Kruskal Constrained MST [10])*
1. *Start the tree by including the least-cost link.*
2. *Add the next least-cost remaining link that does not form a loop or violate any constraints.*
3. *Repeat step 2 until all links are connected; the result is the heuristic CMST.*

As an example consider the terminals and candidate links of Fig. 9.11a, where the external flows and link costs are shown. Node 1 is assumed to be the central site. If there are no constraints it is easily verified that the links are selected in the order l_{52}, l_{32}, l_{42}, and l_{23} to form the MST at a total cost of 11. However, all the flow must go over l_{31} to reach the central site. This may be undesirable for at least two reasons. A flow of 10 units, such as kbps, may require an excessive link capacity that is not available. There

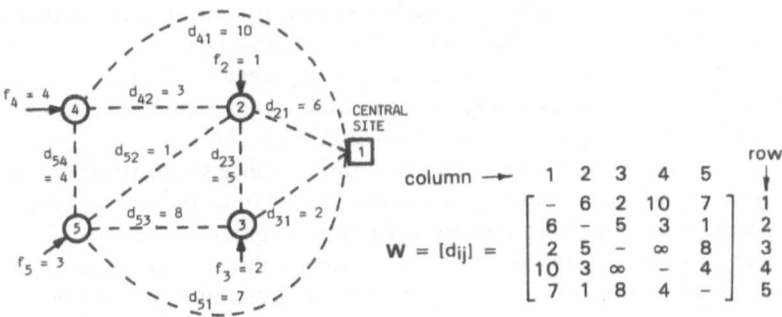

$$W = [d_{ij}] = \begin{bmatrix} - & 6 & 2 & 10 & 7 \\ 6 & - & 5 & 3 & 1 \\ 2 & 5 & - & \infty & 8 \\ 10 & 3 & \infty & - & 4 \\ 7 & 1 & 8 & 4 & - \end{bmatrix} \begin{matrix} 1 \\ 2 \\ 3 \\ 4 \\ 5 \end{matrix}$$

(a)   Candidate Tree Links and Link Cost Matrix

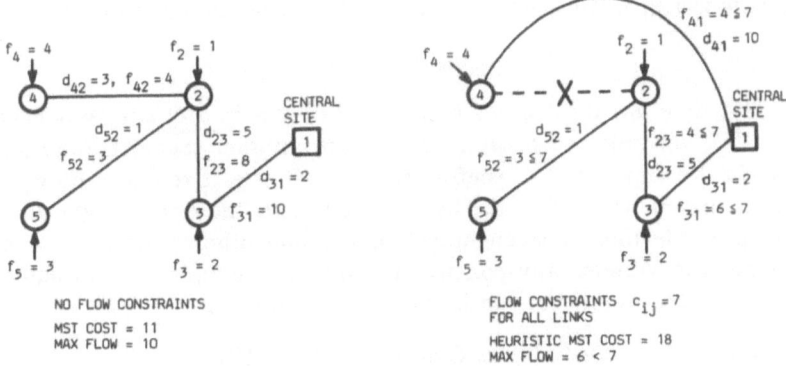

NO FLOW CONSTRAINTS
MST COST = 11
MAX FLOW = 10

FLOW CONSTRAINTS $c_{ij} = 7$
FOR ALL LINKS
HEURISTIC MST COST = 18
MAX FLOW = 6 < 7

(b)   Solution Networks using Kruskal Algorithm

Figure 9.11. Kruskal algorithm design.

are also operational considerations. Since there is only one segment, if either l_{31} or node 3 fails then all network terminals are down, so reliability is a factor. Furthermore, the central site must handle all terminals through one port and, if line control is done through polling, there may be unacceptable delays at the terminals.

Consequently, assume now that we limit the maximum flow on each link to 7 and redesign the network. Again the first selection is l_{52} followed by l_{31}. Next we attempt to add l_{42}, but this makes the total flow into node 2 equal 8, which exceeds the capacity of any outgoing link. Thus we skip l_{42} and next try l_{23}. This try works since the total flow into nodes 3 and 2 does not exceed 7, but the outgoing link l_{31} is now operating almost at capacity. The only remaining node is n_4, which must then be connected directly to the central site at a relatively large cost of 10. The total cost is thus 18. The resulting network, as well as the unconstrained MST, is shown in Fig. 9.11b.

The Kruskal algorithm is straightforward in concept and amenable to computer implementation in practice. It is not difficult to see that additional constraints on the number of terminals per segment can be easily incorporated, since it is necessary only to keep track of the cumulative number of terminals in each segment as the algorithm progresses.

The Esau–Williams heuristic algorithm is more sophisticated than the previous ones, but it usually gives results closer to optimal. The basic idea is to start with each terminal in a separate segment connected in a star topology to the central site, then reconnect nodes in different segments whenever there is a reduction in cost and no constraints are violated. Thus many small segments are progressively consolidated into fewer large ones. This is done by calculating *trade-off values* t_{ij} that represent the change in cost upon reconnecting node i to node j. To determine each t_{ij} we define the *node weight* w_i for each node $i > 1$, then base the decision on

$$t_{ij} = d_{ij} - w_i$$

Initially, $w_i = d_{i1}$, the cost of connecting to the central site. Whenever a node i^* is reconnected to some node j^* in another segment, then for all nodes in the segment of i^* (before reconnection) w_i is replaced by w_{j^*} and t_{ij} recalculated. Thus all nodes in a segment have the same weight. If there is no possible link between node i and some other node k, or if such reconnection violates any constraints, then we set $t_{ik} = \infty$. Formally the Esau–Williams algorithm can be stated as follows:

Algorithm 9.4 *(Esau-Williams Constrained MST [9])*
  1. *Start with all nodes $i > 1$ connected directly to the central site node 1, set node weights $w_i = d_{i1}$, and calculate all trade-offs $t_{ij} = d_{ij} - w_i$; set t_{ij} to ∞ if $d_{ij} = \infty$ or if reconnecting i to j will violate any constraints.*

2. Find $t_{i*j*} = min \{t_{ij} < 0|i$ and j in different segments$\}$; if this cannot be done then stop; the existing tree is the heuristic CMST.

3. If node $i*$ can be reconnected to node $j*$ without violating any constraints do so, then set $w_i = w_{j*}$ for all nodes i in the segment of $i*$ (before reconnection) and recalculate the corresponding trade-offs t_{ij}; otherwise set $t_{i*j*} = \infty$.

4. Go to step 2.

In step 2, there is no cost savings unless the trade-off is negative; if there is no further possible cost savings the algorithm terminates. This may not be the true optimal CMST, but it is the best the heuristic algorithm can do. In step 3 the constraint test will, of course, depend on the nature of each constraint. For example, link capacity constraints can be checked by summing all external flows upstream of each link, and the number of (polled) terminals per segment can be checked by simply keeping a running count.

Two examples will help clarify the algorithm. First, we work the preceding example with all link capacities again 7, then we rework the same problem but allow no more than two terminals per segment. Thus consider first the example shown in Fig. 9.11a with cost matrix **W**. It is convenient to display the trade-off calculations in tabular from as shown in Fig. 9.12. The order of iteration is indicated by the circled numbers. At iteration 1 $t_{i*j*} = t_{42}$ so node 4 is reconnected to 2 at a savings of 7 units. This, of course, eliminates the possibility of reconnecting 4 to either 3 or 5, and also the reverse connection of 2 to 4. In steps 3, 4, and 6 the total flow into the proposed segment would be greater than capacity. The only other feasible reconnection is node 2 to 3, leaving 3 and 5 still connected directly to the central site. The result is shown with a total cost of 17. Note particularly that this is different, and slightly better, than the Kruskal algorithm network of Fig. 9.11b, but there is no guarantee that it is actually optimal.

Now suppose an additional constraint of *no more than two terminals per segment* is added. With this additional constraint to check, the algorithm proceeds as before until step 7. If node 2 is reconnected to 3, there would be a three-terminal segment, so $t_{23} = \infty$. There are then no further $t_{ij} < 0$ terms; therefore nodes 3 and 5 remain directly connected to the central site and the total cost becomes 19 with a maximum flow of only 5 units.

Although we illustrated multiple constraints only in the last example, it should be apparent that the Prim and Kruskal algorithms are also applicable to the multiple constraint case. It is only necessary to devise appropriate tests for each type of constraint. All are straightforward to program for computer solution, and data structures containing the adjacency matrix **M** or other topological descriptions can be easily generated.

LINK COST		FIRST ITERATION				SECOND ITERATION			
ij	d_{ij}	w_i	t_{ij}	step	comment	w_i	t_{ij}	step	comment
node 2									
23	5	6	−1			6	−1	⑦	$t_{i*j*} = t_{23}$
24	3	6	−3	⑤	eliminated by ①	6	—		same segment
25	1	6	−5	④	flow constraint	6			> 0
node 3									
32	5	2	+3		> 0	2	+3		> 0
34	∞	2	∞		> 0	2	∞		> 0
35	8	2	+6		> 0	2	+6		> 0
node 4									
42	3	10	−7	①	$t_{i*j*} = t_{42}$	6	—		same segment
43	∞	10	∞		> 0	6	∞		> 0
45	4	10	−6	②	eliminated by ①	6	—		eliminated by ①
node 5									
52	1	7	−6	③	flow constraint	7	∞		> 0
53	8	7	+1		> 0	7	+1		> 0
54	4	7	−3	⑥	flow constraint	7	∞		> 0

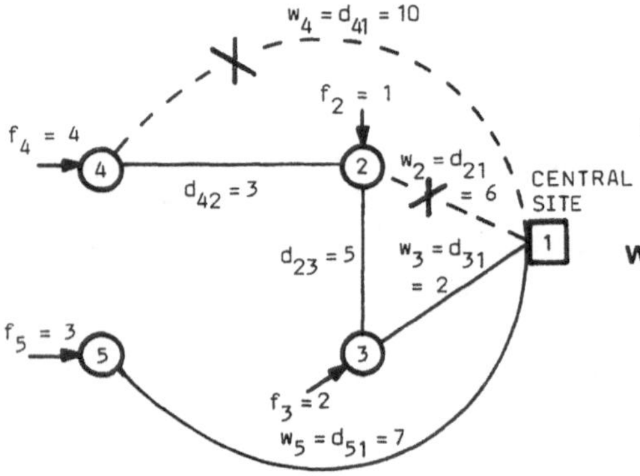

$$W = [d_{ij}] = \begin{bmatrix} - & 6 & 2 & 10 & 7 \\ 6 & - & 5 & 3 & 1 \\ 2 & 5 & - & \infty & 8 \\ 10 & 3 & \infty & - & 4 \\ 7 & 1 & 8 & 4 & - \end{bmatrix} \begin{matrix} 1 \\ 2 \\ 3 \\ 4 \\ 5 \end{matrix}$$

FLOW CONSTRAINT $c_{ij} = 7$
FOR ALL LINKS

HEURISTIC MST COST = 17
MAX FLOW = 7 ≤ 7

Figure 9.12. Esau–Williams algorithm design.

The tree configurations we have considered so far have not included concentrators per se, but rather assumed that any multidrop lines were controlled from the single central site, e.g., by polling. However, in many design situations it is much more efficient to use multiplexers to concentrate data remotely before it is sent to the central site. For example, a common situation is where there are many different terminal locations in each of several cities, all of which must communicate with a single central site. There can be large savings in leased line costs if the traffic of each city is multiplexed over one or a few high-speed lines. Common design issues in such cases include: how many concentrators are needed (can some cities' traffic be combined?); where should each concentrator be located (alone or at a terminal?); is more than one level of concentration appropriate (multiplexed inputs to other multiplexers?); how should the terminals be connected to the concentrators (which terminals and what topology?). In the remainder of this section we consider this problem. It is first formulated as 0–1 integer program, and then two computationally efficient heuristic algorithms are presented.

In approaching the design of tree networks containing concentrators we shall restrict our attention to only *two levels of concentration* (at the central site and remote sites) for simplicity. Topologies will be either a *tree* with multidrop links or a *star* with a single point-to-point link from each terminal to a concentrator. The latter arrangement is more appropriate to networks with conventional multiplexers (FDMPX or TDMPX) which do not normally provide multidrop polling. For more modern and complex networks with intelligent remote concentrators or statistical multiplexers, concentrator polling is often available and polled multidrop lines may result in significant savings. Once the terminals have been assigned to their respective concentrators, then the selection of connecting links is a relatively straightforward constrained tree design problem.

Thus we shall distinguish four basic subproblems to be considered: (1) the selection of potential concentrator sites, (2) the assignment of concentrators to some sites, (3) the assignment of each terminal to either a remote concentrator or the central site, and finally (4) the connection of terminals to these sites with a tree or star topology. Note that there is a considerable degree of interaction among them. For example, it may be preferable to connect no terminals to some initial concentrator site, or perhaps initial terminal assignments must be later changed to another concentrator because of constraint violations. Obviously, multiplexers will have inherent limitations such as the number of available ports, the number of drops that can be polled, and the link capacity to the central site.

Although the actual final concentrator locations will not generally be known initially, it is often the case that all potential sites are known. As an extreme case, for example, every terminal location can be considered for a possible concentrator site. In such cases, the network design problem can be formulated as an integer program and solved optimally provided sufficient computing resources are available.

To set up the program for the case of point-to-point links [11], let $\{T_1, T_2, \ldots, T_n\}$ denote the set of all n terminals and $\{S_0, S_1, S_2, \ldots, S_m\}$ be the $m + 1$ concentrator sites to be considered, with S_0 denoting the central site. Also let d_{ij} be the cost of connecting T_i to S_j (note that the direct connection of T_i and T_j is not allowed here), and indicate such a connection by setting $e_{ij} = 1$ if T_i connects to S_j and $e_{ij} = 0$ otherwise. For each potential remote concentrator site S_i let d_j be the fixed cost; e.g., the unit itself plus the link to S_0, and indicate the presence or absence of a concentrator at S_j by setting the binary variable y_j to 1 or 0, respectively. Finally, let the capacity of concentrator at S_j be k_j point-to-point terminal links. Then we have the following 0–1 integer program to minimize the total cost:

$$\text{minimize } z = \sum_{j=0}^{m} \sum_{i=1}^{n} d_{ij}e_{ij} + \sum_{j=1}^{m} d_j y_j$$

$$\text{subject to } \sum_{j=1}^{m} e_{ij} = 1, \qquad i = 1, 2, \dots, n$$

$$\sum_{i=1}^{n} e_{ij} \leq k_j, \qquad j = 1, 2, \dots, m$$

$$e_{ij} \text{ and } y_j = 0, 1, \qquad i = 1, 2, \dots, n; \qquad j = 1, 2, \dots, m$$

The first constraint requires each terminal to connect to exactly one concentrator and the second ensures that there are not more terminals than ports at each concentrator. In general, such integer programs can be formally solved for the optimal network, but for larger networks the computational cost may be excessive. For the special case where the concentrators are already installed initially, the problem reduces to finding the least-cost feasible assignment of n terminals to m concentrators. This is a form of the standard warehouse location problem in operations research.

With the nature of the concentrator problem established by the optimal formulation, we now turn to several heuristic methods that greatly reduce the computational requirements and improve the design flexibility at the expense of suboptimality. The first method is known as the *ADD algorithm*, which was originally proposed in the context of the classical warehouse location problem. Beginning with knowledge of all terminal and potential concentrator sites, this algorithm reassigns terminals from the central site to various concentrators in star topologies. Since the concentrators are also in a star relative to the central site, this is sometimes called a *star–star* configuration.

The basic idea of the ADD algorithm is to iteratively find the best conditional solution with one remote concentrator, then use this result to find the best conditional solution with two, and similarly for three, etc. Intuitively, if concentrators are advantageous at all, then there should be a significant savings in line costs as the initial concentrators are installed. Eventually, however, the fixed cost of adding still more concentrators will outweigh the line savings. If the minimal cost with k concentrators is denoted z_k^*, then the minimum of the sequence z_0^*, z_1^*, z_2^*, ..., z_m^* is the heuristic solution. At each iteration of the ADD algorithm one concentrator is added to those already selected, which reduces the amount of computation enormously compared to that of trying all possible combinations.

Like the Esau–Williams algorithm, trade-off values t_{ij} are used to decide which terminals to reconnect. There are several variations of the ADD algorithm, but the following is representative and straightforward. It is assumed that no more than k_j terminals can be connected to a concentrator at site S_j.

Algorithm 9.5 *(ADD Star Concentrator Network)*
1. *Start with all terminals T_i connected to the central site S_0 at cost $z_0^* = \sum_{i=1}^n d_{i0}$; set $j_0^* = 0$.*
2. *For each remaining site $S_j, j \neq j_0^*, j_1^*, \ldots, j_{k-1}^*$, after $k-1$ assignments calculate the trade-off values $t_{ij} = d_{ij} - d_{ij*}$, where T_i is connected to S_{j*}; if for some S_j all $t_{ij} > 0$ then skip S_j, and if all remaining S_j are skipped or no sites remain, then stop.*
3. *For each unassigned S_j in step 2 calculate the cost change Δz_k from putting the kth concentrator at S_j, $\Delta z_k^j = \sum$ (most negative $t_{ij} < 0$) $+ d_j$, and find $\Delta z_k^* = \min \{\Delta z_k^j\}$ with corresponding index j_k^*.*
4. *Calculate the cost with k concentrators as $z_k^* = z_{k-1}^* + \Delta z_k^*$; if $z_k^* \leqslant z_{k-1}^*$ install a concentrator at $S_{j_k^*}$ and connect the terminals used in calculating $\Delta z_k^{j_k^*}$; if $z_k^* > z_{k-1}^*$ assume z_{k-1}^* is the solution and stop.*
5. *Go to step 2.*

Most practical versions include some type of final "Bump and Shift" routine that effects any obvious improvements to further reduce the cost. For example, two underutilized concentrators located close together may be combined, or a concentrator may be removed if all previously assigned terminals were later reassigned by the algorithm. The algorithm may halt as soon as the costs z_k^* begin to increase, as in step 4, or it may be designed to continue on through all m iterations in the event that later values of z_k^* begin decreasing again.

To illustrate the ADD algorithm, consider the example of Fig. 9.13a. The cost matrix gives all d_{ij} and d_j values. In the first iteration (b) one concentrator at S_2 gives the largest decrease in cost for four connected terminals but can only accommodate three. Either T_4 or T_5 can be connected, but not both. At the second iteration a second concentrator is added at S_3 and T_4 is reconnected from S_2 to S_3. A third iteration provides no further improvement so the algorithm terminates. However, a casual inspection of the heuristic results indicates that reconnecting T_5 from S_3 to S_2 (which now has a spare port from T_4) further decreases the cost from 24 to 23. Note that site s_1 was never used, and that T_3 remained connected to the central site throughout.

The ADD algorithm is intuitively appealing since it does not depend on the order in which the sites are considered. It is most appropriate where the final network requires relatively few concentrators. A related DROP

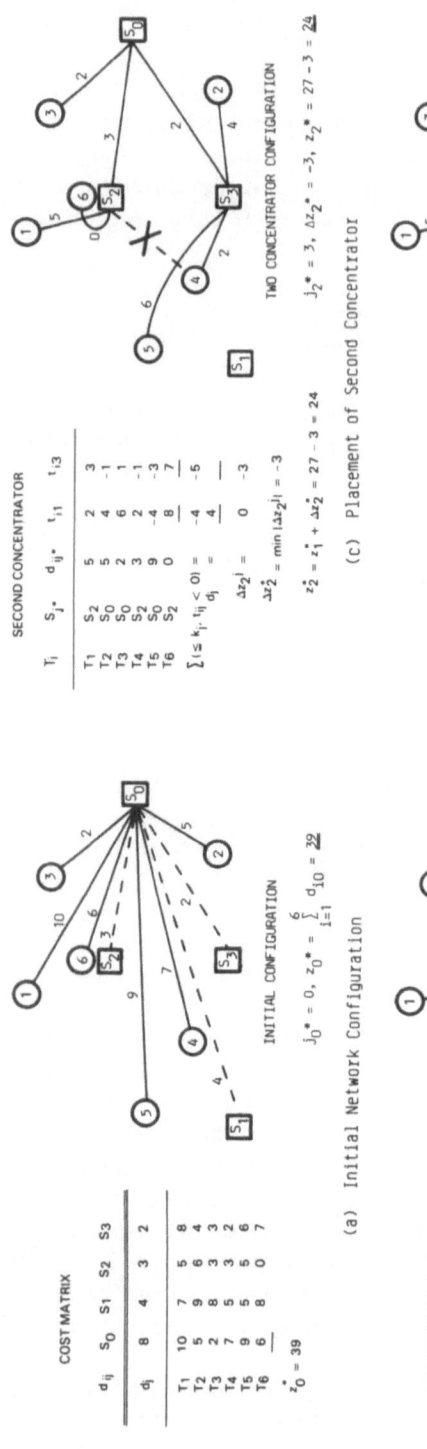

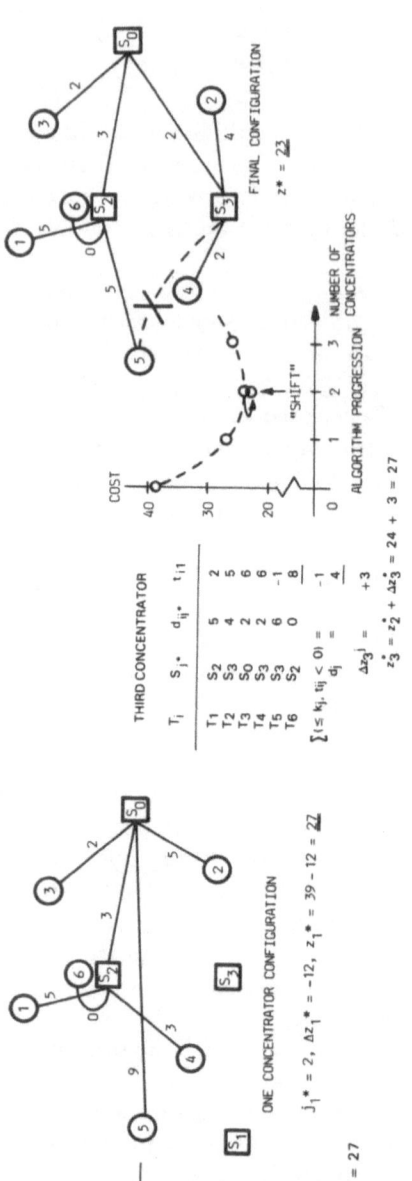

(a) Initial Network Configuration

(b) Placement of First Concentrator

(c) Placement of Second Concentrator

(d) Final Network after "Bump and Shift"

Figure 9.13. ADD algorithm design.

algorithm starts with concentrators at all sites and drops one per iteration. Clearly the DROP approach is more efficient when there will be concentrators at most of the sites. This, of course, may not be known a priori.

The second heuristic approach to designing networks with concentrators is more general than both the integer programming and ADD approaches, in that it selects the potential concentrator sites, assigns the concentrators and terminals, and allows tree topologies around each concentrator. The basic approach of this NEWCLUST algorithm [12] is to determine first those terminal sites which are near the largest number of other terminals, then collocate the concentrators with those terminals. A preliminary assignment of each terminal to the nearest concentrator is refined so as to satisfy various constraints. Next the terminal clusters are connected to their assigned concentrators using any convenient tree network design algorithm, and finally all concentrators are connected to the central site. Further refinements, such as combining underutilized concentrators, are possible at each stage of the algorithm. The final topology is then a *tree–star* for the terminals and concentrators, respectively.

The entire NEWCLUST algorithm as given in the preceding reference is rather involved, so we shall focus here on the aspects of concentrator siting and terminal assignments. For a total of n terminals we assume that for each concentrator there can be a maximum of $k_j = k_0$ connected terminals, either directly or multidropped. The limited algorithm can now be formally stated:

Algorithm 9.6 *(NEWCLUST Tree Concentrator Network)*
1. *From the cost matrix W determine the $k_0 - 1$ nearest neighbors of each terminal; then determine the frequency x_i with which each terminal T_i occurs.*
2. *For each value j of n_i, $0 \leqslant j \leqslant F$ determine the list L_j of terminals with frequency of occurrence j and order the lists $L_1, L_2, \ldots, L_F$; then calculate the weighted integer mean $m_x = 1 + [(1/n)\sum_{i=1}^{F} ix_i]$.*
3. *Starting with concentrator sites S_j collocated with the terminals T_j in L_F make the lowest cost terminal assignments to each concentrator; if more sites are required add them successively from the lists $L_{F-1}, L_{F-2}, \ldots, L_{m_x}$.*
4. *For each concentrator site S_j where the constraints are not satisfied, reconnect to S_{j*} those excess T_i having the smallest trade-off costs $t_i = d_{ij*} - d_{ij}$, where $j^* \neq j$ indexes the concentrator site (other than S_j) with smallest connection cost for T_i.*
5. *Connect each concentrator directly to the central site and refine the terminal assignments to allow combining underutilized concentrator sites and assigning terminals directly to the central site when less costly.*

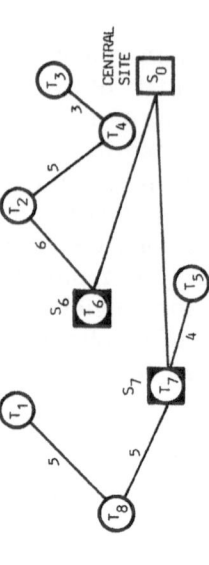

Counting T_6 there are now $5 > k_6 = 4$ terminals assigned to site S_6 so one terminal must be reassigned to another site in C, and here S_7 is the only viable choice. The tradeoffs for each terminal assigned to S_6 are given by $t_i = d_{ij^*}-d_{ij}$, where j^* indexes the lowest-valued d_{ij} for all S_j in C, $S_j \neq S_6$, for $i = 1, 2, 3, 4$:

$$t_1 = d_{1j^*}-d_{16} = d_{17}-d_{16} = 6-5 = 1$$
$$t_2 = d_{2j^*}-d_{26} = d_{27}-d_{26} = 9-6 = 3$$
$$t_3 = d_{3j^*}-d_{36} = d_{37}-d_{36} = 12-9 = 3$$
$$t_4 = d_{4j^*}-d_{46} = d_{47}-d_{46} = 10-7 = 3$$

Thus we reconnect T_1 to S_1 to obtain the least costly choice. The resulting concentrator locations and terminal assignments are now as shown below:

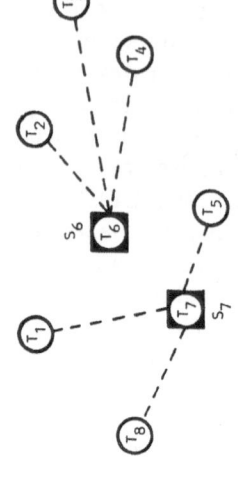

$w = |d_{ij}| =$

	1	2	3	4	5	6	7	8
	-	8	13	11	9	5	6	5
	8	-	5	5	3	6	9	12
	13	5	-	3	10	9	12	16
	11	5	3	-	6	7	10	15
	9	3	10	6	-	4	4	9
	5	6	9	7	4	-	4	8
	6	9	12	10	4	4	-	5
	5	12	16	15	9	8	5	-

(iii) Calculate the weighted average of the terminal assignment frequencies:

$$m_x = \left[\frac{1}{F}\sum_{i=1}^{F} x_i\right] + 1$$

$$= \left\{\frac{1}{8}[0+0+4\times3+2\times4+1\times5+0+1\times7]\right\} + 1$$

$$= \left[\frac{32}{8}\right] = 5 \leq J \leq F = 7$$

(ii) Calculate k-Nearest Neighbor (NN) Table for $k_Q = 4$ terminals/concentrator

T_i	k_Q-NN				freq x_i	freq list L_j for T_i
T1	1	6	8	7	3	$L_1 = 0$
T2	2	3	4	6	3	$L_2 = 0$
T3	3	4	2	6	3	$L_3 = |T1,T2,T3,T8|$
T4	4	3	2	5	4	$L_4 = |T4,T5|$
T5	5	6	7	1	4	$L_5 = |T7|$
T6	6	5	7	1	7*	$L_6 = 0$
T7	7	5	6	8	5	$L_7 = |T6|$
T8	8	1	7	6	3	$L_8 = 0$

*F = 7

Thus we shall only consider $L_5 - L_7$, so the potential concentrator sites are C = $\{S_6, S_7\}$ at T_6, T_7. Initially the closest terminals are assigned to concentrator sites S_6 and S_7 collocated with T_6 and T_7, respectively, as shown below:

S6	S7
T6	T7
T1	T5
T2	T8
T3	
T4	

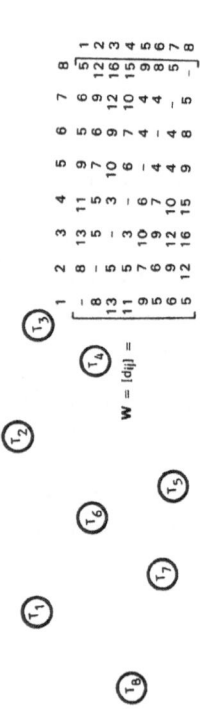

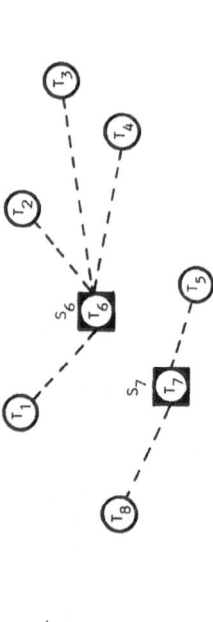

(a) Initial Concentrator Locations and Terminal Assignments

(b) Revised Terminal Assignment Satisfying the Constraints

Figure 9.14. NEWCLUST algorithm design.

6. *For each remaining concentrator site locate a concentrator and connect all assigned terminals with a tree network satisfying all constraints; e.g., link capacity, number of drops.*

A few comments and an example should help clarify the algorithm. In step 6 any design algorithm for a CMST can be used; e.g., Kruskal, Esau–Williams, etc. In step 3 there may be an insufficient number of concentrator sites in L_F to handle all terminals, in which case less centralized sites are allowed in an ordered manner. Note that once these potential remote concentrator sites have been selected, another algorithm such as ADD could be used instead of NEWCLUST.

Now consider the example of Fig. 9.14a consisting of eight terminals with the cost matrix **W**. For $k_0 = 4$ the three nearest neighbors of each terminal are located, the frequencies of occurrence totaled up, and the terminal frequency lists determined with $F = 7$. Then m_x is calculated to be 5. Consequently we first try L_7, but the $k_0 = 4$ constraint can never be satisfied. Next L_5 is added (L_6 is empty) and the closest terminal assignments are made to S_6 and S_7, as indicated by the dashed lines. Since S_6 got five terminals, one must be reassigned to S_7. The trade-off calculations are shown in Fig. 9.14b, along with the resulting MST connections at each concentrator. The two concentrators are finally connected directly to the central site S_0.

This section has presented and illustrated a selection of practical techniques for tree network design under various common constraints. Point-to-point and multidrop links are viable alternatives, with and without remote concentrators. Although the approaches that we have considered are both practical and representative, there are, of course, many other heuristic ways to design tree networks. For example, clever techniques for trading off delay and the cost of link capacity have been developed. At this point it will be instructive to look back at the more concrete material of the last chapter and relate it to the more theoretical developments of the past two sections. With this background of tree networks and graph theory, we now turn to the more general subject of multiply connected network design in the remainder of the chapter.

9.4. Link Capacity and Network Flow

In this section we consider the three key network issues of delay, link capacity, and maximal traffic flow. They are clearly interrelated since, for example, adding more capacity can increase the rate of traffic flow between any pair of nodes, or can decrease the buffer queueing delay at intermediate nodes. Although these three issues all apply to tree networks, our focus here is on the more general (and interesting) application to multiply connected, or mesh, networks. Thus we are now beginning to consider decentralized backbone networks as opposed to the centralized local access networks of the preceding section. We shall first look at delay and capacity assignment, then at network flow.

Delay in computer data communications networks arises in several places. We shall consider packet switching here, but the same comments apply to message switching as well. There is a finite *propagation delay* on each link which, though normally small, can be significant on satellite or even transcontinental terrestrial links. Nodal processors introduce a finite amount of *processing delay* in handling packet headers and error checksums, and calculating routing, etc. However, the major source of delay is nearly always due to buffering at nodes of incoming and, particularly, outgoing packets. This *buffer delay* is well modeled by conventional queueing theory, and we shall approach it from this standpoint. The Appendix contains an adequate coverage of probability and queueing theory for our introductory treatment. There may also be delay from various other sources including error control retransmissions and preemption by higher-priority traffic, but we shall ignore them here and focus on the more basic buffer delay only. Packets arriving at a node are immediately assigned to an outgoing link l_i and placed in the corresponding link buffer. They are then sent out over the link which has finite capacity c_i bps. The packet *arrivals* at the buffer are assumed random, described by the Poisson probability distribution with an average of λ_i packets per unit time interval. Thus, as shown in the Appendix, the *interarrival time* is exponentially distributed.

The link buffer is assumed large enough so that arriving packets are always accepted, hence is modeled as *infinite*. Packets work their way through the queue on a *first-in-first-out* (FIFO) basis and eventually go to the line I/O port and are transmitted. Thus the digital line and corresponding I/O processing constitute the queueing server. If the average packet length is $1/\mu_i$ bits per packet, then the time to send an average message is $1/\mu_i c_i$ sec and hence the *service rate* is $\mu_i c_i$ packets per second.† Now we assume the packet lengths vary, with many relatively short transaction and control packets intermixed with fewer long message packets. Then an *exponential service time* distribution is a plausible model and results in a tractable $M/M/1$ queueing system to which all of the $M/M/1$ queue results in the Appendix apply.

In particular, the traffic intensity is $\rho_i = \lambda_i/\mu_i c_i$. Consequently, for $\rho < 1$ the server can "keep up" with the arrivals and the average number of packets in the queueing system is

$$L_i = \frac{\rho_i}{1 - \rho_i} = \frac{\lambda_i}{\mu_i c_i - \lambda_i} \qquad \text{average queue length}$$

† Here the notation $\mu_i c_i$ used in the same manner as μ alone is used in the Appendix.

Then the average waiting time, or *packet delay*, through the queue and server is $\bar{T}_i = L_i / \lambda_i$ or

$$\bar{T}_i = \frac{\rho_i}{\lambda_i(1 - \rho_i)} = \frac{1}{\mu_i c_i - \lambda_i} \qquad \text{average wait in queueing system}$$

To extend this result to networks we need several additional properties of queues. First is the fact that the departures from an $M/M/1$ queue are also Poisson so that $M/M/1$ queues connected in tandem continue to operate as $M/M/1$ queues. Since the sum of two independent Poisson random variables is also Poisson, the preceding property extends directly to tree networks. In fact, a more fundamental result [13] extends this property to a large class of mesh networks provided primarily that the service times are all independent. This latter provision is not strictly true, since service time depends on packet length and is thus closely related at each node a packet visits. However, for reasonably large multiply connected networks, a typical outgoing link receives a mix of packets from multiple incoming links so the service times are much less correlated and independence is plausible. Practical experience and simulation support this model. With this conventional *independence assumption* we can now model and analyze fairly general networks with simple queueing techniques. We shall first obtain a widely used expression for average network delay $\bar{T}$, which is then used to obtain an instructive expression for the optimal link capacities [14].

For an arbitrary (tree or mesh) packet network with n nodes and l links, let each outgoing link along with its buffer and processor be modeled by an $M/M/1$ queueing system. The end-to-end traffic flow† in packets per second from node j to k over a fixed route is γ_{jk}, so the average total *external* flow into (or out of) the network, i.e., the *throughput*, is

$$\gamma = \sum_{j=1}^{n} \sum_{k=1}^{n} \gamma_{jk}$$

Also let λ_i be the average packet flow on link i so that the average total *internal* traffic in packets per second is

$$\lambda = \sum_{i=1}^{l} \lambda_i$$

Note that $\gamma \leqslant \lambda$ since each γ_{jk} will be counted on each link traversed between nodes j and k to get λ. In fact, the average number of such "hops" per message is just $\bar{n} = \lambda / \gamma$.

† In our prior flow notation we have $f_i = \lambda_i / \mu_i$ for link i in *bps*; similarly γ_{jk} can be converted into an external flow into node j in bps.

Now for the entire network the total average delay, represented by $\gamma \bar{T}$, should intuitively be equal to the sum of the delays on each link; hence we define

$$\bar{T} = \frac{1}{\gamma} \sum_{i=1}^{l} \lambda_i \bar{T}_i = \sum_{i=1}^{l} \frac{\lambda_i}{\gamma} \left[\frac{1}{\mu_i c_i - \lambda_i} \right]$$

where the average packet time $\bar{T}_i$ on link i is given by the previously derived expression. $\bar{T}$ is the average time a packet is in the network and is a widely used indicator of network performance. It can be easily modified to include control packet flow, link propagation delay, and nodal processing time [14]. The underlying assumptions are that each $M/M/1$ queue can be treated independently and the traffic and link capacities are known or can be determined.

Next consider the theoretical problem of finding the link capacities that minimize $\bar{T}$ subject to a total link capacity cost D. The network topology and traffic are assumed known. Furthermore, we assume that available link capacity is continuously variable with cost consisting of fixed and proportional components. Thus the cost for link i is $d_{i0} + d_i c_i$ and the total link capacity cost is

$$D = \sum_{i=1}^{l} (d_{i0} + d_i c_i) = d_0 + \sum_{i=1}^{l} d_i c_i$$

The resulting optimization is formally stated:

minimize $\bar{T}$

subject to $D = d_0 + \sum_{i=1}^{l} d_i c_i$

Since $\bar{T}$ is a nonlinear function of the variables c_i this is not a linear program. However, it can be easily converted to an unconstrained optimization by minimizing the *Lagrangian* function

$$L(c_i, c_2, \ldots, c_l, \xi) = \bar{T}(c_1, c_2, \ldots, c_l) + \xi \left(D - d_0 - \sum_{i=1}^{l} d_i c_i \right)$$

The new variable ξ is called a Lagrange multiplier. Setting the gradient to zero and solving the resulting set of simultaneous equations easily yields

$$c_i^* = \frac{\lambda_i}{\mu_i} + \left[\frac{D - d_0 - \sum\limits_{j=1}^{l} d_i \lambda_j / \mu_j}{\sum\limits_{j=1}^{l} (\lambda_j d_j / \mu_j)^{1/2}} \right] \left(\frac{\lambda_i}{\mu_i d_i} \right)$$

$$= \frac{\lambda_i}{\mu_i} + K \left(\frac{\lambda_i / \mu_i}{d_i} \right)^{1/2}$$

where K denotes the constant quantity in brackets. This optimal solution first assigns the absolute minimum required capacity λ_i / μ_i to each link, then apportions out the weighted remainder K in a manner proportional to the square root of the minimum capacity and inversely proportional to the square root of the incremental capacity cost for each link. This is certainly a reasonable solution.

There are several instructive comments and variations on the preceding result. First, the minimal delay $\bar{T}^*$ corresponding to the solution c_i^* is

$$\bar{T}^* = \frac{1}{\gamma K} \sum_{i=1}^{l} \left(\frac{\lambda_i d_i}{\mu_i} \right)^{1/2}$$

which depends inversely on the weighted excess capacity K. As this excess decreases, i.e., the links are designed for near minimal capacities, K becomes small and the average delay increases dramatically. This is the same phenomenon for the network that occurs in a single queue as the traffic intensity approaches unity. In fact, this limiting needs to occur in only one internal queue to greatly affect $\bar{T}^*$. To decrease this sensitivity of $\bar{T}$ to individual $\bar{T}_i$ values, similar optimization approaches have resulted in other allocations of the excess capacity, most notably by assigning each link either an equal or a directly proportional share. Although in practice capacity is available only in discrete amounts, e.g., available modem rates, this optimal result can provide both a good starting point and important intuitions for practical designs.

Having characterized network delay and related it to link capacity assignment, we now turn to the related subject of traffic flow. The primary issue of interest in this section is how much traffic can be sent either between one given pair of nodes or between any pair of nodes. Closely related to, but distinct from, maximal flow is the question of how traffic should be routed. Routing is considered in the following section. The fundamental result on network flows is the classical *max-flow min-cut* theorem, which not only determines the maximal flow but also provides an algorithm for finding it.

We begin by assuming the network topology and link capacities are known, and we define the general problem by formulating it as a linear program. Thus consider the external flow f_s from *source node s* to *destination node t*, so $f_t = -f_s$, with no other external flows. Within the network this flow may split (bifurcate) and go over many different paths, but on each link between nodes i and j the flow f_{ij} is constrained by the corresponding link capacity, $0 \leqslant f_{ij} \leqslant c_{ij}$. If we define $c_{ij} = 0$, and hence $f_{ij} = 0$, when there is no link from i to j, then the LP for maximal flow is

$$\text{maximize } f_s$$

$$\text{subject to } \sum_{\substack{j=1 \\ (j \neq s)}}^{n} (f_{sj} - f_{js}) = f_s$$

$$\sum_{\substack{j=1 \\ (j \neq t)}}^{n} (f_{tj} - f_{jt}) = -f_s$$

$$\sum_{\substack{j=1 \\ (j \neq s, t, i)}}^{n} (f_{ij} - f_{ji}) = 0, \qquad i = 1, 2, \ldots, n; \qquad i \neq s, t$$

$$0 \leqslant f_{ij} \leqslant c_{ij}, \qquad i, j = 1, 2, \ldots, n; \; i \neq j$$

The first three constraints are just conservation of flow at nodes s, t, and all others, respectively. One of these equations is redundant and can be omitted.[†] The last constraint is the link capacity. All this can be put into standard form and solved directly if there is sufficient computing capability. This approach is not very efficient, however, since the dimensionality is large and the coefficient (incidence) matrix $\mathbf{A}$ has a high proportion of zero components. A more efficient method of solution is based on the max-flow min-cut theorem.

To establish this theorem, recall that a cut-set, or just *cut*, is a minimal set of links that disconnect the network if removed. For two distinct nodes s and t, an *s-t cut-set* has node s in one component S and node t in the other component $\bar{S}$. We denote the cut-set by $(S, \bar{S})$, and define its *capacity* $c(S, \bar{S})$ as the sum of the capacities of all links of the cut-set with flow directed from S to $\bar{S}$. Note that for any nodes s and t there may be many *s-t cut-sets* with different capacities. Since all traffic flow from s to t must go through these cut-sets, we would expect that the one with minimum

† In electrical networks the flow constraints are the Kirchhoff current law equations, and the redundant equation corresponds to the electrical ground node.

capacity would limit the maximum possible flow. This intuitive result can be stated formally:

Theorem 9.1 *(Max-Flow Min-Cut [15].) For any network the maximal flow from node s to t equals the minimum capacity of all the cut-sets separating s and t.*

Thus the maximal flow determination is reduced to a search for a minimum capacity cut-set.

To illustrate the theorem consider the example of Fig. 9.15a. Note that all links except l_{52} are FDX, but may have different capacities in each direction. The capacity value c_{ij} and direction i to j will always be indicated as coming *out of* each node i. For $s = 5$ and $t = 3$, all s-t cut-sets are easily found by inspection. Then the capacity of each is determined by simply adding up the member link capacities directed from each component S containing the source node to the corresponding $\bar{S}$ containing the sink node. The minimum capacity is 7 across the cut-set of l_{23} and l_{43}, hence this is the maximum possible flow from node 5 to node 3.

The major problem in extending this simple example to larger networks is, of course, finding all of the cut-sets. The conventional way of circumventing this impediment is to incrementally augment the network flow on all directed paths from s to t until no further increase is possible. Fortunately, the problem of finding all directed paths can be approached systematically. Before doing this we need to distinguish between forward and reverse flows. Thus let $f(S, \bar{S})$ denote the *forward* flow across *cut-set* $(S, \bar{S})$ from component S to $\bar{S}$, *i.e.*, s to t, and similarly $f(\bar{S}, S)$ the *reverse* flow from $\bar{S}$ to S. Then for any feasible flow f_s and any s-t cut-set $(S, \bar{S})$ we have

$$f_s = f(S, \bar{S}) - f(\bar{S}, S) \le c(S, \bar{S})$$

Furthermore, if f_s is maximal then $f(S, \bar{S}) = c(S, \bar{S})$, so $f(\bar{S}, S) = 0$ and there is no reverse flow. Intuitively, reverse flow indicates packets are being (suboptimally) looped back through a cut-set on different paths. When all

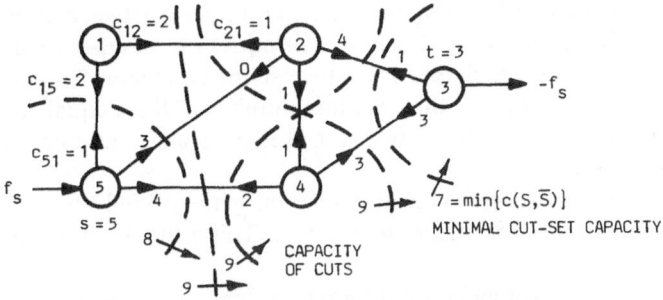

Figure 9.15. Maximum flow across minimum cut-set.

traffic is between only one pair of terminals, the opposing flows can be canceled. We can now state an algorithm for finding the maximal flow between any given terminal pair s and t:

Algorithm 9.7 *(Maximum Flow Between Two Terminals)*†
  1. *Find any directed path from s to t having positive capacity remaining on each link; if no such path exists, the flow is optimal, so stop.*
  2. *For the path of step 1 find the smallest unused link capacity ε and*
     a. *subtract ε from all remaining path link capacity oriented in the forward direction (from s to t);*
     b. *add ε to all remaining path link capacity oriented in the reverse direction (from t to s)*

The algorithm terminates in step 1 when no more source-to-destination capacity exists, i.e., the minimum cut-set has no remaining unused capacity. The maximum flow then equals the original capacity of this cut-set.

To illustrate this, consider again the example of Fig. 9.15. The first path selected (arbitrarily) is via nodes (5, 1, 2, 4, 3) for which $\varepsilon = 1$ from l_{51}. We then decrement all forward capacities and increment all reverse capacities along this path as Fig. 9.16a shows. The procedure is repeated in (b), (c), and (d) until no path can be found. The minimum cut-set is obviously links l_{23} and l_{43} and the capacity of 7 is the maximum flow. Note that this is also the sum of the incremental capacities.

The major drawback of Algorithm 9.7 is that it may be very difficult to find paths with positive capacity by inspection in large networks. A systematic approach that always finds an existing path is to start at node s and progressively connect to an adjacent node if the link has positive forward capacity. Either the sink node t will eventually be reached, or else there is no path and the maximum flow is found. More efficient versions of this approach involve labeling the nodes with the unused path capacities as they are added [15].

There are several interesting ramifications of the max-flow min-cut theorem. It can be used to find the *cohesion* of a network by setting all link capacities to unity and solving for the maximal flow between all node pairs. For any such pair s and t, this flow will equal the number of links in the minimal s-t cut-set, which then is the fewest links that must fail to disconnect those nodes. Furthermore, this minimal number of links equals the number of link-disjoint paths from node s to t. It is possible to find connectivity in a similar manner.

It is also interesting to consider the dual of the primal LP we formulated earlier to maximize flow from node s to t. This primal LP maximizes flow

---

† It is theoretically possible for the algorithm to fail if flow values are irrational numbers [15], but this can always be avoided in practice by a slight rounding and/or rescaling.

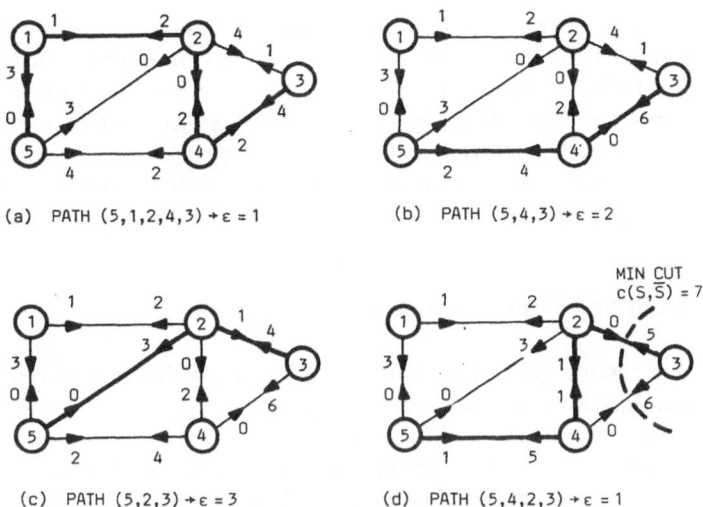

Figure 9.16. Maximum flow algorithm.

subject to flow constraints at nodes and on links, and thereby finds the maximum flow through the minimum cut. On the other hand, the dual LP minimizes the total network link capacity subject to constraints on the links used to carry the flow (resources used). At the dual basic feasible solutions, the dual variables turn out to be either 0 or 1, from which the links of the minimum $s-t$ cut-set are identified. Therefore, the dual finds the *minimum cut* while the primal determines the *maximum flow*, and the equality of these solutions is exactly what the max-flow min-cut theorem states!

From a broader network perspective it may be desirable to know the maximum flow between any given pair of terminals (but still only one pair at a time). This *multiterminal maximal flow* problem could, of course, be solved by applying the preceding method to all $n(n - 1)/2$ possible node pairs. However, there are more efficient techniques.

The network flow problems that we have considered so far are known as *single commodity* problems, where in our context a commodity is the traffic of one communications session. In general, of course, a data communications network will support many simultaneous sessions, with all the resulting packets (commodities) being multiplexed over the network links. This is the *multicommodity flow problem*. It is much more difficult computationally than the single commodity case, but can still be formulated as a linear program. A number of specific cases of practical interest have tractable solutions [16]. It is also possible to approach the problem by transforming it into a single commodity problem under appropriate conditions.

At this point it should be apparent that delay, capacity, and flow in networks are closely related. In this section we have characterized and modeled each, and presented some solution techniques. There are many more interesting theoretical details that could be presented, but we have covered all the fundamentals needed for the level of this text. Consequently, we now shift our focus from the network *design* issues of this section to the related network *operational* issues of routing and flow control in the next section. Finally, in the last section we shall return to the general mesh network design problem.

9.5. Routing and Flow Control

In this section we begin with the assumption that the physical data communications network has already been designed with appropriate nodal switches and link capacities. Our next concerns are how to *route* traffic flow efficiently between various communicating node combinations simultaneously and what type of *flow control* will be necessary to continuously keep the traffic evenly distributed over the network links so as to avoid overloading any nodes or links. These two issues are actually as much a matter of protocol design as of physical network design, and will thus also be considered in the next chapter for specific network architectures. Here we will cover the subjects much more broadly in the context of selecting appropriate techniques during network design.

The relationship of routing and flow control to network design is obvious from our introductory Discount Dog example. For example, there should be sufficient network connectivity to allow efficient routes to be determined for the anticipated traffic loads. There should also be sufficient link capacity and routing flexibility so that failures or traffic overloads do not lead to internal network *congestion*. This latter condition is defined as the undesirable situation where further increase in the overall network input results in an actual *decrease* in the overall output, as Fig. 9.17 shows. Intuitively, conditions like full buffers in parts of the network result in the increasing use of internal bandwidth for repeated retransmissions of discarded or unacknowledged packets. It is even possible for this condition to "snowball' to the point that absolutely nothing gets through, in which case the entire network is *deadlocked*. Some type of emergency *congestion control* must then be implemented. As Fig. 9.17 shows, good flow control should manage overloads so that the maximum network capacity is asymptotically approached without congestion. This inevitably results in some other performance degradation, e.g., delay of throughput. With this brief introduction, we now look at a variety of practical techniques for routing and flow control on computer data communications networks.

There are many ways to approach routing in data networks, with the most appropriate method dependent on the nature and objectives of the

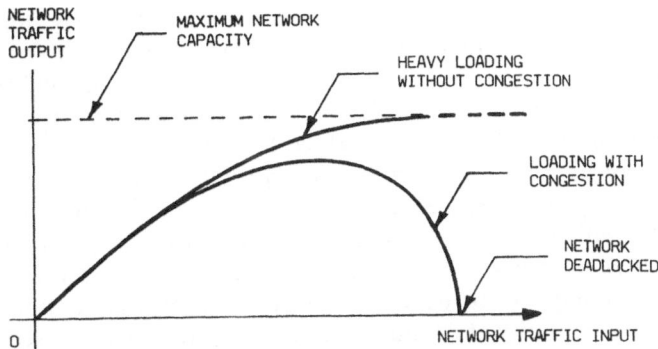

Figure 9.17. Network throughput and congestion.

network. Typical routing criteria include minimum cost, maximum through-put, minimum peak or average delay, high reliability, maximum survivabil-ity, and fast adaptation to structural network changes. These may be either routing performance objectives or constraints, with the selected combination depending on the particular network and design philosophy.

In order to appreciate the rich variety of different routing techniques available, it will be helpful to look at some of the common ways routing is classified. Perhaps the most basic is the *permanence of the routes.* This ranges all the way from permanent and switched *dedicated channels* assigned to each communicating terminal pair, to the connectionless *datagram* routing where each packet is self-contained and routed completely independently of any other one. In between there are *virtual circuits* where each user's traffic is multiplexed over a sequence of nodes and links that remain fixed either permanently or for some interval such as the duration of a call, a session, a message, or a packet.

Another classification is the *location of routing control.* This ranges from one central *routing control center,* to regional or local control centers, to completely decentralized control with each node making its own routing decisions. A related classification is the *degree of operator intervention.* This may vary from intervention only in the event of grave network failures, through various lesser situations, to frequent intervention for routing updates.

At each node routing information is typically stored in some type of *routing table* indicating the outgoing link for the packet destination. At one extreme there may be an extensive table listing primary and alternative links for each possible destination. At the other extreme there may be no table at all, but rather just an algorithm for selecting the outgoing link for each arriving packet. The *update time period* of the routing table or algorithm can range from essentially never to annually, monthly, daily, hourly, and down to less than a second. Updating is often considered either *fixed* (static), or *adaptive* (dynamic) to accommodate changing network conditions. The *scope of routing and update information* is either the entire network (global), the adjacent or near-adjacent nodes (local), or only the one node (isolated). Routing may be either *deterministic* or *stochastic* with the route chosen probabilistically, e.g., by generating a random number to select one of several alternative links at each node. There may be a *unique route* between users or *alternate routes*. A few other routing classifications are also possible, but this taxonomy will be adequate for our level of presentation.

The great majority of routing techniques in practice are based on tables (directory routing), and the most straightforward approach is to calculate for each node a table of *shortest paths*† to all other nodes. This table not only determines the "best" outgoing link, but can allow the originating node to specify an *explicit route* all the way through the network. Shortest path routing tables may be generated centrally or locally, updated frequently or seldom, and may be based on collected or estimated information.

As with any network graph, the criterion for path "distance" is not necessarily the physical length. Common alternate criteria are delay, number of hops, economic cost, and queue lengths. Since the criterion for path distance may vary over time with traffic patterns and topological changes, it is inevitable that exact current routing information is never available. Consequently, current routing tables end up based on past history or on estimates of the present. There is a trade-off between routing accuracy and the overhead required to obtain it, and this clearly depends on how dynamic the network is and how routing information is conveyed.

The shortest route problem can be formulated as a linear program in terms of network flows. For clarity we initially formulate the problem in terms of one source node s and one destination node t, although the extension to find the shortest-path tree to all nodes is simple. It is, in fact, a variant of the well-known transshipment problem of operations research. For all existing links l_{ij} we let d_{ij} be the corresponding length, and set the capacity to unity. If there is no link from i to j, we set d_{ij} and $c_{ij} = 0$. Then minimizing the cost of a unit of flow from s to t will force it to follow the shortest path. The LP statement is then

$$\text{minimize} \sum_{i=1}^{n} \sum_{j=1}^{n} d_{ij} f_{ij}$$

$$\text{subject to} \sum_{\substack{j=1 \\ (j \neq s)}}^{n} (f_{sj} - f_{js}) = 1$$

† Note that the shortest path is not the same as the MST in general.

$$\sum_{\substack{j=1 \\ (j \neq t)}}^{n} (f_{tj} - f_{jt}) = -1$$

$$\sum_{\substack{j=1 \\ (j \neq s,\, t,\, i)}}^{n} (f_{ij} - f_{ji}) = 0, \qquad i = 1, 2, \ldots, n; \qquad i \neq s, t$$

$$0 \leq f_{ij} \leq c_{ij},\ c_{ij} = 1 \text{ or } 0, \qquad i, j = 1, 2, \ldots, n; \qquad i \neq j$$

This form is, of course, very similar to the maximum flow problem formulations and can be solved by the simplex algorithm. However, as is usually the case, there are much more efficient alternatives to the direct linear programming solution.

Of the various shortest-path algorithms perhaps the most representative are the Dijkstra algorithm for a single originating node, and the Floyd algorithm [17], which finds the shortest paths among all node pairs. Intuitively, the latter would be appropriate for a central routing control center, while the former might be executed by each node. We shall look at the Dijkstra algorithm here, since it is both simpler and more instructive. There are, of course, many other viable shortest-path algorithms.

The Dijkstra "tree growing" algorithm is based on the simple but very fundamental *optimality principle* that the shortest path to any final node also contains the shortest paths to all intermediate nodes along the path. Thus we can progress through the network, permanently selecting the next link to provide the shortest path from the source node s until all nodes have become permanent. The algorithm does this formally by *temporarily* labeling each node with a node weight equaling the shortest distance through those nodes that have been selected (labeled *permanent*) at the current iteration. It is stated for node s as follows:

Algorithm 9.8 *(Dijkstra Shortest-Path Tree* [18]*)*

1. *Set $d_{si} = \infty$ if there is no s to i link and set $w_i = d_{si}$; assign node s the temporary label $[-, 0]$, and temporarily label all other nodes i with $[s, w_i]$.*
2. *Among all temporary nodes find one j^* with the smallest weight and make it permanent; if all nodes to which the shortest path is desired are permanently labeled, then stop.*
3. *For all temporary nodes i adjacent to permanent node j^* just labeled in step 2 calculate $w_{j^*} + d_{j^*i}$; if $w_{j^*} + d_{j^*i} < w_i$ update the node label to $[j^*, w_{j^*} + d_{j^*i} \rightarrow w_i]$.*
4. *Go to step 2.*

If the shortest-path tree to all nodes is desired, we simply let the algorithm run until all nodes are permanently labeled. Note that the shortest path may not be unique and will generally be different for each starting node.

As an example consider the network of Fig. 9.18a, which shows the initial labeling for node 1 and also the labeling sequence table indicating the order of labeling. The asterisk indicates a permanent label. In part (b) the resulting shortest-path tree is shown along with the routing table for node 1. A similar result for node 4 is shown in part (c). Note that the final labels give not only the shortest distance but also the predecessor node so that the actual path can always be established.

The Dijkstra algorithm is always optimal for the normal case where the "distance" criterion is nonnegative. In many operational networks the distance criterion is taken to be delay, which tends to vary with time. To maintain good routing information for routing table updates, the delay information may be exchanged in special packets or perhaps estimated from

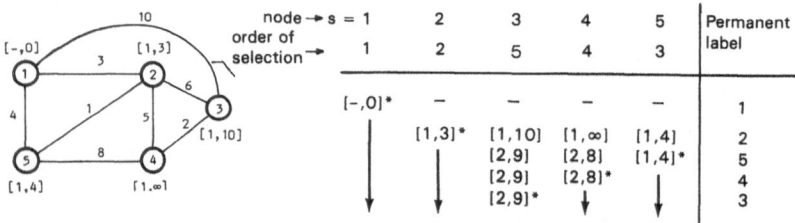

node → s = 1	2	3	4	5	Permanent label
order of selection → 1	2	5	4	3	
[-,0]*	—	—	—	—	1
	[1,3]*	[1,10]	[1,∞]	[1,4]	2
		[2,9]	[2,8]	[1,4]*	5
		[2,9]	[2,8]*		4
		[2,9]*			3

(a) Initial Labeling and Table of Labeling Sequence

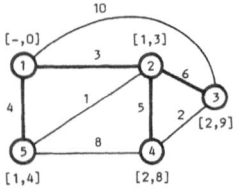

NODE 1 ROUTING TABLE

Ultimate destination	Next node
1	-
2	2
3	2
4	2
5	5

(b) Final Labeling and Routing Table for Node 1

NODE 4 ROUTING TABLE

Ultimate destination	Next node
1	2
2	2
3	3
4	-
5	2

(c) Final Labeling and Routing Table for Node 4

Figure 9.18. Shortest-path routing tables.

traffic flow or buffer queue lengths. The routing tables may then be computed from global information and distributed from a central site, or they may be produced by each node from local or global delay estimates.

An extension of the single-path table routing just considered that is usually desirable in larger networks is to have *alternative routes*. This, for example, helps even the load and reduce delay during long file transfers, and provides uninterrupted operation in the event of network failures. Routing can be further extended to provide *disjoint alternate routes* when high reliability is necessary. Another variation of directory routing is *hierarchial routing*, which is particularly attractive for very large networks. For node-to-node routing tables there can be thousands of entries which can be inefficient to use, much less to keep updated. A solution is to divide the network into regions and send all traffic for a region to one node of that region for further intraregional routing. An entirely different routing method may then be used inside each region.

In addition to directory routing, there are a number of interesting schemes that do not require tables. These schemes tend to trade-off the high overhead and slow response of near-optimal routing for simplicity and fast adaptation. The simplest such scheme, known as *flooding*, simply retransmits any received packet on all outgoing links. With no constraints this approach can quickly saturate the network with duplicate packets. However, it becomes viable if, for example, duplicate packets are discarded at each node, packets are flooded only in the direction of the destination, and packets automatically "self-destruct" after some fixed packet lifetime. Flooding is particularly attractive for special situations which occur infrequently but require high survivability, e.g., an attack on a military network.

Another simple routing approach is to put each received packet at a node on the shortest outgoing link queue, regardless of direction. Known in the vernacular as *hot-potato routing*, this approach at least keeps things moving even if in the wrong direction! Like flooding, it can be constrained to obtain more effective routing. *Proportional routing* assigns outgoing traffic to primary and alternative routes probabilistically, usually by basing the route selection on a random number. This method can be very effective with almost no overhead. Of course, the possible routes must somehow be determined.

Another interesting approach is known as *backwards learning routing*. Here a node estimates the delay of outgoing paths by observing the delay of received packets. This is akin to estimating commuter traffic on a freeway by observing only one direction. Backwards learning obviously depends heavily on how evenly the traffic is distributed among the network links, and can be counterproductive for very uneven distributions, e.g., the morning and evening freeway rush hours.

There is one more routing technique that is very important from a theoretical standpoint, since it solves the optimal routing for the multicommodity flow problem. As with the capacity assignment problem of the last section, we minimize the average network delay $\bar{T}$. However, now the variables are the link flows λ_i with the capacity assignments assumed known. The problem is nonlinear but convex, and an optimal solution technique known as the *flow-deviation method* [14] provides the optimal routing. It is interesting to note that the shortest-path problem is solved once in each iteration of this algorithm. Unfortunately the computation is rather involved relative to other viable suboptimal approaches.

Having discussed routing at some length, we now turn to the related subject of *flow control*. Given the network capacities and routing procedure, the objective of flow control is to regulate the external and internal traffic so that the network capacity is not exceeded and performance remains acceptable to the users. Good routing can mitigate the flow control requirements, but routing algorithms alone are seldom able to respond fast enough to overcome sudden changes in traffic loads. The only short-term recourse is to adjust the traffic flow, both on internal links and end-to-end circuits. Good flow control does this with minimal performance degradation and lost traffic.

We have already studied several flow control techniques in previous chapters. For example the XOFF statistical multiplexer command allows traffic from either end of a channel to be temporarily stopped when the multiplexer buffers become full. The BSC WACK and SDLC RNR commands are similar flow control mechanisms at the data link level. *Sliding windows*, which were used in SDLC for data link error control, can be used effectively for end-to-end flow control by dynamically reducing the window size as buffers become full. Figure 9.19a shows the sliding window operation for window length 5. Note the window is advanced only when an acknowledgment is received. For a one-packet window operation becomes HDX, and flow can be stopped entirely by setting the window size to zero.

In large networks with many users, simple end-to-end flow control will not generally be sufficient to prevent congestion. For example, it is possible that traffic from a number of different channels could be simultaneously routed through the same node, thereby quickly filling all its buffers. Traffic could be lost or, if node-to-node flow control is possible, then the traffic jam could spread to adjacent nodes. In the extreme, two or more nodes with traffic for each other might be unable to send or receive any further traffic, a condition known as *store-and-forward deadlock*, or lockup. To clear the condition, the nodes may have to be manually reset with a resulting loss of packets. A somewhat better solution is to provide a *small emergency buffer* in each node where every incoming packet can be inspected. In this way each node can at least receive acknowledgments and control packets

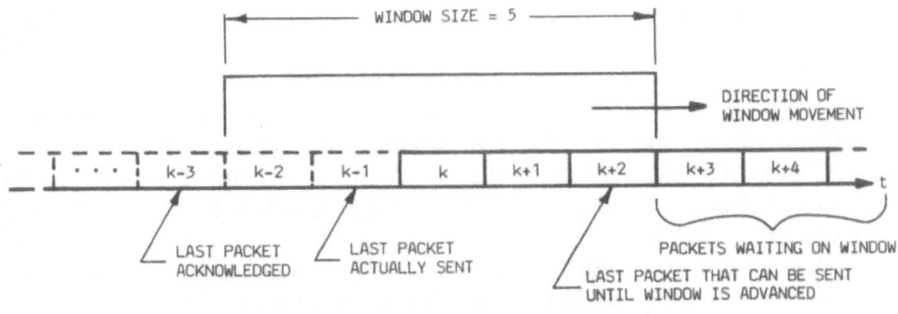

(a) Sliding Window Flow Control Operation

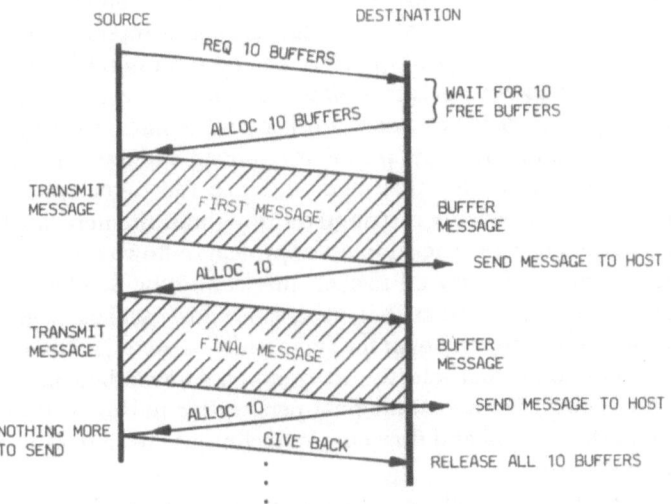

(b) Buffer Allocation Flow Control

Figure 9.19. Flow control techniques.

from higher protocol levels indicating how the deadlock is to be cleared. Internal nodes may also be allowed to flow control sources by sending a control message called a *choke packet* requiring the source to reduce its input rate.

A similar condition can occur end-to-end with the reassembly of packets into messages. If the PAD buffer space becomes full of incompletely assembled messages, then they can be neither delivered nor completed! The resulting deadlock is known as *reassembly lockup,* and again packets may be lost in resolving it. One common solution is to require the originating end to request a *buffer allocation* for the entire message prior to sending it. For multiple messages or long files, the allocation may be automatically renewed each time a message is successfully delivered, as shown by Fig. 9.19b.

Other lockups can occur with protocol uncertainties, for example, when two users try to access the same network resource and there is no rule for resolving the tie. This is a classical problem with file access in distributed data bases. Similarly, link or circuit failures in one direction only may cause *protocol lockups.* Good protocol design and appropriate timeouts can prevent or resolve such eventualities.

Congestion is initiated by excessive traffic entering the network, and then compounded by additional internal traffic because of retransmissions and control packets. One flow control approach, known as *rate flow control,* dynamically limits the rate at which each source can send packets into the network. A more exact control on the number of internal packets is through *token (isarithmic) flow control.* Here a fixed total number of token and user packets are allowed. Before a node can introduce a user packet into the network, it must first "capture" a token and replace it with the new packet. At the destination, the user packet is removed and a token reinserted. There is an obvious price for this method, since the tokens use up bandwidth and must be kept evenly distributed throughout the network if all users are to have equal access. The simplest approach to flow control is for congested nodes simply to *discard* all packets that cannot be handled. This just tends to postpone the problem, however, since lost packets will eventually be retransmitted, generally over the same route. Also, it is obvious that discarding must be somewhat selective to exempt acknowledgments and emergency control packets. To add some final perspective to this section we shall now consider the routing and flow control techniques used in several operational networks [20].

ARPANET initially provided a testbed for developing packet routing algorithms. Packets are routed individually with up to eight packets per message. The original distributed routing algorithm was based on local delay information obtained by exchanging delay vector packets at less than one second intervals. Delay was estimated by the average queue size for outgoing links. Each node then built its own routing table, which in effect contained the estimated shortest path *from* all other nodes. This approach proved to have undesirable oscillation and time lags, and after various modifications, was eventually replaced as discussed in the previous chapter. The new algorithm provides a global delay information update every ten seconds, from which each node calculates a shortest-path tree *to* all other nodes by using a modified Dijkstra algorithm. Global updates are distributed using flooding to route the delay packets among all nodes, and are only around 1% of the network overhead. Flow control uses an end-to-end message acknowledgment scheme known as *request for next message* (RFNM). To overcome reassembly lockup with multiple-packet messages, the originating node is required to request and receive a buffer allocation sufficient for the message prior to sending it. Once the message is reassembled and delivered, another allocation is automatically returned with the RFNM packet. This process then continues until there are no further messages, at which time the buffers are released in response to a "give back" message.

TELENET was based to some extent on the ARPANET technology and consequently also uses the RFNM acknowledgment. Only one message of up to eight packets was originally allowed to be outstanding between source and destination. Routing uses a two-level hierarchial approach, with area routing through the backbone network. Routing tables are calculated at a central network control center with primary and secondary line selections, then downline loaded to each node. The routing criterion is based on unused line capacity rather than delay, which provides a more uniform traffic distribution across the whole network. This is, of course, a prime objective of flow control and consequently there is little congestion. The network control center also reruns the network design program every few months to maintain a near-optimal configuration.

TYMNET uses centralized routing and decentralized flow control. A central control node called SAM (Supervisor in Active Mode) calculates all routes with a modified Floyd shortest-path algorithm. The route cost criterion is based on line loading, with the number of channels per link and per node limited. This cost is increased or decreased depending on the delays reported to SAM by the terminal nodes. SAM assigns a route when a user logs on to TELENET. Flow control is based on what is called back-pressure. Every half second adjacent nodes exchange back-pressure vectors indicating primarily the buffer size for each virtual circuit through the nodes. When a buffer size exceeds a predetermined threshold, the corresponding adjacent node input is stopped until the back-pressure returns to an acceptable value. This dynamic scheme has worked well, with no end-to-end flow control required and no significant deadlock problems.

These examples have hopefully illustrated the practicality of the routing and flow control techniques presented in this section. Many other operational networks worldwide employ one of the standard network architectures, and consequently use the particular routing and flow control techniques specified by the appropriate protocol of that architecture. Since these two functions are integrally interrelated with all the other network architecture functions, we shall defer their consideration until the next chapter.

9.6. Comprehensive Network Design Approaches

In the preceding sections of this chapter we have studied a number of design approaches to various subproblems of general network design. These included tree network designs, connectivity, capacity assignment, routing, and flow control. In all but the first problem we assumed that the topology of a mesh network was known and fixed. In this section we shall address the general mesh topological design problem. However, it is extremely difficult as an optimization problem and, as usual, heuristic algorithms are currently the only viable general approach. Furthermore, there is extensive interaction among the network topology and all the other network parameters, e.g., delay and reliability.

We can formulate the comprehensive network design problem in a rather general way as follows:

minimize	network cost
or	
maximize	network performance
with respect to	topology, link capacity, multicommodity flow
subject to	constraints on reliability, connectivity, delay, throughput, concentrator locations, component costs, available resources, etc.
given	terminal and node locations, anticipated traffic, cost data between node pairs.

This statement obviously leaves room for plenty of variation, such as the functional dependence of the cost function on items listed as constraints. Instead of trying to solve the comprehensive problem head on, we shall return to our piecemeal approach and now look at the topological design subproblem, which is difficult enough in itself. There are two well-documented approaches that we shall consider, namely, the branch-exchange and cut-saturation methods. Both are heuristic and simple in concept, but are highly iterative and invariably must be executed as a computer algorithm for practical network design.

The *branch-exchange algorithm* was originally used in the design of the ARPANET. It essentially operates by first finding any feasible† starting solution. It then replaces selected links with other links, typically in pairs connecting the same nodes. If there is any improvement the new links are kept, otherwise the original links are restored and another link pertubation is tried. When all such pertubations are exhausted a local optimal solution is obtained. Since there can in general be many such local optima, the algorithm is rerun from a new initial feasible solution to get another local optimum. This iterative process is continued until either an acceptable local optimum is found or we run out of money! The formal statement is as follows:

---

† Feasible again means that the solution satisfies all constraints.

Algorithm 9.9 *(Branch-Exchange* [21]*)*

1. *Generate a starting network that is feasible, i.e., satisfies constraints on delay, traffic, connectivity, etc.*
2. *Perturb the network by replacing a set of existing links with a set of new links connecting the same nodes; if no further transformations are possible go to step 5.*
3. *Assign capacity and determine traffic flow.*
4. *If the result is either infeasible or there is no improvement in the performance criterion, replace the set of links removed in the perturbations and go back to step 2; otherwise keep the feasible perturbed network and go to step 2.*
5. *Record the local optimum, compare with previous optima, and if within tolerance of optimality then stop; otherwise go to step 1.*

There are several key stages of the algorithm. First, there must be some way of generating different starting solutions that are feasible. A common technique is to generate links randomly and test each until feasibility is obtained. Next is the selection of the sets of links to exchange. These are normally two links each, connecting four nodes, but other combinations may be desired. A systematic selection criterion such as the highest-cost links or lowest-degree nodes can improve algorithm performance. The heart of the algorithm is step 3, which entails most of the computation (and design cost). Simple schemes such as equal link capacities and shortest-path routing are highly desirable whenever possible. Finally, in the last step some type of stopping rule must be selected. Figure 9.20 indicates some possible

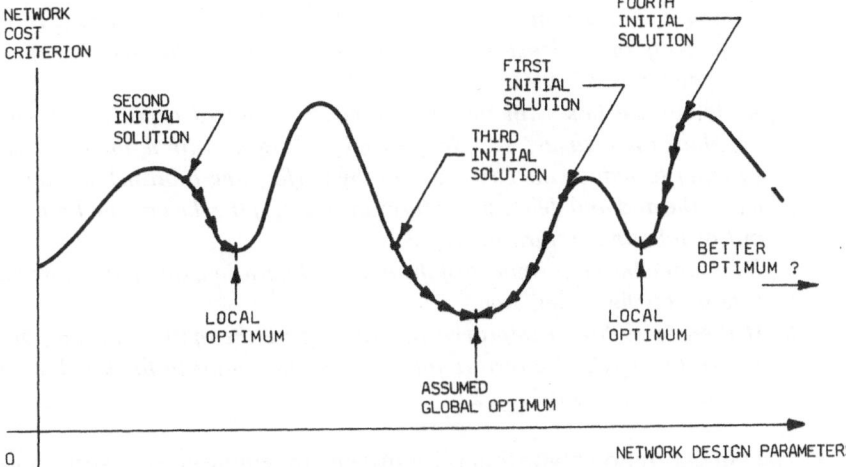

Figure 9.20. Branch-exchange solutions.

sequences of perturbation solutions, with each leading to a local minimum. A logical approach is to select the best existing local solution if no better solution has been found after a predetermined number of complete iterations since that solution. After all, the advantage of a heuristic algorithm is lost if it is turned into an exhaustive search for the global optimum.

The *cut-saturation algorithm* is more sophisticated than the branch-exchange, but it is also more computationally efficient and generally gives more optimal results. It starts with some specified throughput and the external traffic flows. The initial part locates a saturated cut-set, around which additional links are added and/or deleted to attain the required throughput. When this is achieved, further additions and deletions are used to reduce the cost criterion while keeping throughput within an acceptable tolerance and meeting all constraints. In practice it is used by TELENET for its periodic network redesign mentioned in the last section. With this intuitive background we can state the algorithm:

Algorithm 9.10. *(Cut-Saturation* [11]*)*
> 1. *Start with some initial topology that is feasible except perhaps for throughput.*
> 2. *Determine the optimal routing (flow) and order each link i according to its utilization f_i/c_i; then find the saturated cut-set by temporarily removing the highest utilization links in succession until the network is disconnected; each such link that would reconnect the network is part of the desired cut-set.*
> 3. *Obtain the specified throughput (within tolerance) by repetitively adding or deleting links, subject to any other constraints, as follows;*
>   > a. add *the most cost-effective link around the saturated cut-set between nodes that are near the "center of traffic" of each network component, or at least two nodes removed from the actual cut-set ("distance-2" rule);*
>   > b. delete *the link with the most expensive unused capacity, i.e., for which the measure $e_i = d_i(c_i - f_i)/c_i$ is largest, with d_i the link cost.*
> 4. *Attempt to reduce the cost criterion by performing a branch exchange using the add and delete procedures in step 3; if the throughput becomes out of tolerance return to step 2.*
> 5. *Determine the performance of the new configuration, and if the stopping rule is satisfied, then stop.*
> 6. *If there is sufficient improvement, recompute the optimal routing and return to step 4; else restore the configuration prior to the last branch exchange and go to step 2.*

The cut-saturation algorithm is intuitively appealing since it adds more capacity at the apparent throughput bottleneck in a systematic and cost-effective manner. Much of the algorithm's effectiveness depends on how good the routing algorithm is; consequently the theoretically optimal flow-deviation method is often used. Routing also accounts for the bulk of the computation, since it must be recalculated at each iteration. Many of the heuristics that make the algorithm effective are based on practical experience; for example, long serial sequences of links passing through the minimum cut-set may often be excluded from the branch-exchange step, or "collapsed" into a single equivalent link during the cut-set determination step. The stopping rule typically involves some minimal incremental improvement with an additional run-time or iteration limit.

These large-scale network design algorithms are invariably realized as proprietary software packages, and hence incorporate a great deal of designer experience and intuition. Consequently, provision for operator intervention in the design process can be highly effective. The algorithms also involve solving various subproblems such as routing, capacity assignment, and connectivity, for which there are many different approaches. Thus the comprehensive network design problem is currently as much an art as a science.

In this chapter we have tried to present enough tools, techniques, and examples to develop at least an appreciation of how to approach a broad range of network design problems. There are certainly other possibilities, but the material of the chapter will at least provide the designer with practical techniques that will work and can be refined and modified to suit each particular application.

In the next chapter we turn to the last of the three major topics in wide area networks—that of network architecture. Once we have the network laid out and the components selected, then some type of operating rules must be established among all the different communicating entities of the network. Such a coherent collection of rules is known as a network architecture.

Appendix: Basic Probability and Queueing Theory

This appendix includes those aspects of probability and queueing essential to the understanding of the material of this chapter. For those already familiar with the subjects it should establish the notation and provide some review. For the remaining readers it provides a concise introduction to these important subjects. Although not mandatory for understanding the chapter, the appendix material will greatly improve the depth of understanding of several of the chapter topics.

Historically, probability theory developed intuitively from gambling. Consider a game of chance, or *experiment*, with some particular outcome, or *event*, of interest A. If there are $n(\mathscr{S})$ total possible outcomes, of which $n(A)$ result in event A, then the probability $P(A)$ of A can be defined as

$$P(A) = \frac{n(A)}{n(\mathscr{S})} = \frac{\text{ways } A \text{ can happen}}{\text{ways experiment can happen}}$$

Implicit in this definition are the assumptions that *all outcomes are equally likely* and that there is *statistical regularity*, i.e., the same results apply over many repetitions of the experiment. This classical definition is generally adequate for gambling games, but in many other applications a more rigorous mathematical model is desirable. For example, there may be an infinite number of outcomes if packet delay is the experiment, or the outcomes may not be equally likely if there are priorities associated with packet transmissions.

Modern probability theory is based on the Kolmogorov probability model proposed in 1933. This set-theoretic model is based on a well-defined experiment with a number (finite or infinite) of possible *elementary* outcomes o_i. These are, intuitively, the most basic results of the experiment. The set of all elementary outcomes is called the universe or *sample space* $\mathscr{S}$. Every time the experiment occurs, one of the elements of $\mathscr{S}$ must occur. For example, if the experiment is to roll two dice, then the elementary outcomes (o_{ij}) are just the numbers showing on each die, e.g., o_{25} occurs if the first die is a 2 and the second a 5. The corresponding sample space is just the set $\mathscr{S} = \{o_{ij} \,|\, i = 1, 2, \dots, 6 \text{ and } j = 1, 2, \dots, 6\}$.

An *event* A is now defined as some set of elementary outcomes, and the event occurs if any one of the corresponding elementary outcomes results from the experiment. Mathematically an event A is a set defined on the sample space $\mathscr{S}$. There are, of course, many possible events for a given experiment, and the collection (class) of all of them is denoted by $\mathscr{E}$. For the dice example some events are

$$A = \{\text{at least one 1}\} = \{o_{ij} \,|\, i = 1 \text{ and/or } j = 1\}$$

$$B = \{\text{sum is 4}\} = \{o_{ij} \,|\, i + j = 4\}$$

$$C = \{\text{sum is greater than 8}\} = \{o_{ij} \,|\, i + j > 8\}$$

$$D = \{\text{pair of 6's}\} = \{o_{66}\}$$

$$E = \{\text{sum less than 15}\} = \{o_{ij} \,|\, i + j < 15\} = \mathscr{S}$$

$$F = \{\text{sum is 1}\} = \{o_{ij} \,|\, i + j = 1\} = \varnothing$$

These events are illustrated conceptually in Fig. 9.21. Note that an event can be empty $\varnothing$ (no o_{ij}), be the entire sample space $\mathscr{S}$ (every o_{ij}), or contain

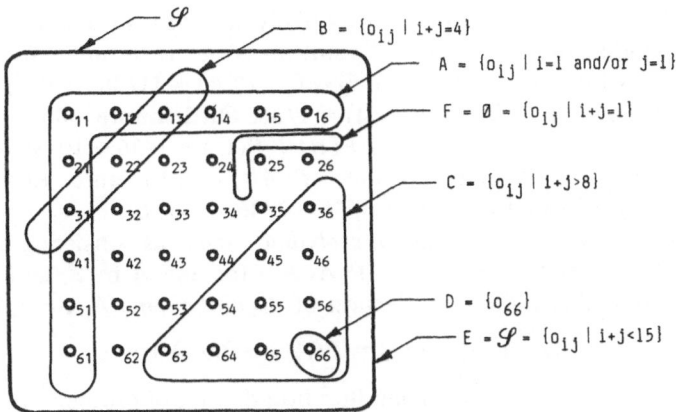

Figure 9.21. Sample space and events.

a single or any combination of elementary events. In our probability model, combinations of events are also events. In set terminology, if A and B are events, then so are their union ($A \cup B$ or just $A + B$), intersection ($A \cap B$ or just AB), and complements ($\bar{A}$, $\bar{B}$). Thus, in the dice example, we could define the additional events

$$G = A \cap B = AB = \{o_{ij} | i + j = 4 \text{ and } i \text{ or } j = 1\}$$

$$H = B \cup C = B + C = \{o_{ij} | i + j = 4 \text{ or } i + j > 8\}$$

$$I = \bar{A} = \{o_{ij} | i \neq 1 \text{ and } j \neq 1\}$$

In fact any finite union or intersection of events will also be an event. Also we have $\bar{\mathcal{S}} = \varnothing$ and vice versa. Two events are said to be *disjoint* if they have no elementary outcome in common, or equivalently, their intersection is empty. In Fig. 9-21, B and C are clearly disjoint and $B \cap C = BC = \varnothing$.

Having defined the sample space $\mathcal{S}$ and class of events $\mathcal{E}$, we next define a *probability function,* or *measure,*† $P(\cdot)$ which assigns to each event A in $\mathcal{E}$ a number between 0 and 1, i.e., $0 \leqslant P(A) \leqslant 1$. In particular, $P(\mathcal{S}) = 1$, $P(\varnothing) = 0$, and for any pair of disjoint event, B and C, $P(B + C) = P(B) + P(C)$. It may be conceptually helpful to think of probability as "mass," with one such unit assigned to $\mathcal{S}$. Then $P(\cdot)$ assigns some fraction of this "mass" to each event. If two events are disjoint, then the combined event is just assigned the sum of each each event's "mass." If two events A and B are not disjoint, then A and B have some common "mass" in their intersection and hence $P(A + B) = P(A) + P(B) - P(AB)$. The last term corrects for the common "mass" being counted twice in $P(A)$ and $P(B)$.

† Formally $P(\cdot)$ is a nonnegative set function that maps sets in $\mathcal{E}$ into the real line interval [0, 1].

For Fig. 9.21 if we assume equally likely elementary outcomes (the die are not loaded), then each has a "point mass" of 1/36 unit. The $P(\mathscr{E}) = P(\mathscr{S}) = 36(1/36) = 1$, $P(F) = P(\varnothing) = 0$, $P(A) = 11/36$, $P(B) = 3/36$, $P(C) = 10/36$, and $P(D) = P(\{o_{66}\}) = 1/36$. Furthermore, since B and C are disjoint, we have $P(B + C) = P(B) + P(C) = 3/36 + 10/36 = 13/36$, but for A and B not disjoint, $P(AB) = 2/36$ and $P(A + B) = P(A) + P(B) - P(AB) = 11/36 + 3/36 - 2/36 = 12/36$.

We can now formally define a *probability space* as consisting of $(\mathscr{S}, \mathscr{E}, P)$ with the properties $P(\mathscr{S}) = 1$, $P(A) \geq 0$ for any A in $\mathscr{E}$, and for any finite (or countably infinite) collection of *disjoint* events A_i, $i = 1, 2, \ldots$

$$P(A_1 + A_2 + \cdots) = P(A_1) + P(A_2) + \cdots$$

This basic model is adequate to handle a broad range of problems of interest in communication networks but we first need to extend our treatment to conditional probability, random variables, and random processes.

Often knowledge of the occurrence of one event will affect the probability of the occurrence of another event, but in other cases it may not. For events A and B, we define the probability of A conditioned on the knowledge of B by $P(A|B) = P(AB)/P(B)$ provided $P(B) \neq 0$. For the previous dice game example we have $P(A) = 11/36$ with no knowledge of B, but if we know B has occurred (sum is 4) then there are only two ways A could also occur (o_{13} or o_{31}). Thus the conditional probability of A given B is $P(A|B) = P(AB)/P(B) = (2/36)/(3/36) = 2/3$, which is a little better than $P(A) = 11/36$. In our probability model conditioning has the effect of restricting the sample space to just those elementary outcomes included in the known event, in our example event B. Note that in general $P(AB) = P(A|B)P(B) = P(B|A)P(A)$, so $P(A|B) = P(B|A)P(A)/P(B)$.

In many situations knowledge of one event has no effect on another. For example, we would not expect the number on one die to influence that on another. In such cases conditioning does not change the unconditioned probability, and the two events are said to be (stochastically) *independent*. In our ongoing example, consider events $U = \{o_{ij}|i = 1\}$ and $V = \{o_{ij}|j = 6\}$. Then $P(U) = P(V) = 1/6$ and $P(UV) = 1/36$ so U and V are not disjoint. However, they are independent because $P(U|V) = (1/36)/(1/6) = 1/6 = P(U)$. Intuitively, when we restrict the sample space to V by conditioning, we restrict the possibilities for U occurring proportionally. Finally, note that if two events are disjoint then the occurrence of one precludes the other, therefore they cannot be independent.

In most practical situations it is desirable to assign numerical values to events in addition to probabilities. For the dice game suppose we will win \$10 if the sum is 7 or 11, and lose \$10 if it is 2, 3, or 12. Otherwise we neither win nor lose, but just play again. Then we could naturally assign +10 to the event $X = \{o_{ij}|i + j = 7 \text{ or } 11\}$, -10 to $Y = \{o_{ij}|i + j = 2, 3 \text{ or } 12\}$, and 0 to $Z = \overline{X + Y}$. A set function $R(\cdot)$ which transfers the corresponding probability "mass" from $\mathscr{S}$ to the real numbers is called a *random variable*. Since $P(X) = 8/36$ and $P(Y) = 4/36$, R here transfers "masses" of $R(X) = 8/36$ to $+10$, $R(Y) = 4/36$ to -10, and the remaining $R(Z) = 24/36$ to 0. This is illustrated in Fig. 9.22a, where the "point masses" are represented mathematically by impulse functions with the areas indicated in parenthesis. If we played a large number of games, we would expect our average winnings to be \$10 (8/36) + \$0 (24/36) − \$10 (4/36) = \$1.11 per game, in other words the *average* or *expected value* of R is 1.11.

Random variables are characterized by their corresponding probability "mass" densities, as described mathematically by the *probability density function* (PDF). For the preceding R, we have the PDF

$$f_R(r) = (4/36)\delta(r + 10) + (24/36)\delta(r) + (8/36)\delta(r - 10)$$

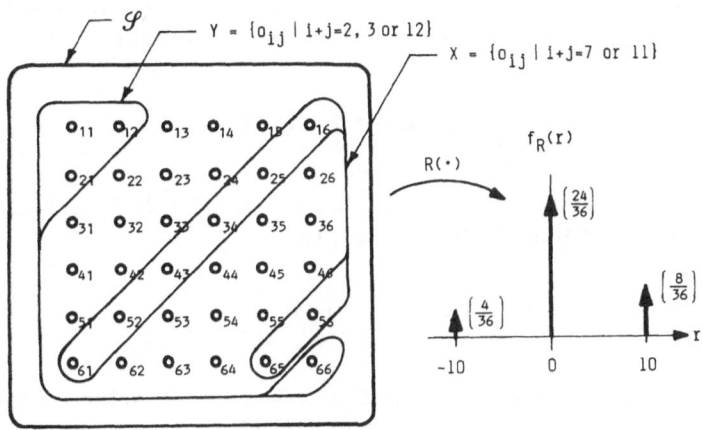

(a) Random Variable and Probability Density Function

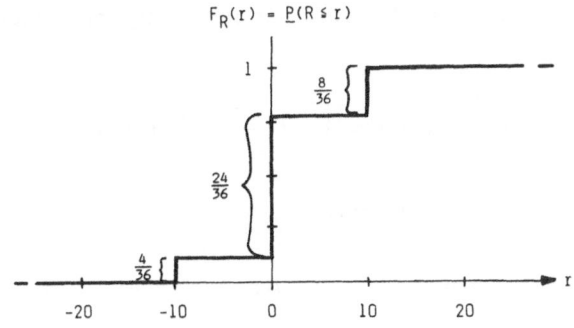

(b) Cumulative Distribution Function of R

Figure 9.22. Random variable characteristics.

as shown in part (a) of the figure. A random variable is discrete or continuous if its PDF is discrete (impulses) or continuous (no impulses). Another approach to the PDF is to define the *cumulative density function* (CDF) of a random variable R by $F_R(r) = P(R \leq r)$. Then $f_R(r)$ and $F_R(r)$ are related by

$$f_R(r) = \frac{d}{dr} F_R(r) \quad \text{and} \quad F_R(r) = \int_{-\infty}^{r} f_R(r) \, dr$$

For the dice game, $F_R(r)$ is shown in Fig. 9.22b. It should be apparent that for any random variable R, $F_R(-\infty) = 0$, $F_R(\infty) = 1$, and $F_R(r)$ never decreases with increasing r.

Once the PDF or CDF of a random variable is known, various averages can be defined which provide simple but very useful descriptions of random events. Consider any random variable T with PDF $f_T(t)$. Then the most basic average is the *mean* or *expected value* of T given by

$$\mu_T = ET = \int_{-\infty}^{\infty} t f_T(t) \, dt$$

Expectation is also sometimes denoted by m_T or $\bar{T}$. If T is discrete with only n possible values t_{ij}, this reduces to

$$ET = \sum_{j=1}^{n} t_j P(T = t_j)$$

For the dice game we have already found the expected value ER intuitively to be our average winnings per game of $1.11.

The mean indicates where the corresponding random variable's probability "mass" is centered, but gives no indication of how it is spread or skewed. A second average, the *variance*, indicating the spread about the mean, is given by

$$\sigma_T^2 = \text{Var } T = \int_{-\infty}^{\infty} (t - \mu_T)^2 f_T(t) \, dt$$

which in the discrete case becomes

$$\text{Var } T = \sum_{j=1}^{n} (t_j - \mu_T)^2 P(T = t_j)$$

Note that the variance is always nonnegative and is independent of where the PDF is centered. In practice it is usually more easily calculated by using the identity $\text{Var } T = E(T - \mu_T)^2 = ET^2 - \mu_T^2$. Finally, the units of the variance are squared; thus we often use the square root or *standard deviation* which has the same units as the mean.

For any pair of random variables R and T we always have $E(R + T) = ER + ET$, but the same property does not hold in general for the variance. Rather, we have $\text{Var}(R + T) = \text{Var } R + \text{Var } T - \text{Cov }(R, T)$, where the last term $\text{Cov }(R, T) = E(R - \mu_R)(T - \mu_T)$ is called the *covariance of R and T*. Analogous to independent events, R and T are *independent random variables* if the PDF of the sum is separable, i.e., if $f_{RT}(r, t) = f_R(r)f_T(t)$. In this case we have $\text{Var }(R + T) = \text{Var } R + \text{Var } T$, and furthermore $E(RT) = (ER)(ET)$ and $\text{Var }(RT) = (\text{Var } R)(\text{Var } T)$. Thus things are much simpler analytically if independence can be assumed and, fortunately, this is often the case.

Having now introduced the idea of a random variable and its general properties, we next consider some specific probability distributions of importance in network design. First, consider the common case of a single experiment that either succeeds (event A) with $P(A) = p$, or else fails with $P(\bar{A}) = q = 1 - p$. This is known as a *Bernoulli trial*; a common example is flipping a coin with A the event of a head. Now suppose the same coin is flipped n times as a new experiment, and that we are interested in the probability of getting k heads. Then the probability of exactly k heads (A) followed by $n - k$ tails $(\bar{A})$ is

$$P(\underbrace{AA\ldots A}_{k}\underbrace{\bar{A}\bar{A}\ldots \bar{A}}_{n-k}) = \underbrace{P(A)P(A)\ldots P(A)}_{k}(\underbrace{P(\bar{A})P(\bar{A})\ldots P(\bar{A})}_{n-k}) = p^k q^{n-k}$$

since the Bernoulli trials are independent. This is only one of many sequences of A and $\bar{A}$ resulting in k heads, and, in fact, there are $\binom{n}{k} = n!/k!(n - k)!$ different combinations. Letting random variable B denote the number of heads in n trials, we have

$$P(B = k) = \binom{n}{k}p^k q^{n-k}, \qquad k = 0, 1, 2, \ldots, n$$

Similarly

$$P(B \leqslant k) = \sum_{m=0}^{k} \binom{n}{m} p^m q^{n-m}$$

and obviously $P(B \leqslant n) = 1$. This is just the well-known binomial expansion for $(p + q)^n$ for the special case of $p + q = 1$. Hence, B is called a *binominal random variable*. Figure 9.23a shows $f_B(k)$ for the case of $n = 5$ and $p = 1/2$, with the discrete probability "mass" again represented by impulses. It is straightforward to show that $EB = np$ and $\text{Var } B = npq$.

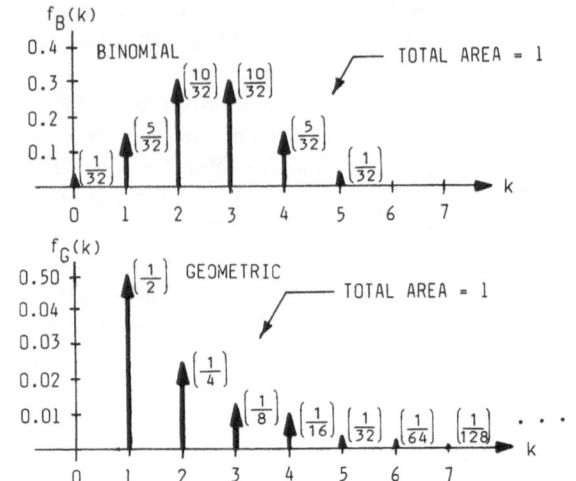

(a) Binomial and Geometric Probability Density Functions for p = ½

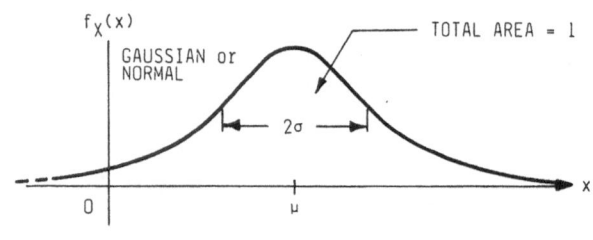

(b) Probability Density of Gaussian Random Variable X

Figure 9.23. Discrete and continuous probability density functions.

Next consider the related experiment of flipping a coin until the first head occurs, and then stopping. The random variable G that counts the total number of these Bernoulli trials has a PDF defined by

$$P(G = k) = q^{k-1}p, \qquad k = 1, 2, 3, \ldots$$

since we must first have $k - 1$ tails prior to the first head. Such a random variable has the *geometric* distribution. The averages of G are $EG = 1/p$ and $\text{Var } G = q/p^2$, and the discrete PDF is also shown in Fig. 9.23a for $p = 1/2$. Notice the number of trials k is (theoretically) unbounded, as there is no certainty of a head after any finite number of trials.

Many of the most important overall random variables have the *Gaussian*, or *normal*, distribution. For a *continuous* random variable X the probability "mass" is distributed according to the PDF

$$f_X(x) = \frac{1}{(2\pi\sigma^2)^{1/2}} e^{-(x-\mu)^2/2\sigma^2}, \qquad -\infty < x < \infty$$

As the notation suggests, $EX = \mu$ and $\text{Var } X = \sigma^2$, with the PDF shown in Fig. 9.22b. Since $F_X(x)$ cannot be expressed in closed form, it is common practice to evaluate Gaussian probabilities using a *standardized* table of $F_Z(z)$ for Z with zero mean and unit variance. $F_X(x)$ is easily converted to Z by noting that

$$F_X(x) = P(X \leq x) = P\left(Z = \frac{X - \mu}{\sigma} \leq \frac{x - \mu}{\sigma}\right) = F_Z\left(\frac{x - \mu}{\sigma}\right)$$

Much of the importance of the Gaussian distribution is due to the fact that the sum of a large number of independent random variables having any individual distributions tends to be Gaussian. The formal statement of this fact is called the Central Limit Theorem. Thus, large population samples and electronic thermal noise are well modeled as Gaussian.

The next distribution of interest describes the pattern of purely random arrivals. To specify exactly what "purely random" means, let λ be the average arrival rate per unit time of "customers," e.g., packets or messages at a network node. Then for a sufficiently small interval $(t, t + \Delta t)$ the probability of one arrival is $\lambda\Delta t + o(\Delta t)$ and of more than one arrival is $o(\Delta t)$, where $o(\Delta t)$ goes to 0 faster than Δt and is thus negligible. Furthermore, an arrival in interval $(t, t + \Delta t)$ is completely *independent* of any other arrivals in other nonoverlapping intervals. Thus nonoverlapping time intervals of equal lengths have the same probability of arrivals, regardless of when they occur. Letting the discrete random variable N_{t_0} be the number of arrivals in an interval of length t_0, the *Poisson* PDF is derived from the binomial [22]

$$P(N_{t_0} = k) = (\lambda t_0)^k e^{-\lambda t_0}/k!$$

For a unit interval, $t_0 = 1$, the mean is $EN_1 = \lambda$ and also $\text{Var } N_1 = \lambda$; the mean and variance are identical! Of course for any t_0 we have $EN_{t_0} = \text{Var } N_{t_0} = \lambda t_0$. Figure 9.24a shows the Poisson PDF for $\lambda = 1, 2, 3$, and 5. Again there is (theoretically) an unbounded number of arrivals possible in any given interval.

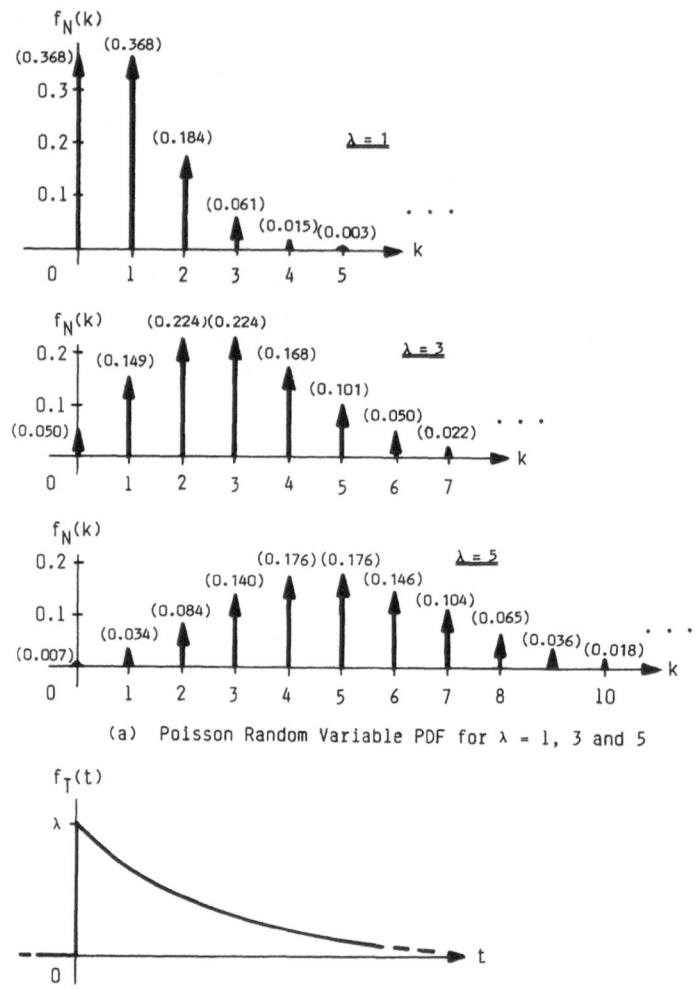

(a) Poisson Random Variable PDF for λ = 1, 3 and 5

(b) Exponential Random Variable PDF

Figure 9.24. Poisson and exponential PDFs.

Closely related to the number of random arrivals in a given interval is the issue of time between successive arrivals, usually called the *interarrival time*. Thus let the continuous random variable T be the interarrival time. The probability of no arrivals in any interval t is just $(\lambda t)^0 e^{-\lambda t}/0! = e^{-\lambda t}$, where λ is the average Poisson arrival rate. Consequently the CDF is

$$F_T(t) = P(T \leq t) = 1 - P(T > t) = 1 - e^{-\lambda t}, \qquad t \geq 0$$

The PDF is then found by simply taking the derivative

$$f_T(t) = \frac{d}{dt} F_T(t) = \lambda\, e^{-\lambda t}, \qquad t \geq 0$$

This is the *exponential distribution*, and is illustrated in Fig. 9.24b. The mean and variance of T are $E\,T = 1/\lambda$ and Var $T = 1/\lambda^2$. Intuitively, the average of T varies inversely with the arrival rate. The larger values of $f_T(t)$ near 0 imply that there are more arrivals close together than far apart. To an observer the arrivals tend to come in clusters.

Both the geometric and exponential distributions have the *memoryless*, or *Markov*, property which rather loosely means that the future values of these random variables depend only on the present and not the past. Mathematically, for the exponential case,

$$P(T > t + s \mid t > s) = P(T > t + s \text{ and } t > s)/P(T > s)$$
$$= e^{-\lambda(t+s)}/e^{-\lambda s} = e^{-\lambda t} = P(T > t)$$

and similarly for the geometric,

$$P(G > m + n \mid m > n) = P(G > m)$$

Intuitively, the probability of the next customer arriving, or of flipping the next head, is the same regardless of when the last customer arrived, or of how many tails have already been flipped! In other words, each separate time increment, or Bernoulli trial, is independent. This is less surprising in view of our definition of purely random arrivals.

With the preceding probability theory established we are finally in a position to consider queues per se. *Queues* are simply waiting lines. They occur in essentially every facet of our daily lives. A queue accepts *customers* (arrivals) according to some *arrival pattern*, stores them temporarily, and then sends them on to one or more *servers* in an order determined by some *queue discipline*. Each server then processes the customer during some *service time*. The served customer then leaves. The entire operation is called a *queueing system*.

Figure 9.25 depicts the key components of a queueing system. A convenient way to indicate these various components is the Kendall notation

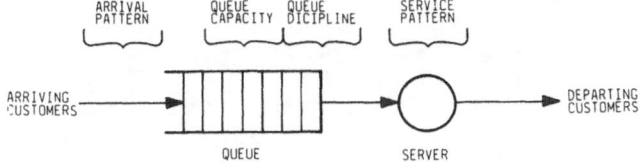

Figure 9.25. Queueing system components.

$a/b/c/d/e$. Here a denotes the arrival pattern; for example, D indicates regular (deterministic) intervals, M indicates Poisson (Markov) random arrivals, and G indicates any general pattern. The b parameter is the service pattern, with categories similar to the arrivals, e.g., M denotes exponential (Markov) service times. The third parameter c is the number of servers, d is the queue capacity, and e indicates the queue discipline such as first-in-first-out (FIFO). In most cases the form $a/b/c$ is used, with the suppressed parameters determined from the context.

In this introductory treatment we shall focus our attention primarily on the $M/M/1$ queue as a model for packet buffers. For this simple but very useful queue the arrivals are random (Poisson) with average rate λ, the service time is exponential with average value $1/\mu$ (so μ is the average service rate), and there is a single server. We will also assume the queue has sufficient capacity to avoid overflow and thus can be considered infinite. The queue discipline will normally be FIFO. Once the $M/M/1$ results are developed it is more or less straightforward to extend them to include multiple servers, finite queue lengths, and priority queue disciplines. The major difficulties come in handling other arrival and service distributions [14].

To analyze the operation of queues we first define the *state* of a queueing system as the number of customers Q in it. This includes those being served as well as those waiting in the queue. The state is obviously nonnegative; it increases by one with each arrival at average rate λ, and decreases by one with each departure at average rate μ. Our main interest is the probability $P(Q = n) = p_n$ that the system is in state n. This certainly depends on λ and μ, and, in general, also on time. However, if we assume λ and μ are constants and that the queue and server parameters are fixed, then the system will reach an equilirium condition so long as the server can "keep up" with the arrivals. Intuitively, we expect that at least $\lambda < \mu$.

To formulate these observations, the state-transition diagram of Fig. 9.26 will be helpful. Each circle represents one state, i.e., one value of the random variable Q, and there is only one arrival or departure at a time. The key insight is that for the system to be in equilibrium it is necessary that, on average, the arrivals and departures at each state must be equal. Consider, for example, state 3. The system will end up with exactly three customers if it is either in state 2 and there is an arrival, or in state 4 and

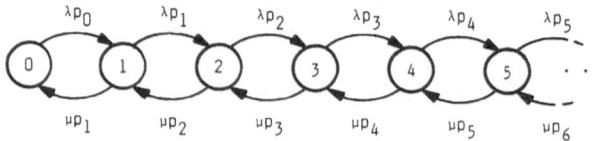

Figure 9.26. State-transition diagram for $M/M/1$ queue.

a departure occurs. Since the proportion of time the system is in state 2 is p_2, the average number of departures from state 2 going to state 3 is just λp_2. Similarly the average number of departures from state 4 to state 3 is μp_4. Consequently, the equilibrium condition for state 3 is $\lambda p_2 + \mu p_4 = \lambda p_3 + \mu p_3$. In general we have

$$\text{state 0: } \lambda p_0 = \mu p_1$$

$$\text{state 1: } (\lambda + \mu)p_1 = \lambda p_0 + \mu p_2$$

$$\cdots$$

$$\text{state } n: (\lambda + \mu)p_n = \lambda p_{n-1} + \mu p_{n+1}$$

$$\cdots$$

Substituting successively gives

$$\lambda p_n = \mu p_{n+1}, \qquad n = 0, 1, 2, \ldots$$

$$p_{n+1} = \left(\frac{\lambda}{\mu}\right)p_n = \left(\frac{\lambda}{\mu}\right)^2 p_{n-1} = \cdots = \left(\frac{\lambda}{\mu}\right)^{n+1} p_0$$

Now the ratio $\lambda/\mu = \rho$, the ratio of arrival to departure rate, is called the *traffic intensity*.† Clearly $\rho < 1$ for the preceding result. Next, note that the system must always be in some state so

$$1 = \sum_{n=0}^{\infty} p_n = p_0 \sum_{n=0}^{\infty} \rho^n = p_0 \left(\frac{1}{1-\rho}\right)$$

hence $p_0 = 1 - \rho$ and so $p_n = \rho^n(1 - \rho)$, $n = 0, 1, 2, \ldots$. This says that the random variable Q has the *geometric* distribution!

With these analytical results were are now in a position to obtain some practical results. Two important such results are the *average number of customers* in the *system* (L) or the *queue* (L_q), and the *average time a customer must wait* in the *system* (W) or the *queue* (W_q). The first result is just the expectation

$$L = EQ = \sum_{n=0}^{\infty} nP(Q = n) = \frac{\rho}{1-\rho} = \frac{\lambda}{\mu - \lambda}$$

† This is also called the *utilization factor* in contexts other than communications.

For the queue itself L_q equals L minus the average number of customers being served $1 - p_0$, thus

$$L_q = L - (1 - p_0) = \frac{\rho}{1 - \rho} - \rho = \frac{\rho^2}{1 - \rho} = \frac{\lambda^2}{\mu(\mu - \lambda)}$$

Notice in Fig. 9.27 that the size of the queue increases rapidly to infinity as the arrival rate approaches the service rate, that is as ρ tends to 1.

To find the waiting time we need a fundamental result known as *Little's formula* [23], which states that the average waiting time in the system is related to the number of customers by $L = \lambda W$. Similarly, for the queue itself $L_q = \lambda W_q$. Intuitively, during a large time interval T there will be an average of λT arrivals, and the total waiting time by all customers is LT. Consequently, the average time per customer is $LT/\lambda T = L/\lambda$. Thus we have the remaining key result for the $M/M/1$ queue,

$$W = \frac{L}{\lambda} = \frac{\rho}{\lambda(1 - \rho)} = \frac{1}{\mu - \lambda}$$

$$W_q = \frac{L_q}{\lambda} = \frac{\rho^2}{\lambda(1 - \rho)} = \frac{\lambda}{\mu(\mu - \lambda)}$$

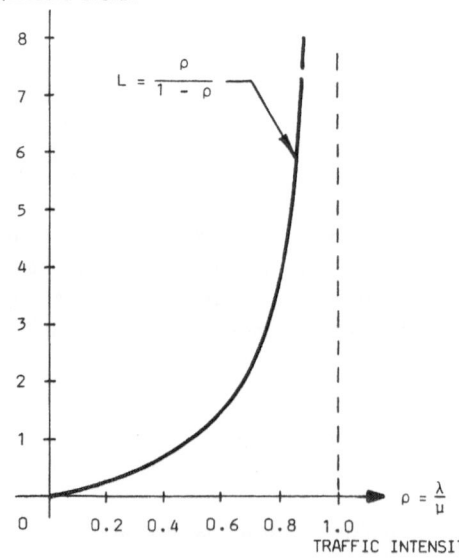

AVERAGE CUSTOMERS
IN QUEUEING SYSTEM

$L = \dfrac{\rho}{1 - \rho}$

$\rho = \dfrac{\lambda}{\mu}$

TRAFFIC INTENSITY

Figure 9.27. Effect of traffic intensity on queueing system length.

Since the average service time is $1/\mu$, it is easily verified that $W = W_q + 1/\mu$. A final result of frequent interest is the probability of a customer having to wait in the system or queue longer than some specified time t. Letting Ω and Ω_q be the corresponding random variables, it can be shown that

$$P(\Omega > t) = e^{-t/W}$$
$$P(\Omega_q > t) = \rho\, e^{-t/W}$$

The four formulas for L, L_q, W, and W_q provide the basic results we shall need for using the $M/M/1$ queue in our treatment of data communications networks. However, there are many other queues of practical importance in more detailed network analyses [22]. Of particular interest are finite length queues $(M/M/1/K)$, multiple server queues $(M/M/s)$, queues with periodic arrivals $(D/M/1)$, and with fixed service times $(M/D/1)$. The latter, for example, is appropriate when all packets have the same length and are processed identically.

To conclude this appendix with an example, consider a packet network nodal processor that buffers incoming packets which arrive randomly (Poisson) at the average rate of 20 per second. The buffer is obviously a queue; and if there is sufficient memory, it is not unreasonable to model it as having infinite capacity. All packets are eventually processed by a single routine and then transmitted on an outgoing link in an average time of $1/30$ sec. Since many packets are relatively short (acknowledgments, control, etc.) and a few are rather long (variable-length messages) with the processing time roughly proportional to the length, an exponential service time distribution is a reasonable compromise between exactness and tractability. Thus we have an $M/M/1$ queueing model with $\rho = \lambda/\mu = 20/30 = 0.667$. The four key parameters are then

$$L = \rho/(1 - \rho) = 2.000 \text{ average packets at node}$$
$$L_q = \rho^2/(1 - \rho) = 1.333 \text{ average packet in buffer}$$
$$W = 1/(\mu - \lambda) = 0.100 \text{ average seconds at node}$$
$$W_q = \lambda/\mu(\mu - \lambda) = 0.067 \text{ average seconds in buffer}$$

Finally, the chance of a packet getting delayed at the node more than one-half second is

$$P(\Omega > 0.50) = e^{-t/W} = e^{-5} = 0.0067$$

Since the arrivals are Poisson, they tend to arrive in clusters. Thus for the average of two packets in the node, 1.33 of them are in the queue itself. Once the server clears out the queue it may remain idle awhile, with the exact proportion of idle time given by $p_0 = 1 - \rho = 1/3$.

References

1. R. C. Prim, Shortest connection networks and some generalizations, *Bell Syst. Tech. J.* **36**(6), 1389–1401 (1957).
2. R. S. Wilkov, Analysis and design of reliable computer networks, *IEEE Trans. Commun.* **20**(6), 660–678 (1972).
3. H. Frank and I. T. Frisch, Network analysis, *Sci. Am.* **223**, 94–103 (1970).
4. F. Boesch, Introduction to basic network problems, *Proc. Symp. Appl. Math.* **26**, 1–29 (1982).
5. D. J. Kleitman, Methods for investigating the connectivity of large graphs, *IEEE Trans. Ckt. Thy.* **16**(5), 232–233 (1969).
6. G. B. Dantzig, Reminiscences about the origins of linear programming, *Oper. Res. Lett.* **1**(2), 43–48 (1982).
7. M. Sakarovitch, *Notes on Linear Programming*, Van Nostrand Reinhold, New York (1971).
8. K. M. Chanady and R. A. Russell, The design of multipoint linkages in a teleprocessing tree network, *IEEE Trans. Comput.* **21**(10), 1062–1066 (1972).
9. A. Kershenbaum and W. Chou, A unified algorithm for designing multidrop teleprocessing networks, *IEEE Trans. Commun.* **22**(11), 1762–1771 (1974).
10. J. B. Kruskal, Jr., On the shortest spanning subtree of a graph and the traveling salesman problem, *Proc. Am. Math. Soc.* **7**, 48–50 (1956).
11. R. R. Boorstyn and H. Frank, Large-scale network topological optimization, *IEEE Trans. Commun.* **25**(1), 29–47 (1977).
12. H. G. Dysart and N. D. Georganas, NEWCLUST: An algorithm for the topological design of two-level, multidrop teleprocessing networks, *IEEE Trans. Commun.* **26**(1), 55–62 (1978).
13. J. R. Jackson, Jobshop-like queueing systems, *Management Sci.* **10**(1), 131–142 (1963).
14. L. Kleinrock, *Queueing Systems*, Vol. II, Wiley, New York, (1976) Vol. II.
15. L. R. Ford and D. R. Fulkerson, *Flows in Networks*, Princeton University Press, Princeton, New Jersey (1962), Chap. 1.
16. H. Frank and I. T. Frisch, *Communication, Transmission, and Transport Networks*, Addison-Wesley, Reading, Massachusetts (1971), Chap. 3.
17. R. W. Floyd, Algorithm 97: Shortest path, *Commun. ACM* **5**(6), 345 (1962).
18. E. W. Dijkstra, A note on two problems in connexion with graphs, *Numerische Math.* **1**, 269–271 (1959).
19. H. Rudin, On routing and "delta routing": A taxonomy and performance comparison of techniques for packet-switched networks, *IEEE Trans. Commun.* **24** (1), 42–59, (1976).
20. M. Schwartz and T. E. Stern, Routing techniques used in computer communication networks, *IEEE Trans. Commun.* **28**(4) (1980), pp. 539–552.
21. H. Frank, I. T. Frisch, and W. Chou, Topological considerations in the design of the ARPA computer network, *Proc. AFIS Spring Joint Computer Conf.* (1970), pp. 581–587.
22. D. R. Cox and W. L. Smith, *Queues*, Wiley, New York (1961).
23. J. C. Little, A proof for the queueing formula: $L = \lambda W$, *Oper. Res.* **9**, 383–387 (1961).

Suggested Readings

G. Chartrand, *Introduction to Graph Theory*, reprinted by Dover, New York (1985).

An inexpensive but very well written and complete treatment of graph theory with a lot of interesting, classical examples.

L. R. Ford and D. R. Fulkerson, *Flows in Networks*, Princeton University Press, Princeton, New Jersey (1962).

The classical work on network flow theory, and the basis for much of the more recent work in the field.

L. Kleinrock, *Queueing Systems* Vols. I and II, Wiley, New York (1975 and 1976).

This two-volume set is an outstanding source for both queueing theory and its application to selected network design problems.

M. Schwartz, *Computer-Communication Network Design and Analysis*, Prentice-Hall, Englewood Cliffs, New Jersey (1977).

An early standard textbook that collects and organizes a large number of existing network design techniques, including numerous examples and exercises.

J. F. Hayes, *Modeling and Analysis of Computer Communications Networks*, Plenum Press, New York (1984).

A very good, current treatment of network design at a considerably more detailed and mathematical level than our one-chapter presentation; a good choice for further study of network design.

Check Your Understanding of Chapter 9—True or False?

1. Most data link delay occurs in the buffers and not during propagation or processing.
2. Removing all branches of a spanning tree will disconnect the network.
3. The Esau–Williams algorithm will always produce an optimal CMST.
4. Both arrivals and departures from an $M/M/1$ queue have the Poisson distribution.
5. Sliding window flow control can be used on links as well as end-to-end.
6. Prim's algorithm will always produce the optimal unconstrained tree network.
7. Maximum flow between two network nodes is always over the MST.
8. It is straightforward, though tedious, to test a network for connectivity.
9. Network routing in packet-switched VANs is usually done according to routing tables.
10. Large-scale network design today is invariably done with heuristic methods.
11. The optimal value of a primal or dual LP must occur at a vertex of the constraint simplex.
12. Shortest path routes will coincide with the MST for multiply connected networks.
13. Good network design algorithms should not allow operator intervention.
14. The least cost network for a set of nodes will always be a spanning tree.
15. Most operational packet networks are the result of optimal computer-aided designs.
16. The ADD algorithm deletes concentrators until the heuristic optimal solution is reached.
17. The minimum number of links terminating on any network node is called the network value.
18. For the same constraints the Kruskal and Prim algorithms give the same solution network.
19. Large, comprehensive network designs are basically done by sophisticated trial-and-error.
20. The NEWCLUST heuristic is to site concentrators near clusters of terminals.
21. Fully connected graphs have a unique minimum spanning tree.
22. An optimal link capacity assignment allocates excess capacity proportionally to all links.
23. Network cohesion can be found using the max-flow min-cut theorem.
24. Congestion can generally be reduced by providing alternate routing in a network.
25. If the traffic intensity is greater than unity, then our network design is in trouble.

10

Network Protocol and Architecture

10.1. Introduction and the OSI Reference Model

This chapter is about protocols, the rules by which digital data networks operate. In the earlier chapters we carefully developed the notion of a *physical data link* composed of transmission lines and modems, and then considered some typical rules of operation for the physical level (RS-232C, RS-449) and for the data link level (BSC, Poll/Select, DDCMP, SDLC). In the preceding two chapters we have analogously developed the notion of a *physical data network* composed of data links and network components. Now in this chapter we shall consider the *rules of operation*, i.e., protocols, for entire networks. However, before delving into the myriad details of the

319

various formal network protocols, it will be helpful first to establish some basic intuition about protocol in general.

Rules of various types permeate every facet of our modern bureaucratic world, and we normally consider protocols as very formal rules—for example, in the context of diplomatic relations. In sophisticated digital networks it is imperative that there be protocols that are thorough and unambiguous so that they can be executed automatically by the various network processors. They ensure a smooth flow of data in normal operation with minimal congestion. They also provide for efficient handling of, and recovery from, abnormal conditions with an absolute minimal amount of human operator intervention. The need for intricate detail, combined with the continual evolution and great variability of network designs, has resulted in essentially all significant network protocols being organized in modular form with each protocol layer providing the necessary support for the layer immediately above it in the protocol hierarchy.

To illustrate this hierarchial layered concept, consider the simple example shown in Fig. 10.1 of a voice telephone call between two business executives. Executive A at Avarice Associates (AA) decides to place a call to her peer Executive B at Big Bucks (BB). It is, of course, unthinkable that such high-level executives would actually dial their own calls;† rather, she asks her secretary to dial the call and let her know when B is on the line and ready to talk. Secretary B then "interfaces" with the telephone and dials the call, thereby activating the entire complex circuit switched process that transpires through the physical telephone network whenever a call is made. There are definite rules for the signaling process from *link-to-link* through the network, and also from *end-to-end* across it. Thus we have protocols for the *physical* hardware *layer*, the *link layer*, and the end-to-end telephone *network layer*. Once the circuit is connected the telephone rings at Big Bucks, which is part of the protocol between the user equipment (telephones). This can be thought of as a *transportation layer* since it provides a voice path to transport whatever type of speech (or other sounds, for that matter) may be desired. Note that it uses the layer immediately below to accomplish its functions.

Once the two secretaries are in communication, there is a conventional way for them to talk, which is sometimes specified by company policy. The basic function of this level of the hierarchy is to set up a session for the ultimate users, hence we have a *session layer* of protocol. In fact, several successive sessions could be set up by the secretaries at this layer if several different executives decided to use the same call without hanging up and

---

† Many telephone company training experts suggest that it is more efficient for everyone to dial and answer their own telephone, but such efficiency is nearly always sacrificed for status whenever a secretary can be obtained.

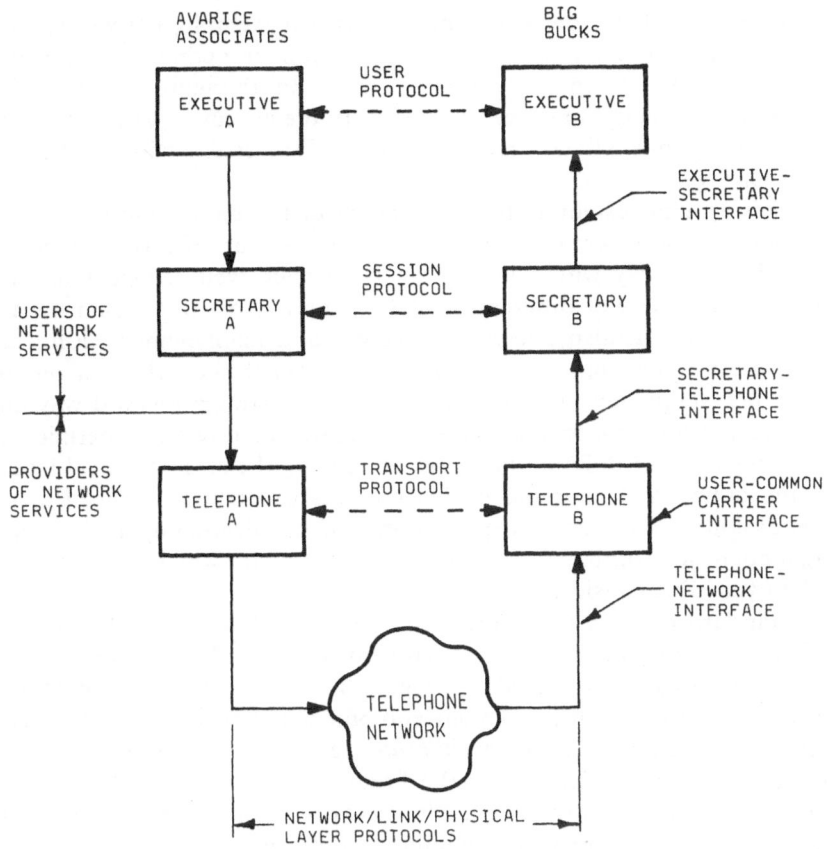

Figure 10.1. Voice network protocol layers.

redialing. Finally the secretaries get everything set and notify the corresponding executives, who then communicate with each other according to those time-honored platitudes that are absolutely essential for success at Avarice Associates, Big Bucks, and all their counterparts in the proverbial military–industrial complex. This *user protocol layer* is the whole purpose of all the lower layers; to the two executive users it appears that there is a direct, or *virtual*, circuit between them alone, and they are totally unaware of all the actions going on in the lower layers. It is helpful to consider the actual telephone system levels as a *transport subnetwork* which simply provides transport service between the telephone instruments of the transportation layer. There is a definite advantage in making the interface to this subnetwork standardized, thus allowing the interchange of user telephone equipment (red telephone set, answering machine, switchboard, etc.), and operation with other underlying subnetworks. For example, a private corporate network, another public telephone system, or perhaps a local area network with voice capability might be used. Furthermore, standardizing the way the secretary interfaces upward with the executive, and downward with the telephone, allows easy substitution of secretaries without affecting the other layers.

This familiar example illustrates much of the underlying philosophy of modern data network protocols. The specification of layered modules which interact only with adjacent layers through well-defined interfaces allows for a variety of different procedures and media to be used for any particular layer. It also allows for changes and maintenance within any layer without affecting other layers, and lets small networks that require only a few simple layers grow in an orderly and straightforward way into larger ones with more protocol layers. Standardized layers also permit easy interconnection of different networks employing the same protocol layers. On the other hand, modular design requires additional overhead in the form of control and addressing functions and duplication of services, so it might not be appropriate for a proprietary network for which high efficiency is the overriding consideration.

Standard network protocols are quite desirable in most situations, as can be seen from the preceding discussion. Historically, most major computer vendors have introduced proprietary *network architectures* which spell out the rules for operation of their equipment in networks. Most notable is the IBM *Systems Network Architecture* (SNA), first introduced in 1974 to provide compatibility among the firm's existing and planned product lines. SNA was originally designed for networks built around a single large main frame, but has since been modified to include multiple hosts in more distributed networks. Soon other vendors announced their own proprietary data network protocols, including the DEC *Digital Network Architecture* (DNA)† and *Burroughs Network Architecture* (BNA). Of these, only SNA was pervasive enough to become much of a de facto standard, and this primarily because of the large number of vendors of IBM-compatible equipment. The data link layer of SNA is SDLC, which has now been adopted and extended by many rival vendors as well as the major data communications standards organizations. Despite SNA, however, there has been a continuing need for public standards for network protocols, and such efforts have been ongoing for over a decade.

To date the most significant international data communications network standardization efforts have come from the CCITT and ISO, which have been working closely together. The ISO in 1977 began to develop a network

† DNA is often referred to as DECNET, but technically this latter term is only the actual software that implements DNA.

protocol model which emerged as the proposed *Open Systems Interconnection (OSI) Reference Model*† in 1979, and became an international standard (ISO 7498) in 1983. It is still subject to future evolution but at present it provides a highly useful guideline for protocol development and has been endorsed by the vast majority of data network providers and users. Most importantly, it has been adopted by CCITT as the X.200 recommendation and has been closely adhered to in the development of the pervasive X.25 protocol standard. The model defines seven layers (levels), and for each layer it specifies the abstract function. ISO goes on to define for each layer the particular services provided (service description), and details of how the layer is to perform those services and functions (protocol specification). Intuitively, users are concerned with services while designers are concerned with protocols.

There will then be a variety of possible specific protocols for each level, and numerous ways then to implement those protocols in either software, firmware, or hardware. Today the OSI model is well defined. In particular, the abstract definition of the Reference Model with its seven layers is essentially completed and provides the basis for the progressively less abstract service descriptions and protocol specifications.

The seven layers not only provide a guide for developing new protocols; they can also be used to compare and classify existing network architectures. Thus these layer definitions are central to any formal effort relating to network protocols. The specific layers defined by ISO are as follows:

Layer 1. The *Physical Layer* indicates the signals, functions, and connectors for accessing and using the physical data path; e.g., wire, coaxial cable, light link, radio, etc.

Layer 2. The *Data Link Layer* provides for the transportation of data across single links, including circuit establishment, error control, sequencing, and maintenance of the link through the use of layer 1. It also defines and delineates the data link frame.

Layer 3. The *Network Layer* provides a functional data path across the network per se, and defines procedures for routing, message sequencing, error recovery, and flow control. This layer is where packet switching and packet multiplexing are done, as well as data transfer between communicating networks (internetworking). Service may be over an initially established path (connected) or separate paths for each packet (connectionless).

Layer 4. The *Transport Layer* furnished a transparent end-to-end path, normally between user equipment, e.g., host-to-host, that ensures the required level of data integrity and is independent of the particular network

---

† Here the word "open" means that the protocols are not proprietary, and heterogeneous networks conforming to them should be able to communicate with each other.

employed. Programs written for the transport layer should then work with any underlying network without modification. When a common carrier is used, this layer is the boundary with the customer equipment. The layer handles packet sequencing, end-to-end error and flow control, and addressing between user processes; and, along with the lower layers, provides a uniform *transport service* to the higher-level protocol layers at the user terminals.

Layer 5. The *Session Layer* is concerned with the setting up (binding) and termination of user sessions between the end-user application programs over the previously established paths, and controlling the dialog flow, e.g., HDX or FDX. It may have provision for error recovery without terminating the session (or disconnecting the data path), and provides much of the synchronization between user processes. The layer adds application-oriented services to the basic transport service of layer 4.

Layer 6. The *Presentation Layer* provides the user application processes with a variety of data transformation services that avoid the necessity of changing the data representations for each new session. Such services are often done by user library routines, and include formating, code conversion, graphics, text compression, encryption, and virtual terminal† coding. This layer is concerned only with the form (syntax) of the data, and not the meaning (semantics).

Layer 7. The *Application Layer* contains the particular user application programs for accomplishing the user's objectives. It may include such services as privacy and authentication, network virtual terminal definition, remote file access, message transfers, logging and accounting, data base access, and the allocation of certain centralized and distributed resources. In many network architectures this layer is left entirely to the user, while in the OSI model only part is left to the user with the remainder specified in the model as a broad collection of highly useful user tasks.

Figure 10.2 shows the typical locations within the network and user terminal equipment of these various protocol layers. Implementation above level 2 has in the past nearly always been done in software because of the complexity and variability. However, as standards for the various layers become fully developed and widely accepted, there will be an increasing availability of board-level and integrated circuit implementations. In the interim, it should be noted that at present most networks do not require the full spectrum of high-level protocol features encompassed by the OSI model, and for those that do, suitable software can either be obtained from vendors or developed in-house.

---

† A *virtual terminal* is a common terminal format defined for the network into which all external (user) terminals must convert their data in order to obtain compatibility among different terminal types using the network.

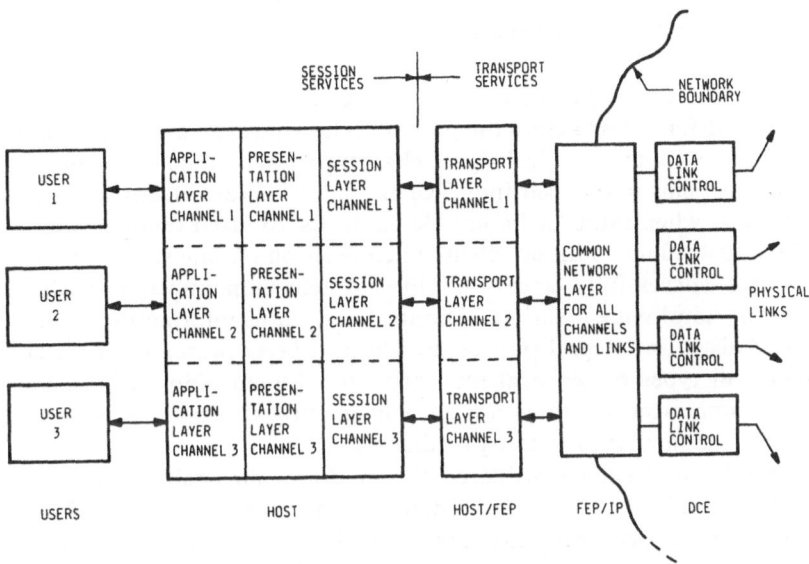

Figure 10.2. Functional relationship of OSI layers.

With the OSI Reference Model the ISO is in the unusual position for a standard of breaking new ground rather than following the historical procedure of accepting (or rejecting) existing standard candidates submitted by vendors. Thus the ISO committees charged with developing the OSI model must create viable new standards that are coherent† and acceptable worldwide in the midst of the highly dynamic and rapidly growing field of computer data communications. This is clearly a formidable task for any organization!

Now that the framework of the OSI Reference Model is available, we can consider specific protocols and network architectures in relation to it. The most pervasive lower-level protocols for data networks are the CCITT X. series, primarily X.25. For the higher levels ISO has defined protocols that correspond to the OSI layers.

In the following sections of this chapter we shall first look at some specific protocols for each of the OSI layers, and then describe two selected proprietary network architectures (SNA and DNA) within the OSI framework. The chapter concludes with a brief introduction to the operation of connected networks, i.e., internetting and interoperability. In the next chapter we will return to the subject of protocols. Then, however, instead of wide area networks we shall consider protocols for the more specialized local area networks.

---

† To appreciate the need for coherence one need only attempt to read and comprehend the actual X.25 standard.

10.2. Physical Layer Protocol

The physical layer of a network architecture is concerned with the actual transfer of bits across the particular transmission medium employed. It involves both the physical signal characteristics and the signal meanings. Physical signals at the user interface are nearly always electrical in nature, since even when radio or lightguide channels are used there is some type of DCE transducer with an electrical connection on one side.† In order to pass bits successfully across a physical telecommunication path, it is necessary to provide a variety of control and status information to the user DTE. This is an integral part of the physical layer protocol. There are, of course, all types of physical interfaces for different applications, but for data communications they can be conveniently categorized according to whether the medium is analog or digital.

We have already considered the predominant analog line interface, which is RS-232C in the United States and V.24 elsewhere. Recall that the V.24 standard covers only the signal meanings (functional description) for the DTE/DCE interchange circuits, while the electrical signal characteristics such as voltage levels and rise times are defined by V.28. Although RS-232C contains both the functional and signal specifications, the newer RS-449 only includes the former with the latter relegated to RS-422 and RS-423. In the past nearly all data transmission over voice grade telecommunications channels was done with modems using RS-232C or V.24/V.28. Because of its widespread acceptance and adequate operation in the majority of applications, there is every indication RS-232C will continue to dominate for many more years before it is significantly phased out by higher performance analog and digital standards. For wideband analog lines the interfaces defined in the CCITT V.35 and V.36 specifications for group band modems are widely used. These interfaces are quite similar to V.24 and V.28, but are capable of handling 48 and 64 kbps, respectively. And, of course, RS-449 operates at up to 2 Mbps. Clearly all of these *analog* physical layer standards are, and will continue to be, most important in practical data networks. However, since they were covered in considerable detail in earlier chapters, we shall not dwell further on them here. Rather, we now turn to the consideration of the physical layer for *all-digital* transmission media.

With the enormous increase in the use of digital terminal equipment, there has been a corresponding increase in the demand for digital communications services. Although conventional modems are now widely used for sending data over analog lines, this is something of a transitory situation, albeit a rather long one, until the widespread availability of all-digital

† This, however, need not always be true; e.g., a fiber optics link could be driven directly by the user DTE, in which case an optical interface would have to be defined for the physical layer.

transmission facilities. There is already a considerable effort in this direction, with the high-speed Local Area Network at the local level and the initial transitions toward the digital Integrated Services Digital Network (ISDN) at the global level. Eventually inherently digital traffic will dominate voice traffic, which, when combined with the overall growth of all types of traffic, will become a strong driving force for all-digital physical transmission channels. The CCITT and other standards organizations and PTTs are well aware of this trend, and consequently much of their communications efforts are now focused on the all-digital aspects of network protocols. With this motivation we shall now consider digital physical layer protocols, mainly the CCITT X.21 standard, which will certainly have significant influence on the future development of any widely accepted all-digital physical interface standard.

Although the X.21 standard has been around since 1972 and is now used for the national digital networks in Japan, Scandanavia, and Germany, there is still considerable room for confusion in understanding it as a physical layer protocol standard. In fact, X.21 is a full protocol for circuit switched digital networks and also includes the data link and network layers. Unfortunately, it was developed well before the OSI layered model, and does not specifically adhere to the same layers. Our concern here, of course, is with those components of X.21 that pertain to the physical level per se, but X.21 also contains procedures for placing, setting up, managing, and clearing circuit switched calls, and for character synchronization of the transmitted digital data. These latter functions are clearly not part of the OSI Physical Level and thus will only be considered very briefly here. It should be noted that X.21 applies to synchronous transmission systems; the asynchronous case is covered in the similar X.20 standard.

The X.21 physical interface contains only eight interchange circuits, two of which are grounds and two of which are clocks. Thus all data and control functions are handled by the remaining four circuits. As shown in Fig. 10.3a, the DTE sends to the DCE on the *Control* (C) and *Transmit* (T) lines, while in the reverse direction the *Indication* (I) and *Receive* (R) are used. The bit clock is provided by the X.21 DCE (just as by the modem DCE in the analog case) over the *Signal Element Timing* (S) line. There is an additional optional line called *Byte Timing* (B) for a character synchronization signal. Before considering the functions of these lines, we shall first establish their electrical signal characteristics.

Just as with RS-449, X.21 specifies *balanced* electrical circuits for all critical signals, with the pertinent standard being X.27.† It allows for data

---

† CCITT Recommendation X.27 is identical to V.11, and is essentially the same as EIA RS-422 (RS-449 Category I circuits); similarly, X.26 and V.10 are identical and basically the same as RS-423 (RS-449 Category II).

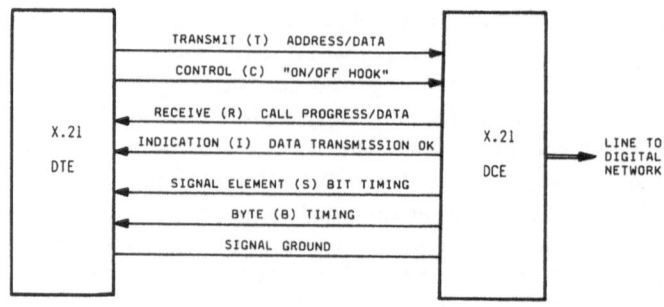

(a) X.21 Interface Circuits

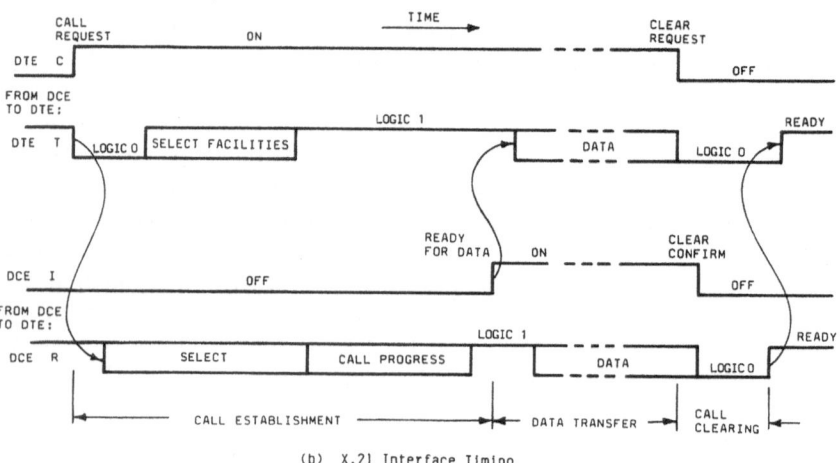

(b) X.21 Interface Timing

Figure 10.3. X.21 physical interface.

rates up to 10 Mbps and distances of up to 1.0 km using balanced line drivers and receivers. Thus all X.27 interchange circuits actually require two wires each, with the result being a 15 pin connector for X.21. The DCE, which interfaces with the network, is always required to use X.27. However, the DTE has the option of using the lower performance X.26 unbalanced interchange circuit whenever data rates are under 9600 bps. Of course, the DTE may use X.27 in all cases if desired. This is the recommended choice, since it ensures high performance and allows any feasible data rate to be accommodated without having to modify the interface drivers and receivers.

Unlike RS-232C, which has separate lines for each interface function, X.21 operates by basically time-multiplexing the control and data information on the T and R lines. Circuits C and I are then used to indicate what the particular information on T and R means,† and are strictly binary in nature. Turning C on is analogous to going off-hook with a conventional telephone, indicating to the DCE that the DTE is requesting a call. In the full circuit switched version of X.21 there is then an interchange of control characters to select the particular facilities desired and to indicate call progress, with each operation (state) preceded by a couple of SYN characters. Upon completion of this signaling the call is set up and the DCE then turns circuit I on to indicate to the DTE that data transfer may begin. Thus I is used somewhat analogously to the RS-232C signal CTS. To signal the end of data transmission the DTE will turn C off, and the DCE turns I off. This entire sequence is shown schematically in Fig. 10.3b. Note in particular that control information cannot be sent during the data transfer phase.

Although the circuit switched function of X.21 is important, our main interest here is in its relation to the CCITT packet switching protocol known as X.25. In 1976 the physical level X.21 was adopted for X.25. For this application the functions of C and I are modified so that when the DTE turns C on, the respective signals on T and R are to comply with the data link and network layer protocol that is specified in X.25. This, of course, replaces the circuit switched X.21 procedures indicated in Fig. 10.3b. The specific packet switching procedures that transpire across the physical interface will be described in the next two sections. X.25 is an extremely important and widely used standard today throughout the world, but there are relatively few networks that provide all-digital physical lines compatible with X.21. Consequently, CCITT has specified as an interim part of X.25 that the physical layer may use X.21 bis, which provides for operation on analog lines with V.-series modems. It is essentially the same as V.24 for data rates up to 9600 bps, and V.35 for 48 kbps, but the various interchange circuit functions are translated into a form that is compatible with the higher layers of X.25. However, regardless of which version of X.21 is used at the physical layer, the data link layer of X.25 will see the same interface to it.

The X.21 physical layer has not been as readily accepted as the other layers of X.25, owing both to the general lack of all-digital lines and to a number of inherent drawbacks. The 15-pin connector is considered by some to be too large, and, when used to connect two DTE devices together directly with no intervening network, the pin connections must be transposed to make each side appear to be connected to a DCE. This, of course, is an old problem with RS-232C which requires a special crossover cable (haywire) adapter. X.21 is also deemed inappropriate for the emerging ISDN, and there are several actual and proposed standards for separate *ISDN physical layer interfaces*. Since the ISDN could potentially include nearly

---

† CCITT Recommendation X.24 defines these functions in a little more detail ... but not much more!

all public data networks eventually carrying both voice and data traffic, the ISDN physical interface may become so pervasive that it would be naturally included in X.25. It is worth noting that just in the United States there has been a huge recent increase in the number of devices that can interface to the AT&T digital network both at T1 rate of 1.544 Mbps and at the lower rates. We shall consider the ISDN interfaces in Chapter 13. The ISO and CCITT have also reconsidered the physical layer and have developed service definitions to fit into the OSI model. Finally, there are well-established physical level standards for local area networks, but these will also be deferred until the next chapter.

Despite an initial appearance of simplicity and the near universal use of RS-232C or V.24, there is considerable change going on with regard to the physical layer for digital data networks, and particularly for packet networks. Although the current X.21 interface is certainly adequate, there are these problem areas with it and a number of alternatives are being considered. The ISDN physical layer specification will undoubtedly have a significant impact on other physical protocols. In the meantime most existing networks will continue to operate with analog lines and modems, allowing time for the new digital networks to grow and develop a suitable physical interface. Regardless of how the physical layer evolves, the boundary with the data link layer should not be affected. We will now consider this layer in the next section.

10.3. Data Link Layer Protocol

The main function of the data link layer is to provide a transparent, error-free digital path across single links. It uses the physical layer to provide timely and efficient services to the network layer. Among these services are typically link initialization, management, and disconnection, along with addressing, error control, sequencing, flow control, and the synchronization of characters and events at each end of the link. It is important to point out that the data link transfers blocks of information, as opposed to the physical layer, which passes only the raw bit stream.

The data link protocol can be considered conceptually as an *envelope* into which information from the network layer is placed for transmission from point to point, as illustrated in Fig. 10.4 for a two-link connection. The user data is appended with various headers from higher-level end-to-end protocols (HLH), and then assigned to an outgoing link according to the network routing procedure. The data link protocol then puts this end-to-end packet into its own protocol *envelope* and transmits it to the next node. Each data link protocol envelope has a *link header* (LH) and *link trailer* (LT) containing whatever information is necessary to provide the specified

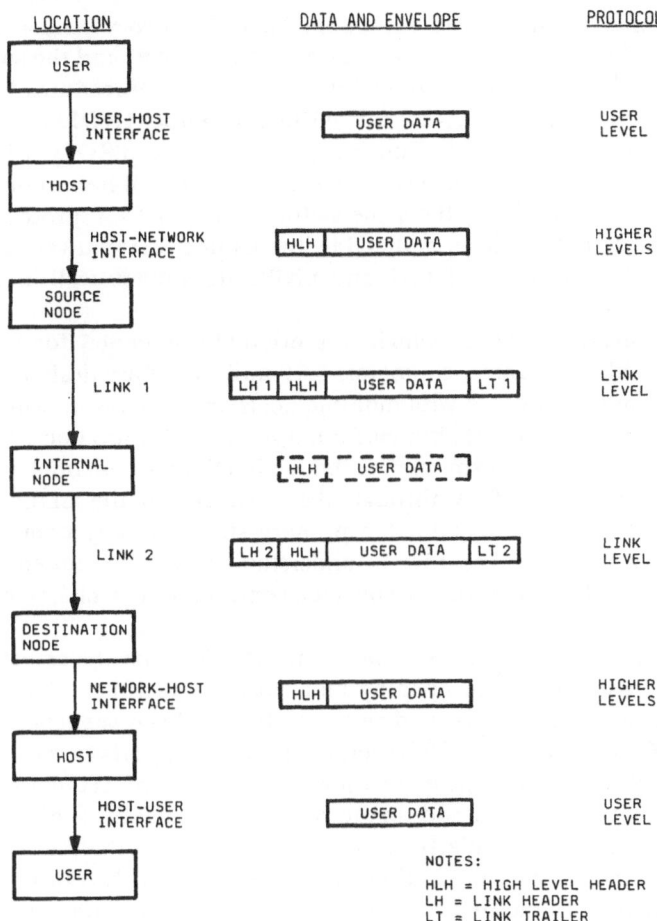

Figure 10.4. Link level protocol "envelopes."

services. Each higher protocol layer also acts as an envelope for yet higher layers. This "telescoping envelopes" analogy again points out the value of the layering approach to protocols; the particular type of envelope used is immaterial so long as it can hold the contents from the next higher layer. This is assured by defining standard interfaces between layers.

Since point-to-point and multidrop links have been around much longer than full digital networks, there is a great variety of feasible data link protocols. For example, we considered four of the most widely used proprietary ones in Chapter 7, and in the preceding section we briefly touched on the data link features of X.21. Most existing network architectures specify a single preferred data link protocol, and the CCITT

X.25 recommendation is no exception. There are two versions which are known as the original X.25 *Link Access Protocol* (LAP) and the subsequent *Balanced Link Access Protocol* (LAPB). These link procedures have been derived from the ISO *High-Level Data Link Control*† (HDLC), which was heavily influenced by SDLC. Thus our previous study of SDLC will provide a good basis for HDLC; however, there are a number of important differences. HDLC is also nearly the same as the ANSI ADCCP protocol, which is the version that has been adopted by the United States government; thus the basic characteristics of LAP and LAPB are already well ingrained in common usage.

In contrast to SDLC, which was originally intended for hierarchial networks with one central computer, LAP and LAPB are designed to work with distributed networks with multiple hosts and processors which do not generally need specific capabilities for loops and multidrop lines. The ability of a link to initiate transmission in both directions is essential to support network layer traffic flow through the network; so the SDLC Normal Response Mode (NRM) with the primary station totally controlling the link is inappropriate. In order to understand how X.25 circumvents this problem, it will be helpful to consider some pertinent characteristics of HDLC.

The basic HDLC frame as shown in Fig. 10.5a has the same structure as SDLC, with the same opening and closing flag (01111110), one-byte address and control fields, and two-byte frame check sequence that uses the same CRC polynomial. However, HDLC also supports *extended addressing* and *extended control fields* to allow option seven-bit sequence numbers ranging from 0 to 127, in addition to the conventional three-bit fields. This latter capability is particularly relevant to satellite links that have extremely long propagation times in relation to the time required to transmit a frame. These extended frame formats are illustrated for *information* (I) frames in Fig. 10.5b and 10.5c. For I frames the P/F bit and sequence numbers operate in a manner similar to SDLC, with the response number indicating the next frame that is expected. The basic *supervisory* (S) frame format is also like SDLC, but in addition to Receive Ready (RR), Reject (REJ), and Receive Not Ready (RNR) there is a fourth type called *Selective Reject* (SREJ). SREJ requests the retransmission of only the single I frame with the sequence number indicated by the N(R) field, but not any following frames. It is useful primarily for satellite links, where retransmission of all frames after an erroneous frame could involve as many as 128 frames with the extended format, and would certainly entail considerable delay in throughput. Some of the *unnumbered* (U) frames of HDLC are also the same as SDLC, although there are none for loop control or testing. For example, *Frame*

---

† HDLC is promulgated in a series of ISO standards issued from 1979 to 1981 [1,2].

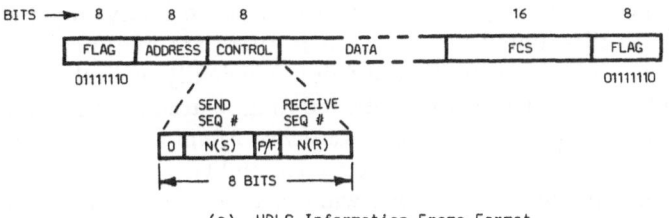

(a) HDLC Information Frame Format

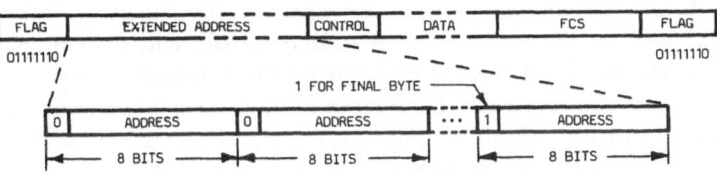

(b) HDLC Extended Addressing Format

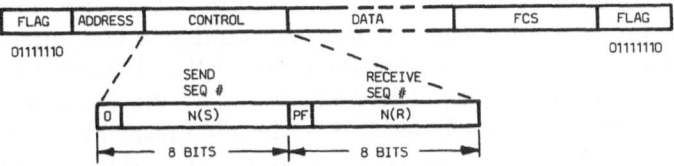

(c) HDLC Extended Control Field for Seven-Bit Sequence Numbers

Figure 10.5. HDLC frame formats.

Reject† (FRMR) is still used to indicate an error condition that cannot be resolved by retransmitting frames, and the corresponding diagnostic information fields are coded the same. There are also Unnumbered Information (UI) and Unnumbered Acknowledgement (UA) frames as in SDLC, plus a new command called *Unnumbered Poll* (UP), which is used to poll secondary stations without affecting the existing sequence number states. Finally, there is a U frame called *Exchange Identification* (XID) used for providing and receiving identification to and from an unknown station.

In addition to the many similarities between SDLC and HDLC, there are also some major differences. These relate mainly to the modes of operation. The only SDLC data transfer mode is the NRM, which was described in Chapter 7. Although supported by HDLC, NRM was designed for centralized networks with the primary station totally controlling the link. Since in this *unbalanced* procedure the secondary station cannot send

† This is called Command Reject (CMDR) in the older versions of both HDLC and SDLC.

unless it is specifically requested to do so by the primary, the NRM is poorly suited to distributed networks where there are multiple peer processors and terminals, and traffic must flow efficiently in both directions across the links. This situation originally led to the introduction of a second active mode called the *Asynchronous Response Mode* (ARM). Here each station has both a primary and secondary capability, therefore either end of the link can initiate transmissions without a specific request from the other.

In the ARM mode the use of the P/F bit is different from that of the NRM. Instead of being used by the primary for soliciting information (polling) and by the secondary for indicating the end of a transmission (final frame), the P bit is used in ARM by the primary simply to solicit a response from the secondary having the F bit set. Since the secondary may transmit asynchronously of the primary, the F response is sent at the earliest opportunity and has no relationship to the end of the secondary transmission. Similarly, a frame with the P bit set is not a poll for information, but rather only a request for a confirming F bit from the secondary. Clearly, for each P bit there must be a corresponding F bit received, and there can be only one such transaction per link outstanding at any time. This, of course, provides a mechanism for check-pointing in both directions, since all frames through the one containing the set P or F bit should have been previously acknowledged. The ARM is set by the command *Set Asynchronous Response Mode* (SARM), which corresponds to Set Normal Response (SNRM) for the NRM of SDLC and HDLC. If the optional seven-bit sequence numbers are to be used, then the corresponding control fields are two bytes long and the appropriate commands are *Set Asynchronous Response Mode Extended* (SARME) and *Set Normal Response Mode Extended* (SNRME).

Although the ARM does mitigate the strict master–slave relationship between the two stations on a link, initial network implementations resulted in certain deadlock conditions which were unacceptable. This resulted in further definition of a strictly *balanced* HDLC procedure which provides a third mode of operation called the *Asynchronous Balanced Mode* (ABM). This is a point-to-point configuration with a *combined station* on each end. Combined stations act as *both* primaries and secondaries, since they send and receive both command and response frames. The *Set Asynchronous Balanced Mode* (SABM) and *Set Asynchronous Balanced Mode Extended* (SABME) U frame commands are used to initiate balanced operation. Here either station can initialize the link, and the appropriate response is a UA frame just as in SDLC. Either station may also disconnect the link with the Disconnect (DISC) command. Use of the P/F bit is essentially the same as with ARM; but since either station can send commands and responses, it is necessary to carefully distinguish whether the bit is used as P or as F.

Table 10.1. HDLC Basic and Optional Functions

	Commands	Responses	Comments
	Basic functions		
	I	I	Information
	RR	RR	Receive Ready
	RNR	RNR	Receive Not Ready
	SNRM/SARM/SABM	—	Set Mode
	DISC	—	Disconnect
	—	UA	Unnumbered Acknowledge
	—	DM	Disconnect Mode
	—	FRMR	Frame Reject
	Optional functions		
1	XID	XID	Exchange Identification
2	REJ	REJ	Reject
3	SREJ	SREJ	Selective Reject
4	UI	UI	Unnumbered Information
5	SIM	RIM	Set/Request Initialization Mode
6	UP	—	Unnumbered Poll
7	—	—	Extended Addressing
8	—	delete I	Command I Frames Only
9	delete I	—	Response I Frames Only
10	SNRME/SARME/SABME	—	Extended Sequence Numbering
11	RSET	—	Mode Reset
12	TEST	TEST	Data Link Test
13	—	RD	Request Disconnect
14	—	—	32-Bit FCS

There are a few other features of HDLC that differ from SDLC, but the above discussion includes all the main features. The entire repertoire of HDLC frames for the three operational modes has been conveniently organized into a combination of basic commands and responses plus a variety of optional ones. A table of this repertoire as defined by HDLC is given as Table 10.1, indicating all the options available. For example, SDLC (without the loop mode) can be considered as equivalent to the NRM mode of HDLC with options 1 and 12. As a further illustration of HDLC operation, Fig. 10.6 shows an example of a link transmission using the ABM. It will be helpful to compare it with the corresponding SDLC example of Fig. 7.10 which uses the NRM.

Notice in Fig. 10.6 that each station transmits whenever it desires, once the link is set up in the ABM. The RR and REJ S frames operate essentially the same as with SDLC, but the addressing convention is significantly different. With NRM and ARM, the frames always contain the address of the *secondary* station; i.e., the primary sends commands and

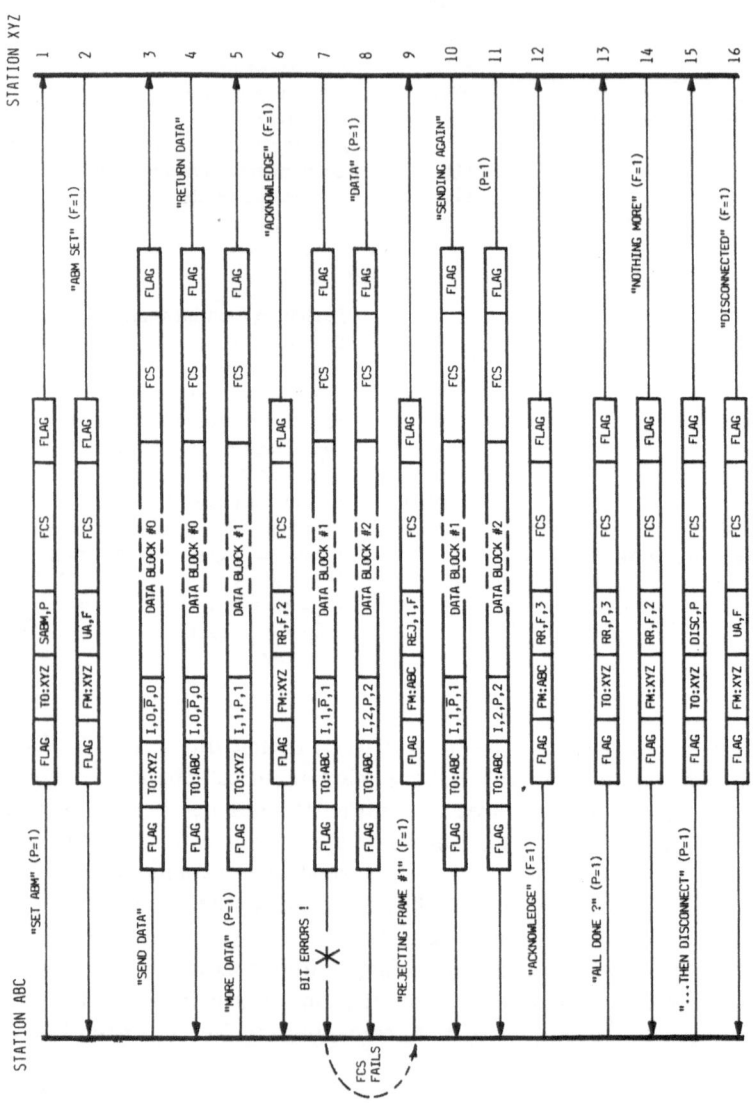

Figure 10.6. HDLC operation in asynchronous balanced mode.

data addressed to the secondary and the secondary always sends its responses and data with its own address. In ABM this convention is ambiguous, since each combined station originates both commands and responses. Thus for the ABM, *command* frames are always sent with the address of the *destination* station (which is acting like a secondary), and *response* frames always have the address of the *originating* station (also acting like a secondary). For example, in Fig. 10.6, the RR frame on line 13 is a response from station ABC and so contains the address ABC, but the next RR frame is a command from ABC and thus has the destination address XYZ. If the addressing conventions are always considered in terms of commands and responses rather primary and secondary stations, there should be no confusion between ABM and the other HDLC modes.

Having now described HDLC, it is straightforward to understand the X.25 data link protocols, LAP and LAPB. It is important to point out one major difference between HDLC and LAP/LAPB at the start. While HDLC is applicable to any pair of nodes, X.25 is intended to operate only across a DTE and DCE interface, which is typically between the user terminal or computer and the actual network interface node. Although X.25 LAP/LAPB can be employed between internal network nodes or between directly connected DTE devices, such applications are beyond the scope of the actual standard. Here we shall adhere to the DTE and DCE convention and not be concerned with the actual internal data link procedures used by the network.

In addition to this fundamental difference with HDLC, X.25 LAP/LAPB also has a number of other variances. Many of these are just additional features such as extended addressing and SREJ of HDLC not included in X.25. For example, there is no NRM in X.25, but LAP and LAPB correspond to the HDLC ARM and ABM, respectively.

Another important difference is the provision in X.25 LAPB for *multilink operation*.† In some situations it may be desirable to use multiple data links in parallel, with the entire group operated as one data link. For example, multiple links can provide additional bandwidth in single-link increments, or can allow link transmission to continue uninterrupted in the event of one or even several single link failures. When sequential frames are transmitted on multiple parallel links it is generally necessary to resequence them at the receiver since different links will have different delays.

The X.25 *multilink procedure* (MLP) only applies to LAPB, and basically adds collective sequencing and flow control on top of the individual LAPB single-link procedures (SLP) already discussed. An additional 16-bit *multilink control field* is inserted immediately after the SLP control field

† This capability was added in the 1984 version of X.25.

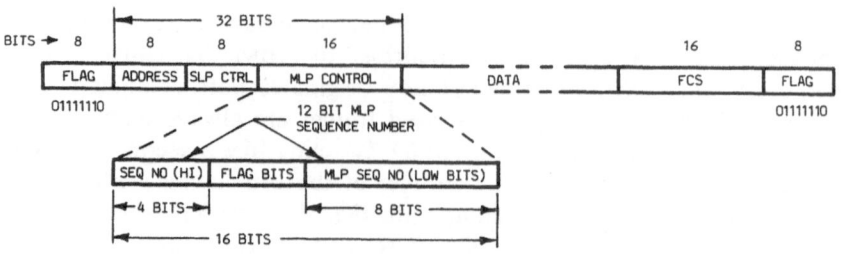

Figure 10.7. Multilink frame format.

with the format shown in Fig. 10.7 for an I frame. This MLP field contains a 12-bit sequence number MN(S) and four flags. Thus multilink sequencing is modulo 4096. Setting the *V flag* voids the multilink sequencing operation, while also setting the *S flag* allows the use of sequence numbers only for duplicate frame detection. The remaining *R and C flags* indicate reset request and reset confirmation, respectively. The MLP allows for multilink resetting without resetting the composite single links. The actual reset involves a thorough handshaking procedure which zeros the sequence counters at both ends of the link.

The X.25 MLP uses the conventional sliding window flow control, with the *transmit window* of width MW being advanced by the SLP acknowledgements. At the destination a similar *receive window* is used. Normally received frames are within this window; but if a frame is lost, frames beyond the window will eventually arrive. X.25 extends the receive window width MW by an additional *guard region* of length MX, within which frames will still be accepted. However, an error message will be generated and the window will not advance. Frames received beyond this guard region will simply be discarded. On each single link the normal LAPB flow control prevails; e.g., RNR indicates a link busy condition to the MLP.

For both SLP and MLP, *addressing* is similar to that of the HDLC balanced mode with the DTE and DCE being the only two stations. LAP and LAPB define the DTE address to be A = 3 or C = 15 and the DCE address as B = 1 or D = 7 for the SLP or MLP, respectively. Hence the DTE sends *commands* TO:A or C and receives *responses* FROM:B or D. This is shown in Fig. 10.8 for both the SLP and MLP cases. Frames with addresses other than A and B, or C and D with MLP, are automatically discarded.

Although a few older networks still support the LAP protocol, practically all present implementations use LAPB, with LAP being gradually phased out. LAPB can also be expected to absorb additional features, from HDLC and elsewhere, as needs arise in the future. This again illustrates

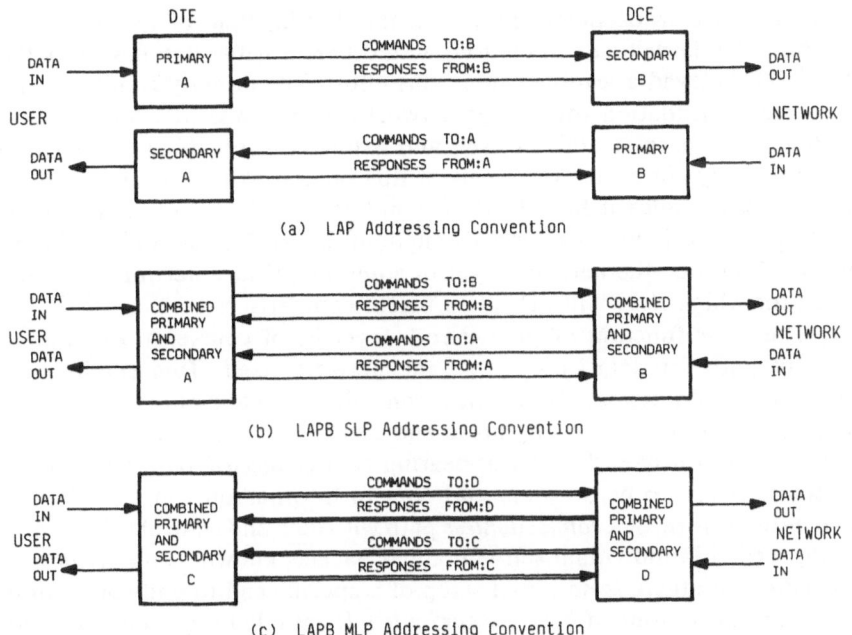

Figure 10.8. X.25 data link addressing conventions.

the evolutionary process that is fundamental to the development of viable complex standards such as X.25.

With our initial introduction of SDLC, its extention to HDLC, and now the specialization of HDLC to X.25 LAP/LAPB, we have a fairly complete treatment of the bit-oriented data link protocols that are so prevalent throughout the modern world. Unlike other layers of the OSI Reference Model, the data link standard is well defined and universally accepted. In the next section we consider the end-to-end network protocol which is built upon data links.

10.4. Network Layer Protocol

The third layer in the OSI Reference Model is the network layer, which is the first protocol layer that operates across the network. A decade ago most networks had point-to-point hierarchial configurations, so that a data link protocol was usually sufficient for normal operations. However, there has been an ever accelerating growth in both the number and the sophistication of data networks, and today many important networks have distributed mesh topologies for which link protocols alone are not sufficient.

This has led to widespread interest in the development of network layer protocol standards that can be used with different network types. Initially, the CCITT played a leading role in this effort. The primary result was the X.25 recommendation for packet networks, which was first approved in 1976 and has been steadily revised and extended ever since† so that it is now relatively complete. The focus of this section will be on the network layer of X.25, since it has clearly become the predominant standard for *public* packet-switched networks throughout the world. Late in the chapter we shall describe the network layer of some *proprietary* network architectures, for which X.25 will provide a useful comparison.

The basic function of the network layer is, of course, to provide an efficient path for data flow across the network itself. This entails such functions as addressing, routing, flow control, packet and message sequencing, and error control on an end-to-end basis. This network path may be in the form of a *virtual circuit* appearing as a *connected* path which must be formally set up before data and control information can flow. It may also be in the form of a *connectionless* path with data and control information being carried in individual self-contained packets known as *datagrams.* In this latter case there is no initial setup of a specific end-to-end connection, but rather the routing of each datagram is left entirely to the network. With packet networks the network layer performs packet-interleaved asynchronous time-division multiplexing on the various data links so as to create either connected or connectionless services for the respective users of the network. Thus a single link may carry many different virtual circuits. Each packet traverses the link inside a frame defined by the data link protocol.

To further clarify the notion of a virtual circuit it may be helpful to consider the software modules that execute the protocol layers as depicted previously in Fig. 10.2. At the data link layer there is a module for each link of a given node, while at the network layer there is only one module per node. The network layer takes information from each of the transport layer modules (one for each virtual circuit), puts it into packets, and sends each packet to the appropriate link according to its routing algorithm. This network protocol function is normally done in an FEP for a large host, or perhaps under the operating system of a small host or terminal.

Virtual circuits can be classified in two types. *Permanent virtual circuits* (PVC) are analogous to leased telephone lines, in that they are assigned to the user when he first subscribes to the network service and remain permanently assigned whether in use or idle. Alternatively, *switched virtual circuits* (SVC), sometimes referred to as *virtual calls*, are set up each time a message sequence is transmitted and then disconnected when the transmission is completed, which is analogous to a dial-up telephone line.

† Revisions of X.25 were approved in the 1980 and 1984 CCITT Plenary Sessions.

Before considering the X.25 network layer, we shall briefly look at the network layer functions of the X.21 circuit-switched protocol. Calls are set up by the DTE by indicating *Call Request* across the DTE/DCE interface, as was shown in Fig. 10.3. At the remote end the DCE sends *Incoming Call* across the interface, to which the normal DTE response is *Call Accepted* if it wants to take the call. Back at the originating end of the network the local DCE now sends a set of characters (SYN SYN + + +...+ +)† meaning *Proceed to Select*. In this context, selection means the DTE selects certain addressing conventions and/or various *facilities*. Examples of typical X.21 facilities include multiple address (conference) calls, call redirection, billing information and closed user groups to effectively create a private subnetwork within a larger X.21 network. Once this *Selection Signal Sequence* is completed the DCE then supplies appropriate *Call Progress* and *Call Connection* status signals, again in the form of control character sequences. The commencement of the data transfer phase is signaled by the DCE turning the I line of the interface ON, and once completed, the call is cleared by the DTE sending *Clear Request* (turning C OFF) and the DCE responding with *Clear Confirmation* (turning I OFF). There are also a few other X.21 commands and responses at the network level, but the above discussion provides a sufficient idea of how the protocol functions with respect to call setup and clearing.

Like X.21, X.25 [3] defines the network layer protocol only across the DTE/DCE interface, and not through the network interior. In fact some major data networks provide X.25 access but use entirely different protocols internally. In the X.25 context we shall normally think of the DTE as user equipment such as an intelligent terminal, host computer, or front-end processor; and we will consider the DCE as the network itself in the form of a nodal processor. Across each DTE/DCE interface there may be up to 4096 *logical channels* which are identified by a *logical channel group number* (0-15) within which there are the *logical channel numbers* (0-225). These logical channels always exist across the interface. They may be idle or they may be assigned to either a switched or permanent virtual circuit or, in earlier versions of X.25, to a datagram channel. Thus, for example, an X.25 virtual circuit will use a separate logical channel at each end of the network plus the necessary path within the network itself.†

The basic X.25 operation with a switched virtual circuit is much like X.21. There is the same general sequence of *call establishment, data transfer,* and *disconnection.* For permanent virtual circuits there is no need for establishment or disconnection, since the virtual circuit is always connected.

---

† The specified character set is the International Alphabet # 5 (IA5), which is essentially the same as ASCII.

‡ This distinction between logical channels and virtual circuits is not so clear in other protocols and the terms *logical* and *virtual* are sometimes used interchangeably.

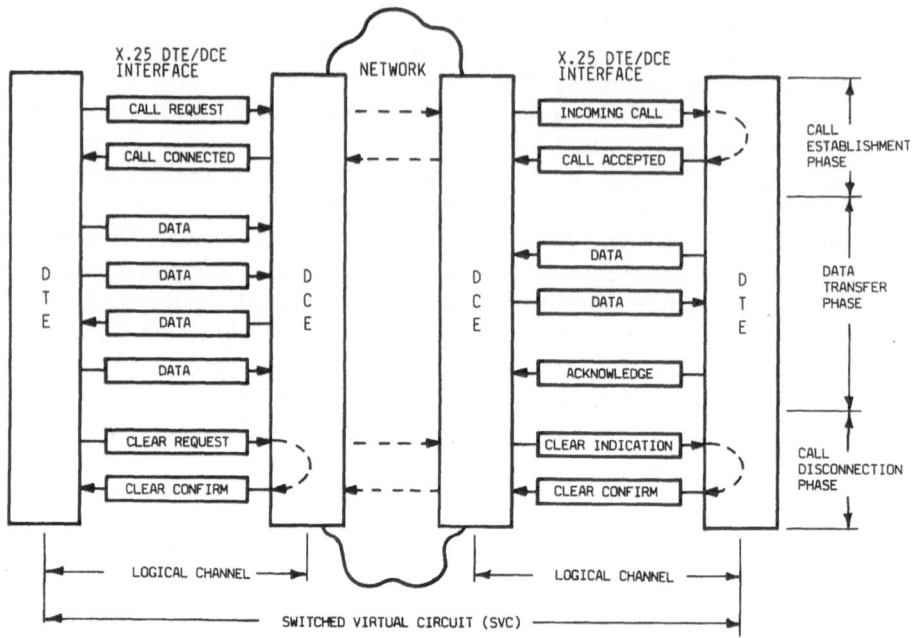

Figure 10.9. Normal X.25 packet sequence.

Figure 10.9 shows the sequence of packets that flow across the DTE/DCE interfaces at both ends of the network. The originating DTE initiates the call by sending a *Call Request* packet to its DCE, which then notifies the destination DCE according to the internal procedure of the particular network. The destination DTE is notified with an *Incoming Call* packet, which is identical to the Call Request packet but flows from DCE to DTE. The remote end indicates acceptance with a *Call Accepted* packet, which is then translated into an identical *Call Connected* packet from the originating DCE to its DTE. This completes the establishment phase and the SVC is now set up and ready for data transfer in both directions.

Once the data transfer phase is complete, either DTE can initiate call clearing by sending a *Clear Request* packet to the respective DCE. Upon receiving this packet the corresponding DCE will return a *Clear Confirmation* packet, and, at the other end of the network, the DCE will send a *Clear Indication* to its DTE. Finally, this DTE sends a *Clear Confirmation* packet back to the network DCE and the disconnection phase is completed. This disconnection sequence is shown in Fig. 10.9. It is also possible for the network DCEs to initiate call clearing by sending a *Clear Indication* to the DTE, which then should respond with *Clear Confirmation.* Our introductory discussion illustrates the basic operation of X.25 at the network level,

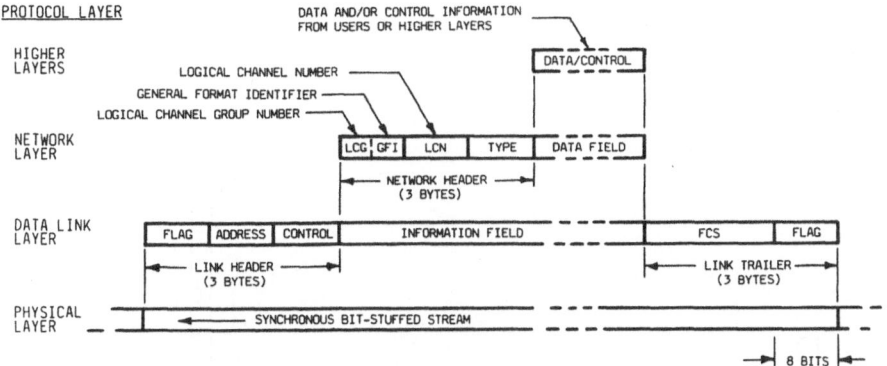

Figure 10.10. X.25 packet and frame relationship.

but there is of course a great deal more complexity. There are packets for control functions as well as for carrying actual data from higher protocol layers and users. Regardless of the type, however, all packets have a header of at least three bytes and are actually transmitted across the DTE/DCE interface as the information field of a data link layer I frame as illustrated in Fig. 10.10.† In general, the third byte always identifies the packet type, while the second byte contains the logical channel number (LCN) and the first byte indicates the logical channel group (LCG).

The *Call Request* (CR) and *Incoming Call* (IC) packets are identical in format as shown in Fig. 10.11a, with the type distinguished by the direction of flow, i.e., DTE to DCE and DCE to DTE, respectively. The DTE addresses are indicated by first providing in byte four a four-bit address length for both the calling and the called DTE, and then providing the actual addresses in the indicated number of bytes beginning with byte five. Coding is in BCD and, should there be an odd number of BCD digits, the final byte is zero-filled. The actual DTE addresses are necessary in this initial packet in order to set up the virtual circuit.

Following the address field comes the optional *facilities field.* There may be one or more facilities requested within this field of up to 63 bytes; or there may be no facilities, in which case the field is omitted. A large number of facilities are available with X.25,‡ dealing with such matters as channel access and direction, closed user groups, reverse charging, flow control windows, acknowledgements, and data rates.§ Additional user

† Note that the packet formats are drawn left-to-right to agree with the frame formats; the X.25 Recommendation indicates packet right-to-left in its illustrations.
‡ CCITT recommendation X.2 lists the facilities for X.25.
§ The X.25 virtual circuit data rate in bps is called the *Throughput Class.*

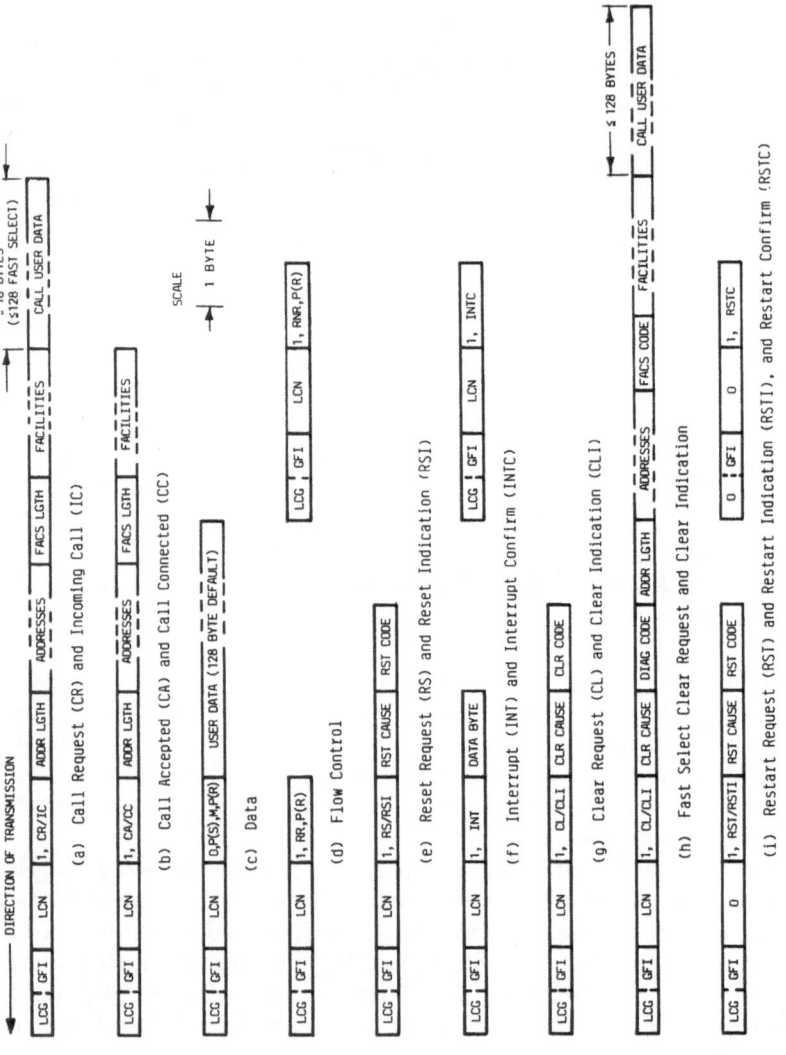

Figure 10.11. X.25 packet formats.

facilities may also be defined by the specific network. The first byte of this field gives the length in bytes, and then each individual facility is indicated successively by an initial byte giving both the *type of facility* and the number of *parameter bytes* following. For example, if a data field length other than the default value of 128 bytes is desired, then the *Packet Size* facility may be used. The first byte identifies this facility and also specifies two following parameter bytes giving a code for the proposed data field size of the called DTE and the calling DTE, respectively.

Another important facility is *Fast Select*, which has a single parameter byte. This allows the DTE to send up to 128 bytes a *call user data* or higher-layer protocol control information along with the initial Call Request and Incoming Call packet. This can save a great deal of time and bandwidth when there is predominantly short, bursty traffic to be sent over an SVC. It also allows the other end of the network to return up to 128 similar bytes in either the Call Accepted and Call Connected packet, or a Clear Request and Clear Indication packet if the transaction is completed. Even if Fast Select is not requested, it is still possible to include up to 16 bytes of data and control information in the optional *call user data field* that concludes the usual Call Request and Incoming Call packet.

The format of the *Call Accepted* (CA) and *Call Connected* (CC) packets is shown in Fig. 10.11b, and again the type is distinguished by the direction of flow. This format is similar to CR/IC but the call user data field is absent except for Fast Select. Some facilities, such as data rate, packet length, and pacing window size, are *negotiable* between the calling and called DTEs. For example, every network will have a predetermined default throughput class value which will be used unless a different class is requested via the corresponding Call Request facility. If the called DTE does not agree with the newly proposed value in the Incoming Call packet, it may indicate a different rate in the facilities field of its Call Accepted reply. However, in any facility negotiation, this counterproposal must have a value *closer* to the default value but can never be beyond it. Thus if the default rate is 2400 bps and the calling DTE proposes 600, the called DTE could counter with 1200 or 2400 but never with 4800 or 300. The absence of a counterproposal in the facilities field implies acceptance of the originally proposed value. It should also be noted that some facilities, such as *Throughput Class Negotiation* and *Reverse Charging*, are subscribed to at the initial network service connection, while others, such as Fast Select and *Window Size*, are determined on a per call basis.

In the data transfer phase of X.25 the *Data* packet is used for transfering data from higher protocol layers. As Fig. 10.11c shows the *type field* (byte 3) is similar to the HDLC control field, in that there is a *Send Sequence Number* P(S) *and a Receive Sequence Number* P(R). These three-bit fields allow sequencing to be done modulo 8, although an extended modulo 128 sequencing option is also possible. †Unlike the data link, however, sequence numbers are used here for *sliding window flow control* rather than error control. The default window size is two; i.e., there can be a maximum of two unacknowledged data packets outstanding at any time in each direction. However, the window size is a negotiable facility, with the negotiations, of course, always being towards two.

There are several flag bits associated with the data packet. First is the *Data/Control flag*, which is the first transmitted bit of byte 3. It will always be 0 for data packets and is set to 1 for all other packet types. The M bit between the P(R) and P(S) fields is called the *More Data* mark, and when set indicates that more data of the current message follows in additional packets. In the final such packet this bit is reset to 0; it thus operates in a manner analogous to the F bit with SDLC, but on an end-to-end basis. In the General Format Identifier (GFI) half of byte one, there are two flags called Q and D in addition to the two bits indicating modulo 8 or 128 sequencing. For normal data traffic Q is set to 0. However, a second level of data transfer may be distinguished by setting Q to 1. A common use of this two-level transmission is to send control information between PADs. The Appendix considers the use of the Q bit with PADs in some detail.

The X.25 sliding window *flow control* normally uses piggyback acknowledgements with FDX operation, but there are also specific *Flow Control* packets. The formats are shown in Fig. 10.11d. *Receive Ready* (RR) packets are used for positive acknowledgements and *Receive Not Ready* (RNR) for indicating a temporary busy condition, again analogous to the data link protocol. An RNR condition is normally cleared with an RR packet. Receiving a value of P(R) allows the sending station to advance the window up to, but not including, the sum of P(R) and the agreed window size. This flow control is illustrated for a window size 3 in Fig. 10.12. There is also a DTE Reject (REJ) packet for requesting retransmission of all packets starting with the sequence number in the P(R) field; however, REJ may be used only when the Packet Retransmission facility is selected. Another flow control mechanism uses the *Delivery Confirmation* (D) bit in the GFI field. When D is set to 1, the values of P(R) are actually obtained from the remote DTE, resulting in end-to-end acknowledgement of packet delivery; otherwise the local DCE is assumed to provide the P(R) values and then is responsible for error-free delivery across the network.

In more drastic situations the flow control mechanism can be reinitialized using a reset sequence of packets with the possible loss of existing packets. Either DTE can send a *Reset Request* (RS) packet to its DCE, which in turn will pass it across the network where it is received as a *Reset*

---

† This is indicated by two bits in the four-bit General Format Identifier, in the first byte of every packet. Code 01 indicates modulo 8 while 10 indicates modulo 128.

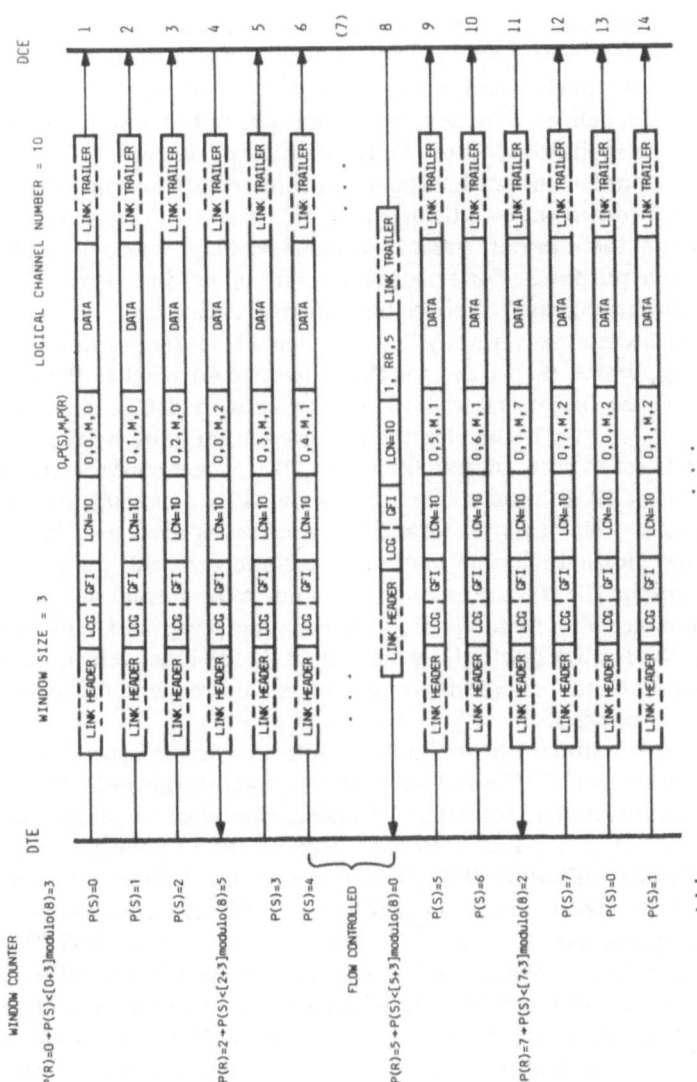

Figure 10.12. X.25 sliding window flow control.

Indication (RSI). Both the local DCE and the remote DTE will then normally respond with a *Reset Confirmation* (RSC) packet. The packet formats are shown in Fig. 10.11e, where the cause and diagnostic bytes give the reason for the resetting. Finally, the flow control can be bypassed entirely through the use of an *Interrupt* (INT) packet, shown in Fig. 10.11f. This is simply a nonsequenced packet containing a *single byte* of data. Interrupts are transmitted and delivered in the normal data stream, but they are not subject to the virtual circuit flow control. Only one such packet can be outstanding at a time, i.e., until acknowledged by a return *Interrupt Confirmation* (INTC) packet. An interrupt allows the transfer of limited information between users outside the normal stream of data flow for various control or administrative purposes. For example, it might be used to clear a deadlock situation without having to reinitialize the virtual circuit.

Virtual circuits are normally cleared when a DTE sends a *Clear Request* (CL) packet, which then traverses the network and reaches the opposite DTE as a *Clear Indication* (CLI) from the network DCE. The format is shown in Fig. 10.11g. If the clearing is initiated normally by the DTE the *cause* byte is set to zero, but otherwise the CLI packet contains the reason for the DTE or DCE initiated clearing followed by a *diagnostic code* byte. In any case, a *Clear Confirmation* (CLC) packet is used to acknowledge the clearing. Normal clearing may also be combined with the Fast Select facility using the Clear Request and Clear Indication format of Fig. 10.11h, which combines the features of both call accepting and call clearing. Following the facilities field there may be up to 128 bytes return user data, after which the virtual circuit disconnection is finalized with the usual Clear Confirmation packets.

Major error situations are handled with the restart procedure, which immediately clears all SVCs and resets all PVCs across the *entire* DTE/DCE interface. Thus restart affects all active logical channels. A restart is initiated either by any DTE sending a *Restart Request* (RST) packet or the DCE sending *Restart Indication* (RSTI). The format for both is shown in Fig. 10.11i, with the cause and diagnostic codes indicating the reason for the reset. The appropriate response is a *Restart Confirmation* (RSTC) packet in either case. After a restart, all SVCs must be reestablished in the normal manner and all PVCs are in their initial state. Obviously, user data may well be lost during a restart. There is also a special X.25 *Diagnostic* packet which can be used to indicate the cause and explanation for various error conditions.

There are a number of additional details of the X.25 network layer protocol such as timeouts, count limits, and state transitions, but the preceding discussion gives a fairly complete introduction to the fundamental characteristics. The operational example in Fig. 10.13 should help to further clarify the uses of the various packet types at both the calling and the called

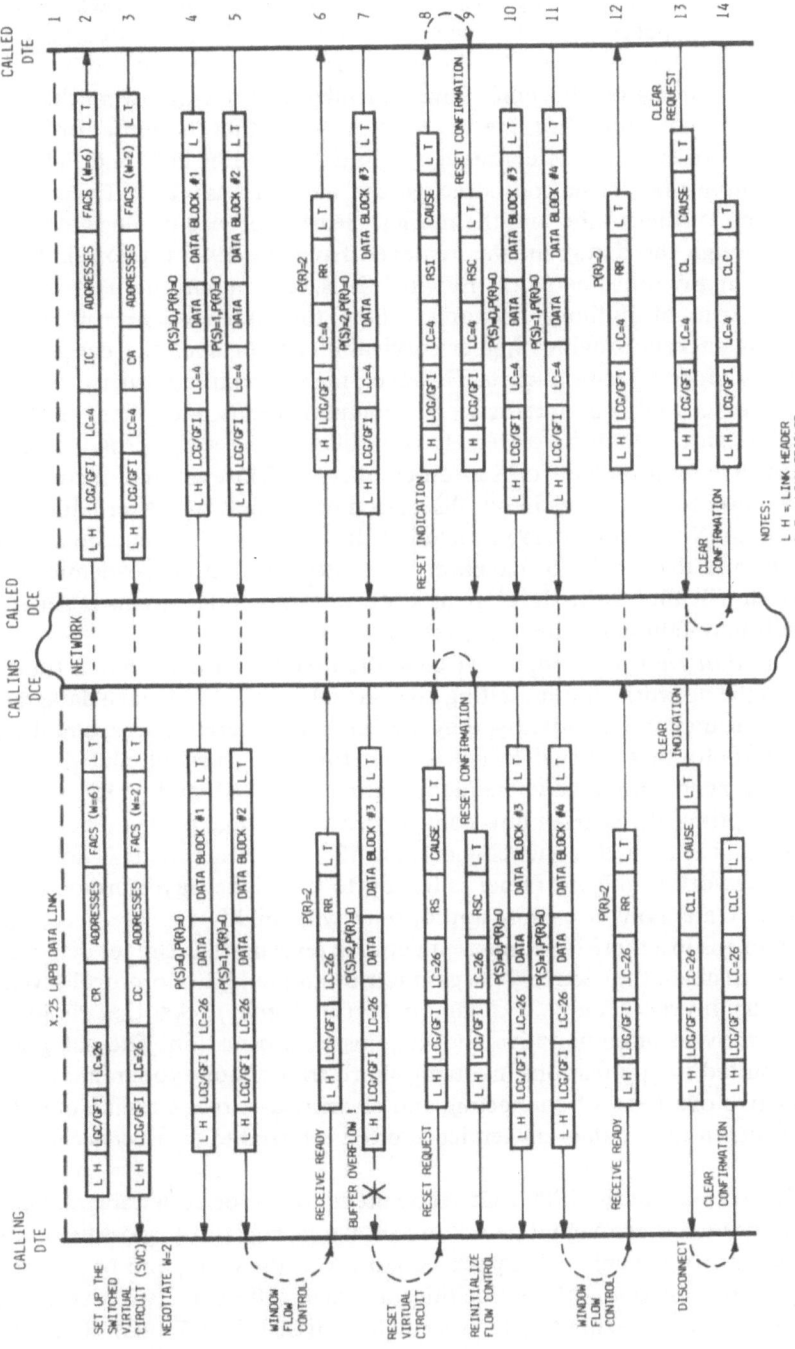

Figure 10.13. X.25 packet interface operation.

ends. The calling end sets up a virtual circuit using logical channels 26 and 4 and also proposes a window size of $W = 6$. This is negotiated back to the default value of $W = 2$ by the called DTE on line 3. As traffic flow progresses the channel becomes flow controlled by the window mechanism after line 5 and again after line 11. Note on line 7 that the buffer overflow results in a reset, with the consequent reinitialization of P(S) and P(R). In this example the lost packet is recognized and retransmitted. Eventually, there is no further traffic and the normal clearing procedure ensues.

Although the datagram was removed from the 1984 version of X.25, there are many other networks in which it is very important. This is particularly true of military networks where the independent routing of datagram packets provides higher survivability than does the contiguous routing of virtual circuit packets. However, many nonmilitary networks use datagrams as well. For internetwork communications datagrams are also attractive, since it can be extremely complicated to set up virtual circuits that traverse multiple networks of different types. Thus we shall include an introduction to datagrams in our discussion of the network layer. Since the versions of X.25 prior to 1984 included the datagram in some detail, we shall discuss the old X.25 datagram here. Although rather academic, this approach will allow us to develop the key properties of datagrams in general and still maintain continuity of presentation in this section.

The *datagram* is a single self-contained packet of data that is routed through the network without setting up a virtual circuit. Since each datagram packet is routed independently, they may go over different paths, and there is no guarantee that they will be delivered in the order transmitted. However, there are send and receive sequence numbers, P(S) and P(R), which, although primarily used for flow control, provide normal packet sequencing modulo 8 or extended sequencing modulo 128. This sequencing is not done by the network itself, but rather is left up to the higher-layer protocols or the user. Datagrams are most appropriate with highly reliable networks, since if one is lost there is no inherent network level mechanism for detecting the loss. Instead, the recovery is again left up to the higher protocol layers. There was, however, an X.25 *datagram Service Signal packet* that could be used to provide administrative and diagnostic information. The datagram is well suited to applications involving short, transaction-type traffic where the relative overhead of connecting and clearing a virtual circuit would be inefficient. In fact, datagram service is often referred to as a *connectionless* service.

The formats of the 1980 X.25 recommendations for both datagram and the datagram Service Signal packet are shown in Fig. 10.14. Most fields are the same as for virtual circuit packets. However, the Q, D, and M flags are always zero and some of the facilities are also different. Two bits of the GFI still code the sequence number field lengths (3 or 7 bits). The final

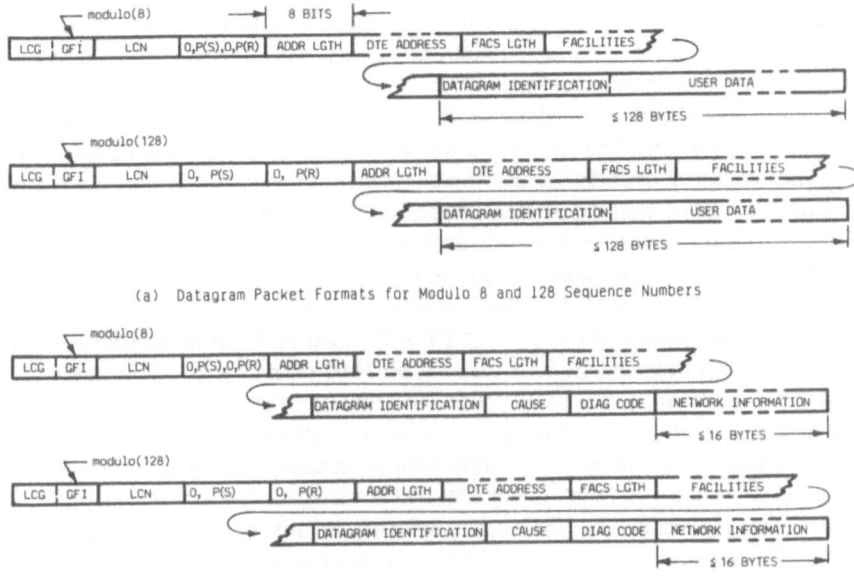

(a)   Datagram Packet Formats for Modulo 8 and 128 Sequence Numbers

(b)   Datagram Service Signal Packet Formats for Modulo 8 and 128 Sequence Numbers

Figure 10.14. Old X.25 datagram packet formats.

field of a datagram contains the user data beginning with an optional two-byte *Datagram Identification* field that is determined by the user. If included, it uses two of the 128 bytes available for data and is returned by the network as diagnostic information in the event of certain error conditions.

The datagram Service Signal packet provides information pertaining either to a *specific* datagram or to the *general* datagram service operation. These two packet types are distinguished in the two-byte Datagram Identification field, which is all zeros for the general case but in the specific case contains the first two bytes of the user data field of the datagram to which it refers. This is followed by single-byte fields containing the cause of the service signal and an amplifying diagnostic code. Finally there is a field of up to 16 bytes for additional information such as invalid fields of the erroneous specific datagram. There are three classes of datagram Service Signals-Specific. *Datagram Rejected* Service Signals indicate that the specific datagram was discarded and any retransmitted version should be corrected in accordance with the diagnostic information supplied. *Datagram Nondelivery* is an optional facility-selected signal that also indicates that the datagram was discarded and should be retransmitted later. Acknowledgements may be provided by Service Signals via the *Datagram Delivery Confirmation,* which is also available only when the corresponding facility has been selected. Datagrams are queued at each logical circuit in the order of arrival and discarded when the queue becomes full. However, Service Signal packets are placed at the head of the queues so that error conditions always receive priority treatment.

The original use of datagrams in X.25 was for short transaction messages. However, the feature was not actually implemented in operational X.25 networks prior to 1984, and the corresponding virtual circuit overhead for single-packet messages was reduced by inclusion of the Fast Select facility. Other non-X.25 network architectures including ARPANET and DNA have used datagrams exclusively.

We have now looked at the lower three layers of the OSI Reference Model for the most important CCITT X.25 standard. In addition to being widely used in practice, X.25 is fairly typical of many other packet network protocols. The network layer provides the means of transferring data across the network itself, and many operational networks provide only these first three layers. They are usually implemented in network equipment such as a nodal processor or FEP. From the user's viewpoint, however, additional protocol is necessary to efficiently perform various applications functions across the network. This is provided by the higher layers of the OSI model, which we consider in the next section.

10.5. High-Level Protocol

In this section we collect the remaining four layers of the OSI References Model. The major emphasis will be on the Transport Layer, which is crucial because it provides the bridge between the *network-dependent* layers (1-3) and the *network-independent* layers (5-7). We then consider the Session, Presentation, and Application layers in less detail. Although the basic versions of these standards have now been developed by OSI and other standards organizations, they have not been widely implemented. These upper three layers normally reside in the users equipment and are more specific to data processing than to the actual data networks which are our primary interest. In effect, we limit our detailed consideration to the transport subsystem which provides a uniform communication service for the user applications.

From a network viewpoint the *quality of service* provided depends on many factors, including the actual network design, network protocols, and the network operation. Thus each network will have a different inherent quality of service to offer the users. However, from the user's viewpoint some users may require better service than the network layer alone can offer. Specific services might include error detection and correction, failure recovery, flow control and sequencing, security, priority, throughput, and delay requirements. It is also very important for user-oriented protocol software to be producible independent of any specific network, since otherwise the entire distributed processing system would be nontransportable and, even worse, subject to any changes made in the network itself! Such mismatches in service quality and network transparency are particularly germane to public data networks, which must accommodate many diverse users.

The main function of the transport layer is to reconcile these inevitable differences between the actual network service and the end user requirements. Consequently, a general transport protocol must be able to provide a broad scope of possible services, many of which overlap with those sometimes found at the network layer. If the network layer is very good or the user does not require a relatively high quality of service, then the transport layer protocol may be quite simple; but if the opposite conditions prevail, then a more sophisticated transport layer is indicated. For example, with datagrams the network protocol is relatively simple, with no sequencing and minimal error control. For this type of network to provide high-quality service to the user session layer, it is necessary for the transport layer to do things like flow control, error recovery, sequencing and buffering, and acknowledgment. Conversely, for a virtual circuit network most of the above functions are done by the network layer, so they need not be repeated again. However, there are still some unique functions that must be done at the transport level. For example, X.25 data packets may be lost when there is a reset or restart. It is then up to the transport layer to keep up with the actual packets at each end and resolve any missing or duplicate packets.

In providing suitable end-to-end quality of service the transport layer also ensures that the session layer sees a uniform transportation system that is independent of the many details of the particular network being used. This is extremely important in practice, since it allows the higher-level protocol software to be written without regard to what network it will be using. The transport software typically runs under the host computer operating system and can be considered as a sophisticated driver routine in that it allows other routines to access a particular network. Thus the same host routines should be able to communicate without any modification over a datagram or virtual circuit packet switched network, a message switched network, or a bus oriented local area network. The transport layer also handles explicit addressing of the destination processes. In fact it is often desirable at this layer to multiplex several channels over a single network layer virtual circuit, thereby more efficiently utilizing the available bandwidth.

The OSI transport layer standard provides a representative example, since it is quite thorough and similar to several other comparable standards.†
The basic data unit, which goes into the data field of a network layer packet,

---

† The OSI standard is closely patterned after the ECMA Standard 72, and is also similar to U.S. NBS FIPS transport protocol. The corresponding CCITT Recommendations are X.214 and X.224.

is called a *Transport Protocol Data Unit* (TPDU). OSI defines ten different TPDUs for setting up, managing, sending data over, and terminating an FDX symmetrical connection between users. There is also a selection of optional and negotiable parameters to define precisely the specific services to be provided.

In the OSI standard there are five different *classes of service* (0–4) for various combinations of networks and users. The class may be negotiated during connection establishment in a manner analogous to that used for X.25 facilities; e.g., negotiations must always be towards the default class. Each class has some mandatory and some optional parameters, with the particular parameters depending on the quality of service to be provided. The five classes defined by OSI are as follows:

Class 0. This *Simple Class* assumes a high-quality network that provides sufficient error control, sequencing, and flow control. Thus it only provides the minimal establishment, data transfer, and termination capabilities. Class 0 is used, for example, with teletex where user requirements are not particularly stringent.

Class 1. The *Basic Error Recovery Class* is similar to Class 0 but assumes the network is failure prone. Thus there is an additional capability to recover from network layer resets. This is done by adding *sequence numbers*† to the TPDUs (modulo 2^7 or 2^{31}) so that lost TPDUs can be detected and retransmitted without the users being aware of the failure. There is also an *expedited data* capability for sending short urgent messages that bypass the normal TPDU sequencing.

Class 2. This *Multiplexing Class* again assumes a high-quality network layer as in Class 0. Here, however, multiple connections can be obtained over a single network layer virtual circuit by *TPDU-interleaved multiplexing.* Related to each such connection is a dynamically adjustable *sliding window flow control mechanism.* The window is advanced by acknowledgment or reject TPDUs which also contain the desired window size, or *credit.* Reducing the credit value decreases the flow of TPDUs on each transport connection individually, and the initial credit values are negotiable.

Class 3. This *Error Recovery and Multiplexing Class* provides for failure recovery while multiplexing. TPDUs lost in a network layer failure are detected by sending the next expected sequence number in a reject TPDU, and then retransmitting all missing numbers to resynchronize the connection. Multiplexing is done as in Class 2.

Class 4. The *Error Detection and Recovery Class* assumes a worst case network with unacceptable error and failure rates. Thus this class provides full error detection and recovery features, including checksums, time-outs

---

† These are, of course, entirely independent of, and different from the packet sequence numbers in the network layer.

and retransmissions, recovery from loss of data and control TPDUs, expedited data, and user notification of nonrecoverable errors. There are, of course, multiplexing and window flow control for each connection. Class 4 is applicable to datagram networks where the conventional network layer functions must be assumed by higher protocol layers.

All five classes together provide a complete selection for quality of service. It is interesting to note that the NBS version of the protocol uses only classes 2 (Basic) and 4 (Enhanced), and includes a connectionless (datagram) capability called *unit data* service.

All TPDUs start with a *Length Indicator* (LI) byte indicating the size of the remaining header. The next byte always identifies the TPDU *type*, and the third byte is the flow control *credit* (window) value where applicable but otherwise is zero. This may then be followed by a number of possible fields, depending on the TPDU type. Figure 10.15 shows the TPDU formats defined by ISO. In addition to DT as shown, the extended 31-bit sequence number formats are also used with ED, AK, EA, and RJ TPDUs. The *source* and *destination references* are just the transport layer connection addresses assigned by each end. Sequence numbers for *Data Transfer* (DT) and *Expedite Data* (ED) TPDUs are either seven bits or, for the extend format, 31 bits. They are preceded by an EOT flag bit, analogous to the X.25 M bit, which is set to 1 to indicate the last TPDU in a message sequence. Since only one ED TPDU can be outstanding at any time, the corresponding EOT is always unity. The *Connection Request* (CR) and *Connection Confirm* (CC) TPDUs contain an additional byte to indicate the preferred class of service, while *Disconnect Request* (DR) and *Error* (ER) include explanation fields.

In addition to the *fixed-length* part of the header there is an optional *variable-length* part that includes the various parameters and, when used, the checksum. Finally, if data is being transferred in DT or ED TPDUs, then this data follows the header subject to any applicable length constraints. Limited amounts of user data can also be carried in the CR, CC, and DR TPDUs for all classes except Class 0.

To illustrate the operation of the transport layer protocol, consider Fig. 10.16, which assumes an unreliable network. The transport connection is set up over the established virtual circuit using CR and CC. The initial flow control window credit is set at 2, Class 4 is specified, and the parameters indicate expedited data is allowed. Then data transfer proceeds until window flow control occurs after line 5. Normal data flow is interrupted for an urgent message using ED and EA on lines 8 and 9. This is followed by a failure of the underlying network layer, and the fourth data TPDU is consequently lost. The virtual circuit reset is eventually signaled to the transport protocol, which then retransmits the unacknowledged TPDU and

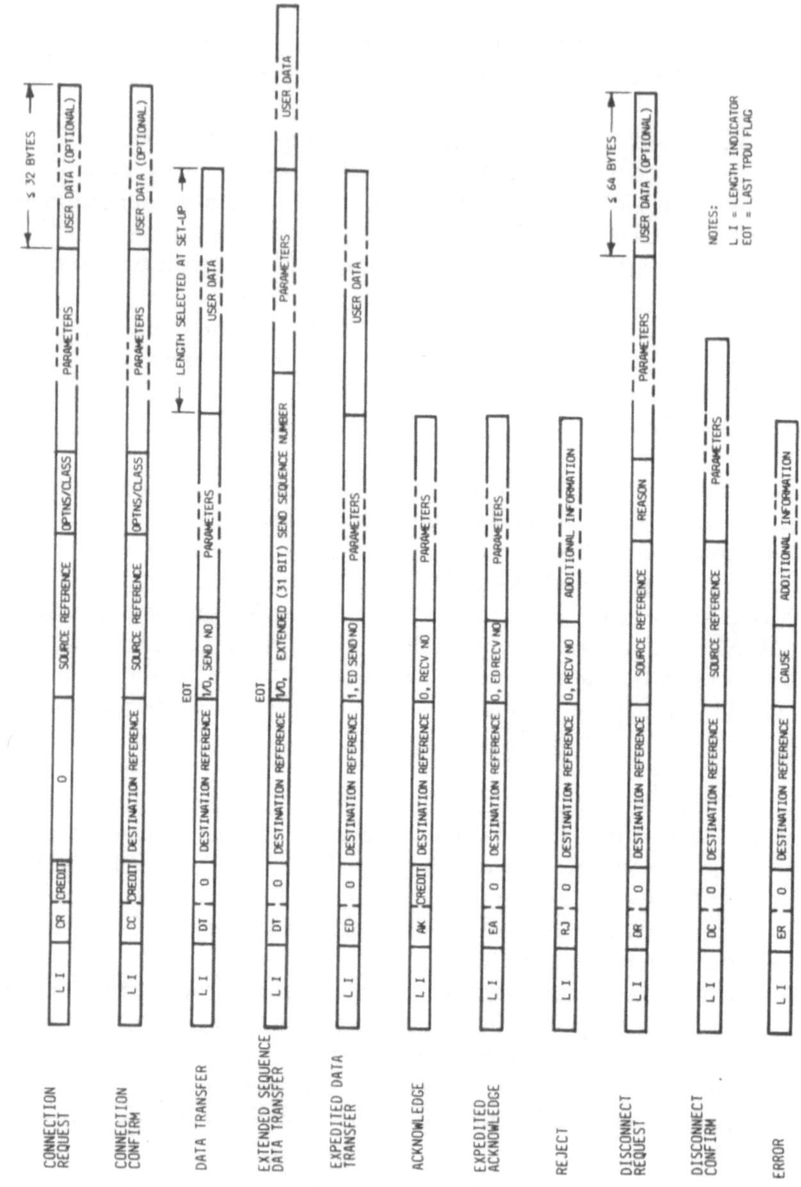

Figure 10.15. OSI transport layer TPDU formats.

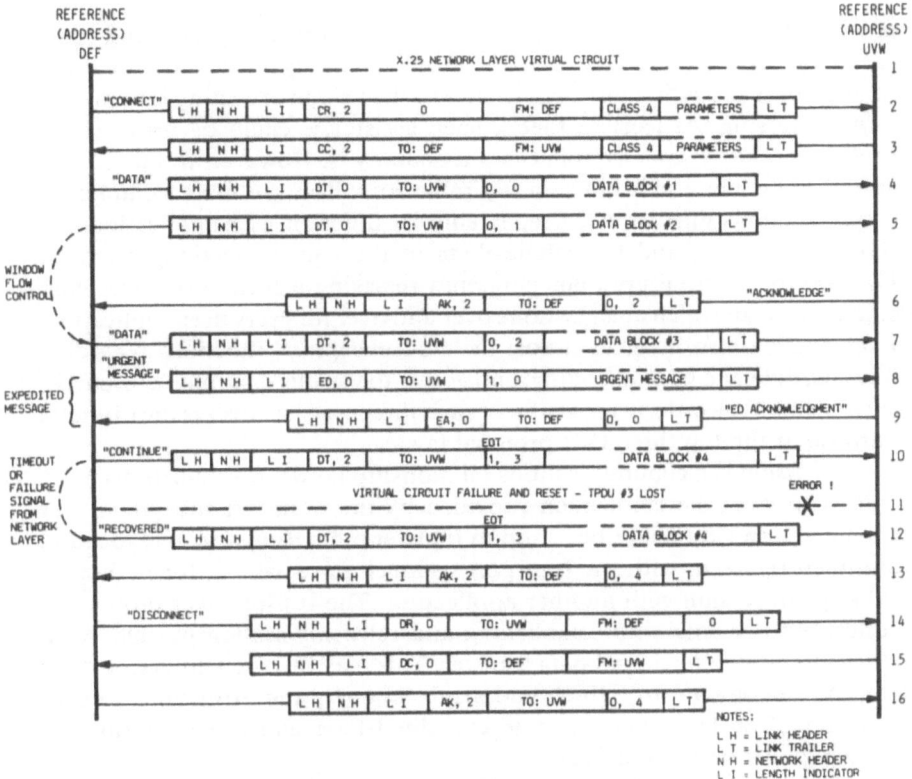

Figure 10.16. Operation of OSI transport layer.

resumes normal operation. The EOT bit in line 12 indicates the end of the TPDU sequence, and the connection is terminated. The final AK provides a positive indication that the disconnection is complete. Despite all the discontinuities in the actual data flow the transport protocol makes the connection appear smooth and continuous to the higher-level users.

With this introduction to the services and OSI protocol of the transport layer, we have completed our treatment of those protocol layers that apply specifically to wide area data communications networks per se. Before going on to the higher-level user layers, it will be helpful to look back over the preceding sections of this chapter and tie together all the different protocols. The envelope analogy should be readily apparent with each layer providing a protocol envelope for the next higher layer. Furthermore, it is now clear that the internal realization of each layer may vary considerably, just so long as the procedures for the interaction between layers, or the *interface*, is consistent and well defined.

Having now established the entire *transport subnetwork* that provides a suitable quality of data communications service, we now turn to those

protocol layers which use the transport services for data processing functions. The Session, Presentation, and Application layers can be thought of
intuitively as providing *direct* assistance to the user in executing his application processing. At least in theory such assistance could be provided by
each user, and in some simple cases it actually is. However, in the general
case it is far more efficient and desirable to collect the more complex of
these functions that are used repeatedly by different user applications, to
standardize them, and to include them in the communications protocol.
For example, setting up, managing, and releasing a session involves considerable detail which must be done over and over for every user application.
Each session also involves a host of "housekeeping" functions regarding
formats, data and control coding, and access to various distributed resources
such as files. The orderly provision of such direct user services is the essential
purpose of the top three OSI protocol layers.

A *session* is a communications relationship between applications users
and exists over the connection provided by the transport subsystem. For
example, a session might be set up to transfer a file from a remote location
and then released, with the transport connection remaining for perhaps a
subsequent session with another application. The session layer protocol is
concerned not only with establishing and releasing session but also with
efficient and uninterrupted data transfer. This may involve addressing, flow
control, dialog control (which end sends when?), and error management
functions. The latter may include expedited data and checkpointing for
unambiguous failure recovery.

There are now several well-defined standards for the session layer,†
with the OSI version again being fairly representative. Thus, without going
into much detail, we shall use it as a guideline for our introductory treatment.
The basic unit of data is called a Session Protocol Data Unit (SPDU),
which of course goes into the protocol envelope of a TPDU.

There are SPDUs appropriate to the various session functions, and in
the OSI case they have been organized into four groups. The *Session Kernel*
includes SPDUs for establishment, data transfer, and disconnection, which
is always implemented. The *Basic Combined Subset* (BCS) adds additional
TPDUs for HDX dialog, expedited data, reporting of exceptional situations,
and negotiated release. A *token* scheme is used to determine which end
gets to transmit, and procedures are defined for acquiring a token under
various conditions.

The *Basic Synchronized Subset* (BSS) primarily adds synchronization
services for the dialog flow. For example, when very long files are transferred
contiguously over conventional networks, there is a reasonable possibility
that eventually there will be a failure that will result in lost data despite

† The ECMA version is defined by ECMA Standard 75.

the best efforts of the transport subsystem. The BSS of the session protocol guards against this potential loss of data by systematically inserting *synchronization marks* in the data stream. These marks then provide a definite checkpoint known to both ends and to which each may return and resume the session dialog without duplication or loss after a failure. The most recent confirmed mark is always taken as the resynchronization point. There is also an SPDU for sending urgent messages outside normal traffic flow.

The most extensive TPDU group is called the *Basic Activity Subset* (BAS), which basically provides management for a collection of *activities* that constitute the overall dialog. Finally, in the event that the session protocol is unable to maintain a session, e.g., a host crash, there is an SPDU for reporting the condition to the users.

The session layer is obviously closely related to the transport layer, adding features to it that make the user's life easier. Some network architectures, e.g., ARPANET, do not have any session layer, while in others, like DNA, it is minimal. The OSI standard includes a selection of subsets to accommodate a range of networks and users, and represents a combination of features from the ECMA and CCITT session standards.

The presentation and application layers are also closely related, both in function and in form. Both consist essentially of collections of direct user services that interact directly with the user to greatly increase the efficiency of running application programs. As the OSI Reference Model indicated, the presentation layer is concerned with the representation, or *syntax*, of the information being transmitted, while the application layer is concerned with its actual meaning, or *semantics*. Existing public standards at these layers have not yet become as broadly accepted as have those for the lower layers, but certain functions such as protocol conversion and remote file transfer are implemented in some form by most sophisticated users.

The *presentation layer* has been perceptively described by one commentator as "a general purpose layer that can do a lot for you"; this is not an understatement. It handles a variety of formatting, character coding, graphics, code conversion, data compression, and other related display services. It also includes, at least conceptually, encryption, videotex, and terminal handling protocols using the virtual terminal and PAD approaches. Encryption and PAD terminal handling are considered in separate sections later, so they will not be developed here.

In order for each end user of a data communications session to understand the *abstract* information being conveyed, it must be represented in *concrete* forms both during transmission and during processing and display. The negotiation and management of precise and appropriate representations for this information is the essential task of the presentation layer protocol. Once agreement is reached, the protocol may just pass traffic between the session and application layers during normal operations. No attempt is made to interpret the meaning of the actual data transferred. This is somewhat like the referee in an Olympic prize fight; once the language and rules are established and the fight (communication ?) begins, he does not interfere so long as everything proceeds normally.

As the diversity of presentation services might suggest, there are many ways to implement the protocol for different types of applications. For example, the OSI standard† defines several presentation *facilities*. The *connection* facility deals with the syntactic negotiations and agreements that are necessary to ensure that a coherent session can take place between applications. A *context* facility includes a collection of common data representation conventions that are widely used in practice and thus incorporated into the protocol for user selection. There is a basic context above which other more specific contexts may be negotiated. Once data transfer begins, the *information-transfer* facility defines special control and management functions, e.g., expedited data. Termination of the presentation connection, both normally and abnormally, is handled by the *termination facility*. The OSI also defines a *standard abstract syntax* for specifying application protocols in a form easily understood by humans, which can then be converted into a concrete syntax for actual machine use.

The use of applications requiring sophisticated presentation services is growing rapidly. For example, *videotex* is now quite common in many countries for distributing text and graphical information over television or telephone channels for ultimate display in the home. There are now several independent graphics standards for videotex, mostly along national lines. In the United States, the version receiving the most attention is descriptively called the North American Presentation-Level Protocol Syntax (NAPLPS). Adopted by ANSI and CCITT, this standard defines several extensive sets of text and graphics symbols, along with the corresponding binary coding for each character. There are also codes for control characters and color selection, plus rules for mixing the primary colors and positioning the resulting text and graphics on a standard grid for display. Using NAPLPS a home or office user can represent all kinds of pictorial and textual information in a standard coded form for display, storage, and transmission. This can be done over any data network capable of supporting NAPLPS at the presentation layer. Other videotex presentation standards, such as Prestel in England, CAPTAIN in Japan, and particularly Telidon in Canada, operate in a similar fashion.

The highest layer in the OSI model is the *application layer*, which is where the user interfaces with the network. This layer directly serves the

---

† The CCITT and OSI standards are essentially the same, but differ significantly from the corresponding ECMA Standard 85.

user's information processing and transmission requirements by providing services necessary for application processes to work together. These services are typically realized as software tasks running directly under the user operating system.

In order for a collection of high-level *application processes* to complete a user task, there must be complete understanding and cooperation among them. This requires agreement on the meaning (semantics) of control and data characters. For example, with a remote file access the file must be addressed, the desired records located, and the resulting information encoded and transmitted in a known format, then converted to the format expected by the user. There must also be protocol for handling unusual situations such as failures, and the file must finally be closed. Both ends recognize the same data characters, e.g., an ASCII end-of-file, as well as the same protocol service instructions.

In the OSI application layer, standard protocol *service elements* are defined that describe the services provided by the application layer. This "bag of services" is divided into two categories. *Common Application Service Elements* (CASE) provide capabilities common to all application processes, including setup and termination of the process associations, interrupts and error recovery, and accounting functions. The other category, called *Specific Application Service Elements* (SASE), includes services for specific application processes such as remote file access and transfer, distributed computing, virtual terminals electronic message handling, and electronic funds transfer. In addition to defining the CASE protocols, OSI has developed specific protocols for remote files called File Transfer, Access, and Management (FTAM); for an extension of remote job entry called Job Transfer and Manipulation (JTM); and for Virtual Terminals (VT).

The VT application layer protocol defines an imaginary terminal with a standard set of characteristics into which every user terminal protocol must be converted in order to access the network resources. The user is generally responsible for this mapping, but only one map is required since the virtual terminal does not change. At the destination terminal or host, communications from all network terminals appear in the same protocol (the VT) regardless of the originating terminal type. This obviously reduces enormously the terminal handling requirements of the host. It also makes adding new terminal types much simpler, since no terminal protocol changes are required except for the new terminal itself.

A virtual terminal involves both the presentation and application layers, since terminals not only display but also produce and communicate information. Much of the early VT development was related to the ARPANET Telnet virtual terminal protocol, but other organizations including ISO now have comparable standards. For example, the ISO version is based on a generic VT model which describes the abstract concepts and services to be provided. From this model several classes of terminal service are developed, including basic, forms, and graphics classes. The *Basic Class* is for the familiar character-oriented terminals operating in a line, scroll, or page mode. Conventions for HDX and FDX operation, various control characteristics, and data transfer are defined. The *Forms Class* provides comparable protocol for form-oriented terminals, where the user enters data by filling out a form displayed on the screen. Other classes will eventually be defined in a similar way.

The application layer protocol is more difficult to define comprehensively than the other layers, since it is essentially a diverse collection of whatever services are appropriate for the users. The implementations are also more complex, since they involve applications data processing as well as repetitive communications tasks. For the more fundamental application layer services, however, suitable protocols standards are now available and new standards will continue to be added to accommodate additional user requirements as they arise.

This section completes our basic study of network architecture. However, there are many more extensions and variations to this broad subject. So far we have only looked at public protocol standards. The next section provides a contrast by considering several important proprietary architectures.

10.6. Proprietary Network Architectures

Historically, network architectures were developed by the major computer vendors. This was a natural course of events, since essentially all digital networks of any size connected computers supplied by a single vendor. It was clearly to the computer vendor's advantage to add some communications capabilities to the existing system software and computer hardware. It also made sense to develop proprietary protocols so that entire product lines could easily communicate between devices. The result of this evolution was the existence of various network architectures that were appropriate for each vendor's particular market and range of capabilities but were incompatible with each other. For example, IBM developed SNA originally for a centralized network with a single mainframe computer, while DEC oriented DNA towards a highly decentralized network of peer minicomputers.

Such situations tended to lock each user into a single vendor. As digital network complexity and pervasiveness has grown, there has been an expanding need for more compatibility and new protocol standards open to all users. As a result, standards organizations like OSI and CCITT are deeply involved in protocol developments such as those of the preceding sections.

In addition, many proprietary network architectures are adding new capabilities to support *interconnection* and *interoperation* with other protocols. This is particularly true of X.25, which is now supported by practically all major network architectures. In fact, many computer vendors have replaced old protocols with the corresponding layers of X.25, and there is every indication that this trend will continue. Many national data communications networks are also now adopting international standard protocols like X.25 and the OSI upper layers, thereby ensuring easy future interconnection with other open networks and a wide availability of compatible equipment.

In this section we will consider in some detail two of the most important proprietary network architectures: the IBM SNA and the DEC DNA. Both of these network architectures are important in their own right, and are widely used in practice. Additionally, they provide a representative indication of complete operational approaches to network architecture and an instructive comparison with the OSI Reference Model. We shall focus here on the network aspects of these architectures, but it should be borne in mind that they are intimately related to the system software of the respective computer equipment that is interconnected by the network.

Digital Equipment Corporation's *Digital Network Architecture* (DNA) was first introduced in 1975 to interconnect DEC minicomputers for real-time operations. It has since gone through several major phases of revision and expansion, and is currently in Phase IV [4]. The actual software is called DECNET, and it consists of various modules corresponding to different protocol layers. The architecture allows for a wide range of operation, e.g., HDX or FDX, synchronous or asynchronous, point-to-point or multipoint, and with any topology. It is highly decentralized, with adaptive routing, and uses datagrams internally. Thus there are no virtual circuits, but the datagram (connectionless) association is called a *logical link*, which should not be confused with a data link or virtual circuit. All nodes have the same (peer) status in the protocol. There is no master–slave type of relationship except on multidrop links, and there is no separate backbone network either. Phases I and II of DNA required point-to-point lines between every pair of user nodes (minicomputers), and were oriented towards real-time data acquisition and control. However, Phase III, announced in 1980, was a true packet network based on datagrams. Phase IV extends the network in several ways, including support for up to 1023 nodes, a gateway capability to IBM SNA networks, and the ability to operate integrally with X.25 networks and with Ethernet Local Area Networks.

The layers of DNA and the corresponding OSI layers are shown in Table 10.2, along with a brief indication of the main DNA functions. At the data link and physical layers are now three separate alternatives. The conventional protocols are DDCMP and either RS-232C or RS-449, but

Table 10.2. DNA Protocol Layers

OSI layer	DNA layer	Typical DNA function
7 Application 6 Presentation	Network Management and Network Application	User interface to network DEC and user software modules Remote file operations (DAP) Network management/testing (NICE) Downline loading/execution (MOP) Virtual terminal capability (NVT) User-to-user communications Gateway functions for X.25 and SNA
5 Session	Session Control	Assistance in logical link setup Logical link addressing functions Transparent during data transfer Some authorization/security functions
4 Transport	End Communications	Creation and termination of logical links End-to-end flow control Integration of DNA and host OS Error control and recovery Message segmentation/reassembly Network Services Protocol (NSP) Data integrity assurance Task-to-task communications
3 Network	Routing	Datagram routing end-to-end Dynamic routing calculation Packet lifetime counting Congestion and buffer control
2 Data Link 1 Physical	Data Link Control Physical Link	Error-free transfer between nodes Multidrop line operation Three alternatives in Phase IV: ○ DDCMP and RS-232C/449 ○ X.25 (HDLC/LAPB and V.21/V.24) ○ ETHERNET LAN

Phase IV has added the options of using either X.25 (HDLC and X.21 or V.24) or replacing the conventional wide area telecommunication network with a high-speed Ethernet LAN. The operation of DDCMP and HDLC have been studied previously, and Ethernet will be covered in detail in the next chapter. The third DNA layer used to be called the Transport Layer, but this was confusing with the OSI layer four, so in Phase IV it is more descriptively called the *Routing Layer*. This layer basically does end-to-end datagram routing based on a least-cost-path algorithm computed at each node using path cost and length information obtained from the exchange of Routine Control Messages among adjacent nodes. The underlying presumption is that the concatination of all locally optimal routes will result in a suitably optimal end-to-end route. The Routing layer also provides congestion control by discarding excess packets to limit the outbound buffer sizes. It detects routing loops by counting the number of nodes each packet traverses and discarding any packet for which this "nodes visited" limit is exceeded. Since this is true datagram routing, there is no guarantee that the datagrams will arrive in sequence.

The fourth layer was formerly called the Network Services Layer, but, again to avoid confusion relative to the OSI terminology, is now called *End Communications*. As with the OSI Transport Layer, this layer provides the required data integrity for the underlying datagram network. It operates over logical links between user tasks and includes such functions as message segmenting, packet sequencing, flow control, error control, and source and destination node addressing. Flow control may or may not be used in DNA. If it is used, then during initialization the receiver indicates to the sender how many segments (the layer four data unit) it will accept. This number is, in effect, the window size for flow control that allows up to this number of unacknowledged segments to be outstanding. The *Network Services Protocol* (NSP) module is the program that executes the layer four protocol. NSP includes data messages for interrupts and data transfer, acknowledgment messages of various types, and control messages for logical link establishment and disconnection. Headers are added by each of the Data Link, Routing, and End Communications layers, as shown in Fig. 10.17. The *Routing Header* contains fields for source and destination *node* addresses and the *number of nodes traversed* count. The *layer four header*, in turn, contains the source and destination *process* addresses, plus a message

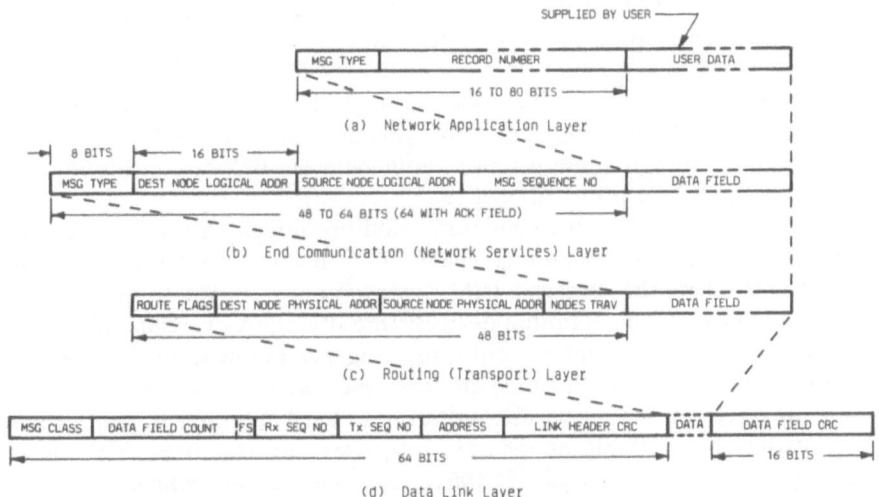

Figure 10.17. Typical DNA protocol layer formats.

sequence number and an optional end-to-end acknowledgment field for message segments. Notice that these three headers alone involve well over 150 bits of overhead.

In DNA the *Session Control Layer* interfaces both with user programs and with higher protocol. It was absent prior to Phase IV and is still rather weak relative to the OSI counterpart, being essentially transparent during data transfer. It performs various name/address translations in addition to the conventional user session establishment and disconnect functions. The remaining *Network Application* and *Network Management Layers* supply the user interface to the network protocol, and consist of a collection of services. The *Data Access Protocol* (DAP) provides remote file transfer and access. Downline program and file loading are handled with the *Maintenance Operation Protocol* (MOP), and the *Network Information and Control Exchange* (NICE) protocol provides various network housekeeping and management services to the user. There are also a *Network Virtual Terminal* (NVT) and gateways for SNA and X.25.

This introduction to DNA illustrates the nature of a rather specialized architecture. Despite being developed prior to the OSI Reference Model, the similarity is striking. This is due both to the influence of DNA on the standards efforts and to the continual evolution of DNA to an increasingly appropriate architecture for modern data communications networks. The overriding advantages of layering are apparent in this evolution.

System Network Architecture (SNA) [5] was introduced by IBM in 1974 in an attempt to standardize the communications methods for their existing and future product lines. Today, with over 20,000 SNA networks installed, it is pervasive. SNA is actually not only a set of protocols, but also includes a collection of associated IBM software and hardware products used in data networking. Although our main interest here is the protocols, it will be necessary to look at the entirety of SNA. Our presentation is further complicated by the fact that, although the concepts are straightforward, the SNA terminology and acronyms are horrendous! Thus we shall initially sort through some essential nomenclature before considering the details of the actual protocols per se.

Like all versatile data communications protocols, SNA has steadily evolved to meet the expanding needs of its users. Initial versions were oriented towards the classical IBM environment of a single main-frame surrounded by various terminal devices. Initially, SNA overcame the problem of only one terminal or multidrop line per application program by adding a software product called an *access method*. Running under the host operating system, an access method allowed each terminal to gain access to a variety of available application programs and resources. The initial such products were called *Advanced Communication Function* for the *Telecommunications Access Method* (ACF/TCAM).

As networks grew larger and more decentralized in the late 1970s, SNA was extended to accommodate multiple hosts, with each host pair being directly connected. Network resources were allocated to each host, and the host plus assigned resources was called a *domain*. Communications within each domain were managed by a SNA program called the *System Services Control Point* (SSCP) located with the host. The SSCP coordinated sessions between users within the domain, and also *cross-domain* sessions with SSCPs in other domains. The current evolution of SNA is toward a more distributed mesh topology with intermediate switching nodes interconnecting multiple host and remote terminals. There is also an increasing capability for SNA to support small computer systems and non-IBM network components, and to provide compatibility with non-SNA protocols such as X.25.

Fundamental to SNA is the concept of *subarea* and *peripheral nodes*. Domains are subdivided into one or more *subareas* for control and routing purposes, with each containing a *subarea node*. Subarea nodes are connected by single or parallel SDLC links, some of which may be operated as a single logical link for better reliability, capacity, etc. Such a related parallel set of data links (multilink) is called a *transmission group*. Peripheral nodes are typically terminals or cluster controllers, with limited networking capability. They communicate through a subarea node, such as a host or a communications controller. An *end user* in SNA is either a person or an application program, and is always represented in the network by a software program called a *Logical Unit* (LU) that executes the necessary high-level protocol functions. Only nodes with end users have LUs, but every node contains a *Physical Unit* (PU) program that monitors the nodes resources under control of the SSCP.

Network addresses are assigned to each LU, PU, and SSCP; consequently they are collectively called *Network Addressable Units* (NAU). Thus every node contains one or more NAUs. A *virtual route* is an FDX logical association between the pair of subarea nodes at each end of a session, with the underlying physical path and components called an *explicit route*. Routing tables are located in the subarea nodes. If the end user is at a peripheral node, then an additional *peripheral link* is required to connect to the virtual route at the corresponding subarea node to obtain a complete end-to-end connection.

Each network address consists of *subarea* and *element addresses*, with peripheral LUs only identified by the latter. Thus subarea nodes must convert from the network address to the *local address* of the peripheral. This is done by a program called the *Boundary Function* (BF). If the end-user is located at the subarea node, then no BF is required. The network of Fig. 10.18 illustrates these basic SNA concepts. There are two domains and several subareas. Note the hierarchial structure of hosts in different domains, then of subarea nodes within a domain, and finally of the local access

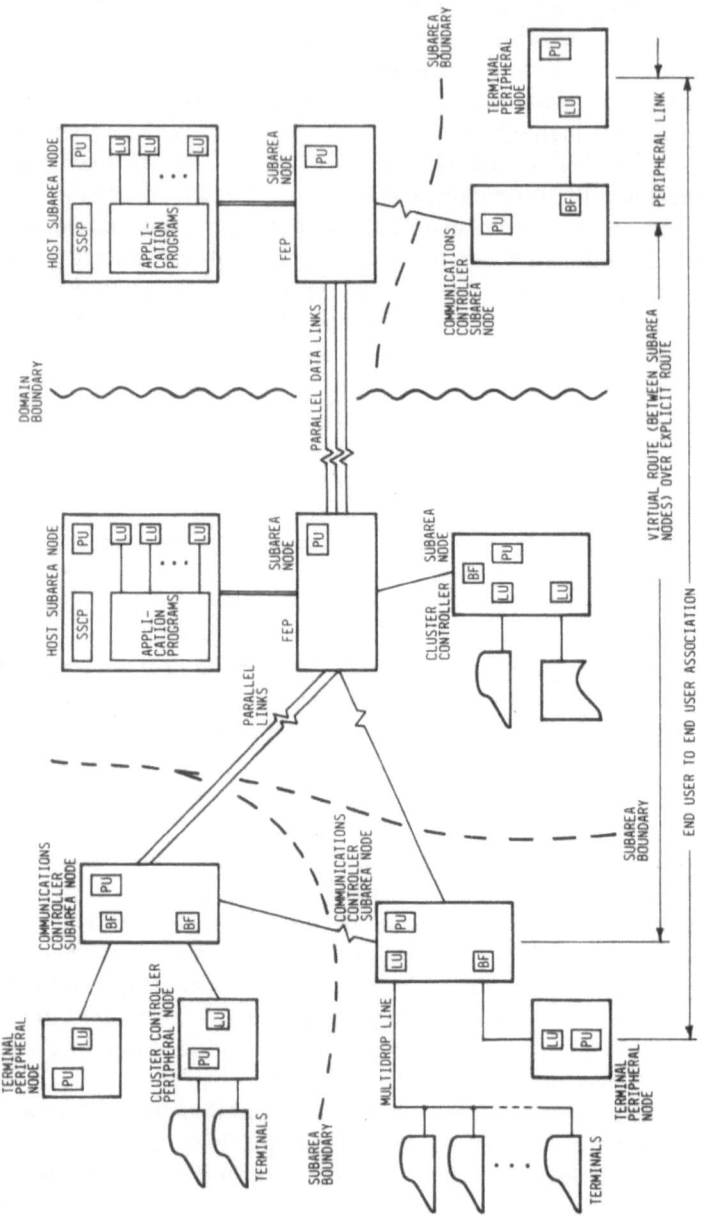

Figure 10.18. SNA network components and configuration.

network or peripheral nodes around subarea nodes. The actual SNA communications program product in the hosts with SSCP is typically either ACF/TCAM or ACF/VTAM (Virtual Telecommunications Access Method). In the other subarea nodes without SSCP, e.g., communications controllers, it is ACF/NCP/VS (Network Control Program). These programs are where functions like SDLC framing, multidrop line polling, buffer control, and various higher-level protocols are executed.

With this terminology established we can now look at the actual SNA protocol. Table 10.3 gives the approximate correspondence between the layers of SNA and the OSI Reference Model. At the physical layer is normally RS-232C, although other interfaces are also possible. The data link layer is, of course, SDLC as described in Chapter 7, including the loop and balanced modes.

Layer three, called *Path Control*, handles the packet (or block) routing and flow control functions. SNA is strictly a virtual circuit protocol, and

Table 10.3. SNA Protocol Layers

OSI layer	SNA layer	Typical SNA functions
7 Application	End User	User interface to networks File access Document interchange
6 Presentation	Presentation Services	Network management, control, and configuration Gateway for X.25 Code and display formatting Code conversion
5 Session	Data Flow Control	Dialog control Chaining Bracketing
4 Transport	Transmission Control	Session setup, operation, and termination Flow control Header construction Encryption
3 Network	Path Control	Address translation Route selection Virtual route operation Flow control Packet sequencing Message segmenting and blocking
2 Data Link	Data Link Control	Error-free data transfer between nodes
1 Physical	Physical	Multidrop line operation SDLC and RS-232C

the Path Control layer is concerned with the establishment, operation, and clearing of these circuits. For example, predetermined static routing tables maintained by the PU at each subarea node are used to determine the ordered sequence of links (transmission groups) and subarea nodes for each virtual route. For a given source and destination there may be several possible choices for the virtual route, depending on factors like existing link loading, type of traffic, and priority level. Since an SNA multilink transmission group can result in packets arriving out of order, they are resequenced by the Path Control protocol *at each subarea node.* Once the packets reach the destination subarea they are either delivered to the end user at that node, or else routed via the BF to the appropriate end user's peripheral node.

Flow control is also a Path Control function. The procedure, known as *virtual-route pacing,* uses a window concept. However, rather than "sliding" as in SDLC, the pacing window "hops." The sending node first sends a *pacing request* message, and normally receives a *pacing response* in reply. This authorizes the sending of a predetermined number, or window, of packets, after which another authorization must be obtained for further packet transmissions. The pacing window size can be dynamically adjusted by congested intermediate subarea nodes through flags in the packet *Trans-mission Header.* Virtual-route pacing is global in nature, since it is affected by other virtual routes passing through intermediate nodes and applies to all sessions carried by a given virtual route.

A final function of the Path Control layer is the *segmenting* of long messages into multiple packets and the *blocking* of multiple short ones into

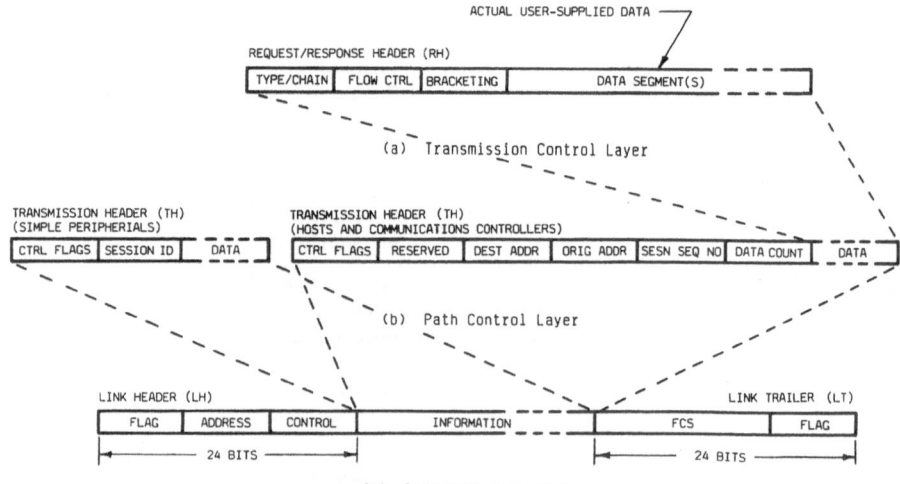

Figure 10.19. Typical SNA protocol formats.

a single packet. Figure 10.19 shows the typical SNA protocol formats, with two forms for the Transmission Header which is added by the Path Control layer. Note the longer address and sequence number fields in the virtual route format.

The remaining SNA upper layers are all oriented towards various aspects of end-user sessions. At each subarea node there is one data link module *per link* and a single path control module *per node*. For the upper layers, however, there is one module *per session*. In fact, these session protocol services are actually provided by the SSCP, PU, LU, and BF software products. Besides being closely interrelated, they do not match the OSI Reference Model as well as the lower layers do.

Transmission Control is the fourth SNA layer. It is concerned with the establishment, operation, and termination of end-user sessions over existing virtual routes. Sessions are set up between a primary and secondary LU via one or more SSCPs that go through a complex handshaking procedure called *binding*. This establishes all necessary parameters for the session, such as character codes, formats, speeds, and flow control parameters. A simplified cross-domain session establishment procedure is shown in Fig. 10.20. Once the session is established this layer builds the *Request/Response Headers*, performs buffering and sequencing at the destination node, provides flow control, and may also encrypt and decrypt traffic. The *session pacing* flow control mechanism is the same as that of virtual-circuit pacing but it controls each session individually with a fixed window size. It basically attempts to keep traffic from one end from overrunning the buffer capacity of the other end, and it may operate in stages via the BF with peripheral nodes.

The *Data Flow Control* layer works more in parallel with Transmission Control than above it to perform the SNA session-related protocol. It is concerned with the integrity of data flow to the session users. This includes session *dialog control*, e.g., who sends and when, and also some coarse flow control. *Chaining* allows messages such as a segmented file to be grouped

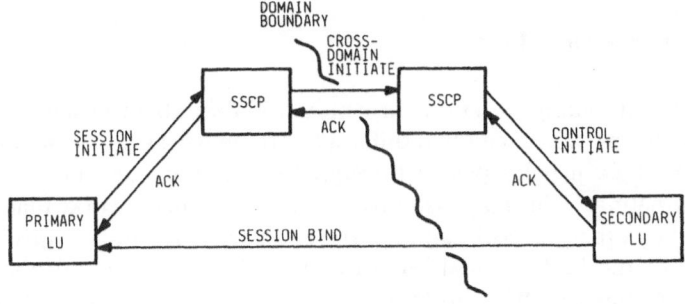

Figure 10.20. Simplified SNA cross-domain nodal operation.

together for uninterrupted transmission and checkpoint error recovery, while *bracketing* ensures that a related series of bidirectional message exchanges will continue to completion without interruption by any other message.

The remaining layers of SNA, namely, *Function Management Data* (FMD) and *NAU Services*, together correspond to the OSI Presentation and Application layers, but the layers are not at all distinct. The Presentation Services sublayer deals with code conversion, text compression, display formatting, and control character translation for different terminal types. These top layers provide a host of user-oriented functions, including remote file access, virtual terminal protocol, data base management, mail and document interchange, network maintenance and control, and accounting services.

SNA is clearly a complex specification, and there are many aspects that we have not covered. However, all the fundamental aspects have at least been mentioned. Despite the unique terminology, SNA fits reasonably well into the OSI Reference Model as well as our network routing and flow control discussions in the preceding chapters. Like other important network architectures, SNA is continually changing to meet the needs of the market, as evidenced by increasing decentralization and interfaces to non-SNA networks. There is also increasing progress towards adapting SNA to operate with separate local networks of small computers and associated resources. For example, in 1982 IBM announced a new LU 6.2 (Advanced Program-to-Program Communications) and PU 2.1 for SNA which allow application programs running in such devices as hosts, workstations, minicomputers, and PC microcomputers to communicate directly without going through the classical hierarchial SNA host.

This section has described two contrasting network architectures in some detail. It will be instructive here not only to compare DNA and SNA, but also to relate them back to the international protocols described in the preceding sections of this chapter.

10.7. Internetwork Protocol

There are many situations in which it is desirable to establish a communications path between two different data networks. For example, international traffic usually passes through two or more networks owned by different nations. This may also involve intermediate networks that are only accessed by other networks and not the end users. Historically this has been the role of the international record carriers for telecommunications. It is becoming increasingly important to be able to interconnect networks that are distinct because of their functions, size limitations, physical locations, protocol differences, security levels, or simply the "territorial imperative" of organizational groups. The military DDN network is a good example of a worldwide connection of distinct military community networks to serve a very general common purpose.

We shall refer to any collection of interconnected networks as an internetwork, or *internet*, and our specific interest here will be in the protocols for internets of heterogeneous packet switched data networks. The internet component networks, sometimes called *subnetworks*, are connected by *gateway* nodes which are common to two or more networks. Effective communication across multiple networks and gateways requires careful protocol design. Such an *internet protocol* function is done theoretically in all protocol layers, but in practice this is predominantly done at the OSI *network* and *transport layers*. At higher layers there is just too much complexity and variation among heterogeneous networks, so high-level protocol headers and data are passed through the internet intact. At the lower layers each network passes data link frames from node to node completely independent of any other network.

The essential internet protocol issues include addressing and routing, flow control, error detection and recovery, reliability, and the general quality of service. However, now these issues must be resolved end-to-end *across the entire internet* as well as across each individual network traversed. These are clearly network and transport layer functions. At the gateway the protocol can just *repackage* a transiting internet packet by imbedding it in the appropriate network packet for the next network. Alternatively, the gateway can actually *replace* the internet packet of one network with a different but equivalent packet in the next network [6].

Another important packet network issue concerns maximum packet length. This may vary considerably among the networks of an internet, and it is quite possible that a large packet may be too long to traverse some intermediate network intact. In such cases there are two basic recourses. The simplest approach is to restrict the maximum length that any originating packet can have to that of the worst case network of either the entire internet or at least those networks involved in any particular call. Alternatively, when packets reach an intermediate network for which they are too long, they can be subdivided, or *fragmented*, into multiple smaller packets that can be accommodated. Once the fragment packets are across this intermediate network, they can either be recombined into the original packet or can continue on to the final destination before recombination. Obviously, with this latter approach there may be further levels of fragmentation, e.g., a 1024-bit packet might first be split into two 512-bit packets, and later each of these fragmented into four 128-bit packets. The final destination reassembly process must then be able to sort out this hierarchy of packet fragment identifiers to recover the original message intact.

We shall look at two widely used internet protocols which represent the two most common approaches. First is the CCITT *X.75* recommendation, which is closely related to X.25 and, similarly, uses the *virtual circuit* approach. In contrast is the ARPANET *Internet Protocol* (IP), which is based on *internet datagrams*. Both of these protocols are applicable to a variety of networking situations. They also are similar to other internet protocols such as the Xerox Internet Transport Protocols.

X.75 [7] defines the protocol for a *split gateway*, with one half in each connected network. Each half-gateway is formally called a *signaling terminal* (STE), and operates over the physical, data link, and network layers. This STE–STE approach is appropriate since it allows each network, e.g., nationality, to develop its own gateway equipment and software so long as it works across the interface. The X.75 internet path consists of a concatenated sequence of network virtual circuits and interconnecting gateways, as illustrated in Fig. 10.21. This sequence is fixed when the call is established, and failure of a gateway or network virtual circuit will break the end-to-end communications.

The X.75 procedures and packets are almost identical with X.25. The major difference is the extension of the user facilities field of the Call Request/Call Connected and Clear Request packets to include a *network utilities* field. The utilities field immediately precedes the facilities field, and contains network-to-network signaling information. Typical utilities include a transit network identification, internet call identifier, flow control window length, maximum packet size, closed user group selection, and fast select. The packet formats are otherwise identical to those shown in Fig. 10.11 for X.25 plus the two-byte multilink field. Extended sequence numbers (modulo 128) are an option.

Flow control is handled at the gateways with the same multilink sliding window as in X.25, with each network also using its own internal procedures. RNR and interrupt packets are also used as in X.25. Acknowledgments are normally on a per link basis, but end-to-end acknowledgment is possible by setting the D bit in the packet header. Error control is handled by discarding and retransmission, using FRMR packets, or by resets and restarts in more drastic situations.

X.75 is oriented towards the interconnection of national X.25 packet networks but is certainly not limited to such networks. It is currently the most prevalent gateway for connecting large public packet networks, and has been used for a number of years by TELENET, DATAPAC, and others.

The ARPANET *Internet Protocol* [8] is a simple datagram procedure for each message. The source node builds *internet datagrams* which are routed independently not only through each network, but also across the entire internet. The datagrams may enter and leave intermediate networks by a number of different gateways, may be fragmented en route, and may

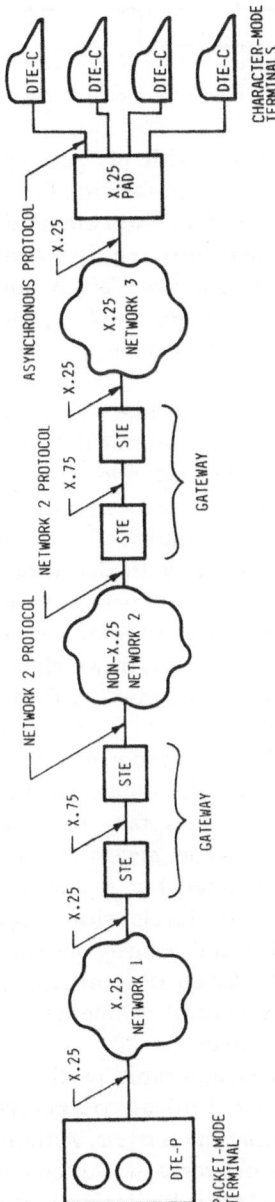

Figure 10.21. X.75 internet structure.

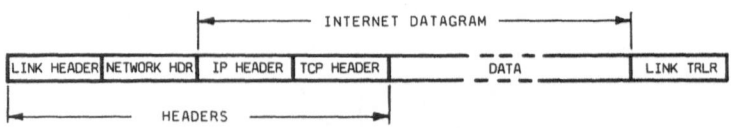

Figure 10.22. Internet protocol frame format.

generally arrive out of sequence with no guarantee of delivery and no acknowledgments returned. However, failure of a network or gateway does not necessarily disconnect the end-to-end communications. IP is clearly an unreliable protocol and consequently requires a strong transport layer to achieve a very high quality of service. This is normally provided by the companion *Transmission control Protocol* (TCP), which operates end-to-end across the internet. The general format of a frame containing an internet packet is shown in Fig. 10.22.

The IP gateway need not be split like X.75, and operates conceptually by removing the internet datagram from the data link and network protocol "envelopes" of the preceding network, and then repackaging it in a corresponding "envelope" of the next network in the route. To the networks the gateway looks like any other host, and is treated accordingly.

IP operates between gateways, with decentralized routing.† A 32-bit *internet address* contains both local and network address subfields, with the latter used to select the next gateway. Datagrams are fragmented as necessary unless a no-fragment option is selected, in which case the datagram will be discarded if it is too long for some network. Congestion is controlled by discarding when buffers are full.

There is also a *Time to Live* IP header field which is decremented as it passes through the internet, and the datagram finally is discarded if not delivered before its time expires. Datagrams may also be *time stamped* with the current time for accounting and measurement functions. There is an error check code for the IP header, but it does not check the data field. The originator can specify service levels with regard to delay, precedence, reliability, throughput, priority, and security classification. The IP datagram format is shown in Fig. 10.23. Since IP is relatively simple, such essential end-to-end functions as flow control, sequencing, and acknowledgments are allocated to the transport layer.

Today there is a continuing and rapid proliferation of all types of data networks, from highly specialized private packet networks to very general national and international common carriers. Although it is logical to build and manage these as separate networks, it is often desirable to communicate with users in other networks. This is becoming particularly true of the class

---

† However, there is a *source routing* option for the originating host to specify a list of gateways that must be placed on the route. This is important, for example, with highly classified traffic.

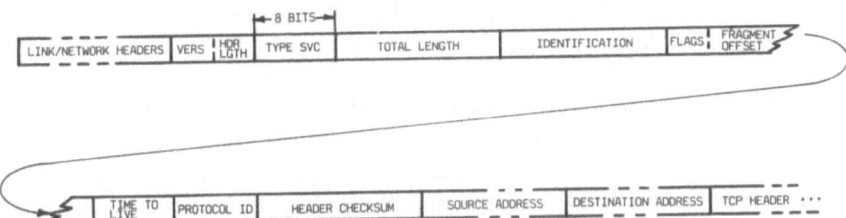

Figure 10.23. IP datagram header format.

of high-performance short-distance local area networks, which is the subject of the next chapter. Thus internetworking should grow in importance and pervasiveness.

This chapter has covered the subjects of network protocol and architecture in a broad but still reasonably detailed manner. The basic concepts have been introduced, with most specific example protocols taken as the international standards that are used in practice worldwide. Theoretically, protocol can be a complex and bottomless subject; but by sticking to established standards we have avoided much of this academic tedium.

With the completion of these last chapters on network components, design, and protocol we now have extended the earlier point-to-point concepts to full-blown wide area networks and internetworks. This background applies to essentially all significant data networks. However, there are several specialized categories of data networks that have additional unique characteristics and capabilities that make them worth further study. In the following chapters we look at a few of these, beginning with local area networks.

Appendix: Packet Assembly and Disassembly

The X.25 standard requires a terminal that is capable of executing the specified packet protocol in order to communicate over an X.25 network. This requires a fairly sophisticated device such as an FEP or even a host computer. There are, however, a huge number of practical situations where it is desirable to use simpler terminals that do not have packet-handling capability without going through an FEP or host. These include transaction-oriented office and reservation systems, information retrieval and data base access, and personal computers. In such cases the network may provide the packet assembly and disassembly capabilities necessary for nonpacket-mode terminals to communicate over packet networks. For example, this approach is used in both ARPANET and TELENET where the conversion devices are known as the Terminal Interface Processor (TIP) and Interactive

Terminal Interface (ITI), respectively. In X.25 the corresponding device is simply called a *Packet Assembler–Disassembler* (PAD).

A PAD can be thought of as a two-faced device that talks to a number of *character-mode* start-stop terminals (DTE-C) at its end of a packet network, and it also talks to a remote *packet-mode* host (DTE-P) located across the network. This type of network service is often highly desirable; for example, it will allow a single mainframe host to be accessed by many remote user terminals located throughout a packet network without the need for the user to provide packet-mode hosts at each remote terminal location. Although the PAD is normally considered to be part of the network, it may also be provided by the user and may even connect to the network via a remote access line. PADs use the first three OSI protocol layers in both directions. Toward the DTE-C it employs whatever protocol the terminal uses, e.g., the RS-232C physical layer and a set of ASCII control and data characters for the setup, operation, data transfer, and clearing of a communications circuit. In the other direction the PAD uses a packet protocol to transfer user data and PAD control information at the network level.

Although various particular networks use proprietary PADs, the major standardization effort is the so-called "triple" X.3/X.28/X.29 set of CCITT Recommendations. In general, X.28 defines how the PAD interacts with the DTE-C and X.3 lists the set of parameters that can be used for this interaction. This communication is done with eight-bit IA5 (ASCII) characters having start and stop bits, and physical access is via either V.21 for analog lines or X.20† for digital lines. In the other direction, X.29 defines how the PAD interacts with the DTE-P, which is typically a host computer that handles packets. The PAD uses conventional X.25 packets for establishing and clearing a virtual circuit and for transferring data in packet form through the network. In addition, however, the PAD must also convey control information between the user DTE-C and the host DTE-P. For example, terminals generally send a long sequence of zeros to indicate a break condition,‡ and it is necessary for the PAD to convey this condition via packets to the DTE-P. X.29 does this by placing the necessary control messages in the data field of X.25 data packets and sending these packets in the usual manner. The DTE-P sends similar packets in return to the PAD. With this dual usage of X.25 data packets it is obviously necessary to somehow distinguish between those containing true user data and those containing PAD control messages. In the latter case, where the data field use is limited, i.e., *qualified*, the packets are distinguished by setting the

---

† CCITT Recommendation X.20 is analogous to X.21, but applies to DTE/DCE interfaces operating in the asynchronous (start–stop) mode.
‡ A break is used to interrupt the normal flow of data by an asynchronous terminal.

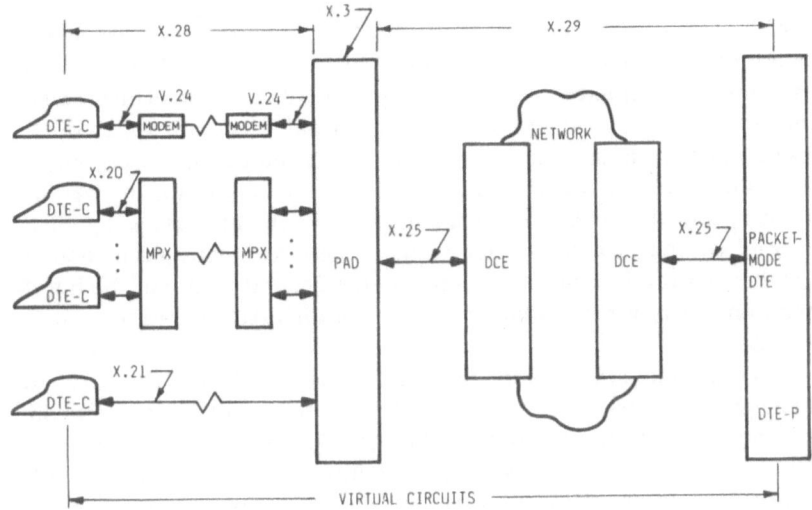

Figure 10.24. X.25 PAD connections.

Qualifier (*Q*) *bit*† of the packet header to unity. Thus the Q bit can be considered intuitively as a one-bit header for the control messages that constitute the network-layer data fields. The above relationships among the PAD, DTE-C, and DTE-P are shown in Fig. 10.24. It is important to note that X.29, like X.25, only defines an interface between peer processes, and says nothing about how the information is handled internal to the network.

Operational details between the DTE-C and PAD are defined by X.28 in terms of a collection of *parameters* given in X.3. This *PAD parameter profile* will be unique for each user terminal and is stored in the PAD during the call. The parameters may be selected individually by the DTE-C during call establishment, or they may be selected as predetermined sets known as *standard profiles.* Typical parameters deal with the familiar terminal options such as character echoing, XON/XOFF flow control, data rate, break handling, line feeds, carriage returns, deletions, and line folding. There are also parameters for shifting out of the data transfer mode (escape), and for indicating to the PAD when to forward a data packet that has been assembled. For example, the *packet forwarding signal* might be a break, the expiration of a timer, or receipt by the PAD of characters such as carriage return, form feed, or ETX.

A DTE-C sends X.28 *PAD commands* to the PAD in order to set or read current parameter values and to establish, reset, interrupt, clear and

† Recall that the Q bit is a flag bit in the X.25 General Format Identifier field in the first byte of the packet header.

obtain the status of the virtual circuit. In return, the PAD sends *PAD service signals* to acknowledge commands, indicate call progress, and provide PAD operational information. PAD commands and service signals are, of course, just strings of characters sent when not in the data mode. Thus the status command STAT requests the status of the virtual circuit, for which the PAD service signal response might be either FREE or ENGAGED. Similarly, the set and read command SET? will set or reset any following parameter numbers, and the PAD will normally reply with the service signal PAR followed by the new parameter values. There are many other details of X.28, but our brief discussion points out the general nature of the interface operation. We now turn to the packet side of the PAD operation, as defined by X.29.

To send traffic through the network, an X.29 PAD uses the X.25 virtual circuit and X.25 packets. A call is set up with a Call Request/Incoming Call packet of the form shown in Fig. 10.25a. The first four bytes of the Call User Data field are reserved for telling the DTE-P what high-level protocol is being used by the PAD to handle the start–stop terminal. For X.29 PADs the first byte is always one with the remaining three bytes all zeros. The remaining part of the Call User Data field may contain up to 12 bytes of user data, which is duly passed on to the DTE-C as user data. Upon receipt of a Call Accepted/Call Connected packet the PAD sends a *Call Connected* (COM) service signal to the DTE-C. The DTE-P may then set and/or read the PAD parameters by exchanging qualified data packets

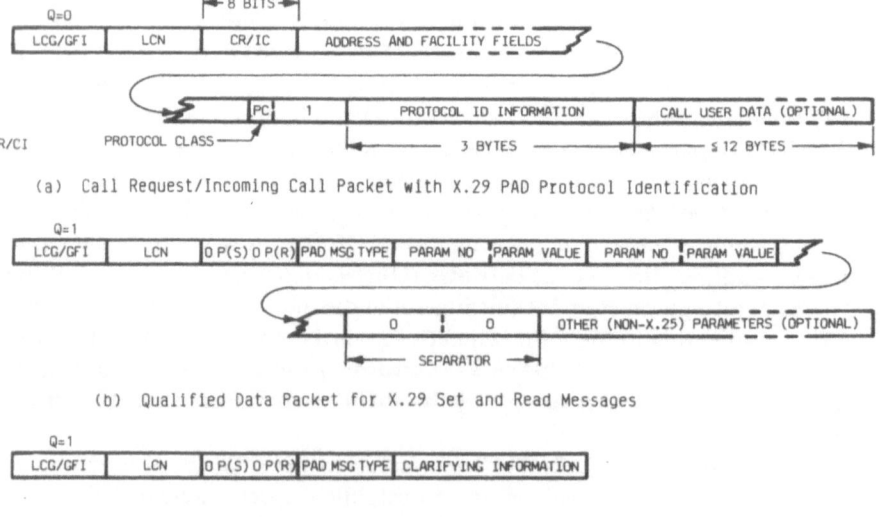

(a) Call Request/Incoming Call Packet with X.29 PAD Protocol Identification

(b) Qualified Data Packet for X.29 Set and Read Messages

(c) Qualified Data Packet for X.29 Break Indication and Error Messages

Figure 10.25. X.29 PAD formats.

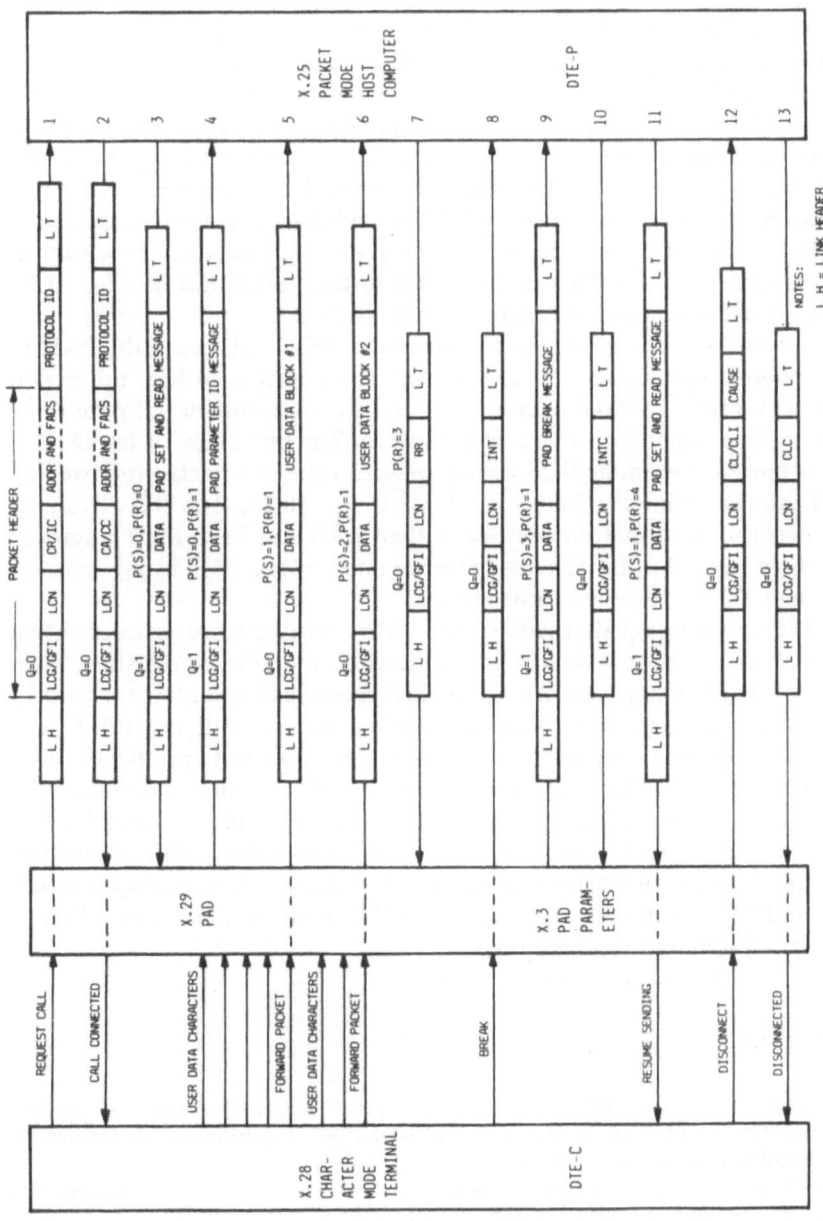

Figure 10.26. Operation of the X.25 PAD.

(Q = 1) with the PAD. There are several types of PAD messages conveyed in these qualified data packets, with all having the format shown in Fig. 10.25b. The first byte of the usual X.25 packet data field indicates the PAD message type, e.g., Set, Read, Set and Read, Parameter Indication, Break Indication, Clear, or Error. The remaining data field bytes indicate the associated details, such as the particular parameters involved or the reason for an error or break. As shown in the figure, any additional network-specific parameters are separated from the conventional X.29 parameters by two all-zero bytes. Once the virtual circuit is established and all parameters are set, data transfer begins using the usual (Q = 0) X.25 data and flow control packets and protocol. This is illustrated in the initial part of Fig. 10.26, which provides a complete example of an X.29 call.

A *break* (or attention) signal from the DTE-C will cause the PAD to take action as specified by the corresponding parameter. In X.25 this action is often to reset the virtual circuit. A less drastic procedure used by TELENET is shown beginning with line 8 of Fig. 10.26. The break key of the DTE-C causes the PAD to immediately send an X.25 Interrupt packet followed by the Break Indication PAD message shown in Fig. 10.25c. The DTE-P returns an Interrupt Confirmation packet and then a Set and Read PAD command to resume normal traffic. The call is terminated by the PAD using the usual X.25 and DTE-C protocol procedures.

The conventional alternative to the PAD approach is the network virtual terminal discussed previously in connection with the application layer protocol. The latter approach is a good deal more flexible for larger networks with many different and changing terminal types. With the PAD, it is necessary to define, and perhaps develop, new parameter profiles for new terminal types. A PAD may also have to handle a terminal differently in order for it to communicate with different hosts or other terminals. This can result in an ever-expanding parameter set as a data network evolves. Thus the conventional wisdom is that the PAD approach is most appropriate for relatively stable networks with a small number of hosts and a limited number of different terminal types.

References

1. *Data Communications—High Level Data Link Control Procedures—Frame Structure* (ISO 3309-1979) and *Elements of Procedure* (ISO 4335-1979), International Organization for Standardization, Geneva (1979).
2. *Data Communications—HDLC Balanced Class of Procedures* (ISO 6256-1981), International Organization for Standardization, Geneva (1981).
3. *Interface Between Data Terminal Equipment (DTE) and Data Circuit-Terminating Equipment (DCE) for Terminals Operating in the Packet Mode on Public Data Networks* (Recommendation X.25), CCITT, Geneva (1984).

4. *DECnet Digital Network Architecture (Phase IV)—General Description* (AA-N149A-TC), Digital Equipment Corporation, Maynard, Massachusetts (1982).
5. *Systems Network Architecture—Concepts and Products* (GC30-3072-0), International Business Machines Corporation, Research Triangle Park, North Carolina (1981).
6. J. B. Postel, Internetwork protocol approaches, *IEEE Trans. Commun.* **28**(4), 604–611 (1980).
7. *Terminal and Transit Call Control Procedures and Data Transfer System on International Circuits Between Packet-Switched Data Networks* (Recommendation X.75), CCITT, Geneva (1980).
8. *Internet Protocol* (MIL-STD-1977), U.S. Defense Communications Agency, Washington, D.C. (1983).

Suggested Readings

P. E. Green, Jr., (ed.), *Computer Network Architectures and Protocols*, Plenum Press, New York (1982).

An extensive technical coverage of the entire gamut of data communications protocol up to 1982 which is very detailed yet still easy to read and understand.

A. S. Tanenbaum, *Computer Networks*, Prentice-Hall, Englewood Cliffs, New Jersey (1981).

An excellent coverage of data network protocol layer by layer at an easy-to-grasp level, but some parts now becoming a bit dated.

B. M. Leiner, R. Cole, J. Postel, and D. Mills, The DARPA Internet Protocol suite, *IEEE Commun. Magazine* **23**(3), 29–34 (1985).

A brief overview of the use of IP and TCP protocols in the U.S. military networks.

D. W. Davies, D. L. A. Barber, W. L. Price, and C. M. Solomonides, *Computer Networks and Their Protocols*, Wiley, New York (1979).

This is a comprehensive text on networks which covers network protocol from the European viewpoint and with a good overall network perspective.

Check Your Understanding of Chapter 10—True or False?

1. The OSI transport layer is concerned only with the quantity of service provided the user.
2. Signaling and data transfer take place over the same X.21 circuits.
3. Videotext is predominantly a function of the OSI session layer.
4. A virtual circuit does not entail any physical paths.
5. TCP/IP packets are intentionally destroyed if not delivered on time.
6. The virtual terminal protocol is an attractive alternative to PADs in large networks.
7. Network layer packets go into frames as data or information.
8. The network layer multiplexes packets statistically and routes them onto data links.
9. The OSI Reference Model is the most important network architecture today.
10. All SNA subarea nodes are addressable via their respective PUs.
11. The X.25 LAPB protocol corresponds to the ARM mode of HDLC.
12. High-level protocols should ideally make link failures transparent to the end users.

13. Qualified X.25 packets are used to carry PAD information rather than data.
14. Most early network architectures were developed by large telephone companies.
15. X.25 layers 2 and 3 both use sliding window and RNR/RR mechanisms for flow control.
16. A restart is a much more drastic error response than a reset in X.25.
17. DNA is a highly distributed, datagram packet network architecture.
18. HDLC allows extensions of the address, control, and FCS fields.
19. Long frames may be fragmented and never recombined while traversing internets.
20. X.25 facilities are negotiated with the Interrupt and Interrupt Confirm packets.
21. Internet communications may use either datagrams or virtual circuits.
22. Digital X.21 is the only acceptable physical interface for X.25.
23. The HDLC command SARME means Send All Risqué Messages Early.
24. SDLC works badly in modern packet networks because it is old and tired.
25. A TPDU credit indicates that the senders have not paid their telephone bills.

Specialized Data Networks

11

Local Area Networks

11.1. Introduction to the LAN

Over the last decade there have been enormous advances in digital electronics, leading to the broad availability of locally accessible computing power. Concurrently, there has been a steady workforce migration from blue-collar to white-collar jobs in the industrialized nations of the world. For this growing hoard of office workers technology has produced word processors, electronic correspondence, data bases, local and remote computing, facsimile, document reproduction, and videoconferencing. In the factory the traditional assembly line worker is steadily being replaced by computer-controlled manufacturing equipment such as programmable controllers, robots, and automatic testing facilities. In both office and factory, as well as in schools, homes, and military units, it is becoming increasingly advantageous for a variety of heterogeneous devices to communicate with each other.

The convergence of such technology and markets has led to the development and growth of the *local area network* (LAN). Intuitively, the LAN can be thought of as an "intelligent cable" or a "distributed packet switch" that can connect all kinds of different terminal devices together locally and allow them to communicate effectively. In direct contrast to telecommunications networks, LANs operate at high speeds over short distances. They supply the bulk of the intelligence required for compatible local communications.

With these basic notions we now define the LAN as a digital communications network having the following characteristics (or at least most of them):

○ high-speed operation (typically Mbps);
○ short distance span (a few miles at most, usually much less);
○ very sophisticated technology (processor-based);
○ support for many different equipment types (different functions and protocols from different vendors);
○ low error rates and high reliability;
○ narrowly owned (usually a single private owner);
○ reasonable cost (overall and per connection).

This definition clearly encompasses a lot of possible networks, and it is not always clearcut where the exact delineation is between other networks such as laboratory instrumentation buses or metropolitan cable TV systems. However, like pornography and the Supreme Court, even if we cannot define the LAN precisely, we shall be able to recognize it when we see it!

Components include some type of cable or *transmission medium* such as wire pairs, coaxial cable, or optical fiber; a connector or *tap* to the transmission medium; a *transceiver* to encode the data and drive the medium at high speed; an intelligent *controller* to perform formatting, addressing, buffering, and error control functions; and finally the actual user terminal equipment, which is called a *station*. The component interrelationship is shown in Fig. 11.1a. Note that the controller and station operate at relatively low speeds, e.g., 2400 bps, while the transceiver and cable are at the high LAN internal speed, perhaps 10 Mbps. Since most LANs provide only a single high-speed bit stream, the bit clock must be recovered from the received data by the transceiver.

Figure 11.1b illustrates a variety of possible LAN stations for office equipment. In addition to computers, display terminals, word processors, printers, and data bases, an LAN allows the sharing among all users of such specialized resources as copiers, facsimile machines, CAD equipment, custom application programs, and electronic mail. Similar examples apply

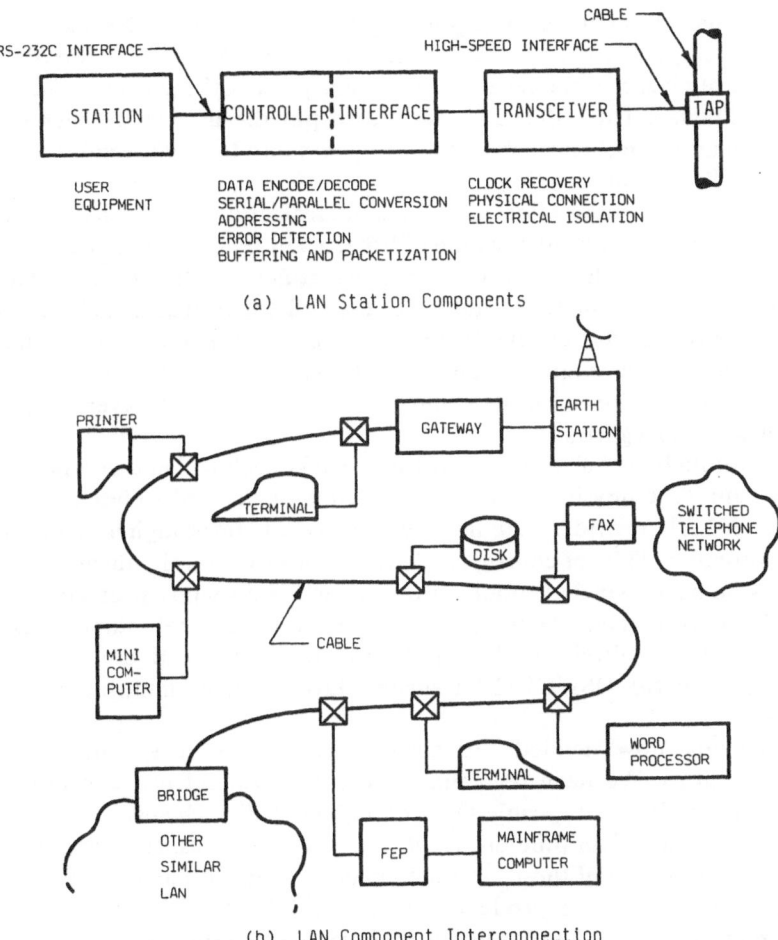

(a) LAN Station Components

(b) LAN Component Interconnection

Figure 11.1. LAN definitions.

to the factory and military environments. LAN users may gain access to another data network via either a *gateway* or a *bridge* station. The latter is used for connecting LANs of the same type, or perhaps segments† of the same LAN, at the data link protocol layer rather than at higher layers. Bridges may reconcile data rates, transmission media, buffer delays, and error performance between two LANs, but do not change the network protocols. Gateways do affect the LAN protocol, and may interface with dissimilar LANs or with various WANs at all protocol layers.

A fundamental assumption underlying LANs is that there is plenty of bandwidth—in the vernacular, "cheap bits." This allows for wide variation

† Segments are connected at the physical layer with *repeaters*.

and sophistication in the corresponding protocols. Early LANs, like WANs, were developed by specific equipment vendors to make their product lines more attractive, and the resulting proprietary networks and protocols were highly diverse. This, of course, tended to put the user at the mercy of a single vendor and hence created a strong demand by small vendors as well as users for standardization.

An initial vendor effort at LAN standardization was led by the Xerox Corporation with the introduction of Ethernet in 1980. Licenses were offered almost free, and this created a large assortment of Ethernet-compatible equipment from numerous vendors. In the U.S. the formal LAN standardization effort was led by the IEEE 802 committee. After a rough start in attempting to develop one universal LAN standard, this group finally decided to develop multiple standards to accommodate several different application categories.

Although the 802 standards are now well developed and accepted, they only define the physical and data link layers of the OSI Reference Model. Thus an LAN provides data link services for whatever higher protocol the user provides. This, of course, can leave the user with a significant problem unless he uses a single vendor's products and proprietary protocol such as DNA. Fortunately, much of the high-level protocol standardization developed for WANs is also applicable to LANs, and this is particularly true of the OSI/CCITT transport layer with its multiple classes of service.

Having now defined the abstract LAN, we next consider some classifications. We have seen that different *transmission media* are used, including various wire, cable, fiber optic, and radio schemes. The medium may be connected in different *topologies*, such as a bus, star, ring, tree, or some combination of these. The data may be transmitted at *baseband*, or it may be modulated to produce a *broadband* LAN with perhaps several different frequency bands. In the unique LAN terminology, broadband means *modulated* rather than a wide bandwidth. To add further confusion, the term "carrier" in the LAN context is used unconventionally to denote the existence of any type of traffic on the LAN medium. This could be a conventional modulated broadband carrier, but carrier also denotes the unmodulated data signal in a baseband LAN.

Perhaps the most important classification from a performance standpoint is the transmission medium *access method*, i.e., the rules for determining which station is allowed to use the network at any particular time. There are many ways to allocate access, including polling, reservations, priority orderings, and pure contention. However, practical LANs tend to use either an *intelligent random access* approach such as waiting until there is no carrier on the medium, or a *token passing* approach where authorization to transmit is rotated among all stations.

In this chapter, we shall organize our study of LANs primarily according to access method, by considering random, token passing, and other access techniques. With the aid of the OSI model, we look at some important LAN protocol issues, including the IEEE 802 standards. Finally, there is a discussion of some nonstandardized LAN access methods and a few representative examples of common vendor LANs. In order to put the LAN in proper perspective it will be instructive to occasionally compare the key points of this chapter with the corresponding ones in the preceding three chapters on wide-area networks.

11.2. Random Access

The simplest way for a station to access an LAN medium is to simply start transmitting. If another station begins transmitting before the first is finished there will be errors which must then be handled by the high-level protocol. This "fire when ready" *multiple access* (MA) approach may work satisfactorily with only a few stations, or even many stations provided they transmit infrequently, but it is extremely inefficient in general.

Multiple access was formally developed for a packet radio network in Hawaii and is consequently called the *pure Aloha* method. It is interesting to note that by requiring all stations to begin packet transmission only at the beginning of synchronous time slots, the theoretical throughput efficiency was doubled from about 18.5% to 37% [1]. This was called *slotted Aloha*, and is a useful approach for satellite-based networks.

On LANs, as well as in personal conversations, MA can be improved by requiring each station to refrain from transmitting until all others are through. However, this *carrier-sense multiple access* (CSMA) approach still does not completely prevent packets from interfering or, in LAN terminology, *colliding*. The problem is the *finite propagation time* between stations. Although measured in microseconds, on long cables at megabit data rates this delay may correspond to many bits of a packet. The two stations that begin transmitting at nearly the same time may still collide and complete short transmissions without realizing it because of this delay. Again error detection must rely on the higher-level protocol. Such situations may easily occur, too, e.g., when all stations that have been waiting immediately attempt to transmit exactly when the current transmission ends.

To resolve this CSMA dilemma, each station can be made to not only "listen *before* talking," but also "listen *while* talking." Each transmitting station now monitors the line signal. If it does not agree with that being transmitted, the station assumes a collision and turns off its transmitter. This is called *CSMA with collision detection* (CSMA/CD). Recurring collisions between the same station pairs can be avoided by incorporating *random backoff* into CSMA/CD. Here each station that detects a collision waits for a random interval before attempting retransmission. The interval is often determined by generating a random number at each detected collision, so each station should backoff by a different amount each time. With a very heavily loaded CSMA/CD network there can still be repeated collisions among a large number of waiting stations, resulting in unlucky stations being locked out for indefinite periods. To mitigate this situation the average backoff interval can be increased with each successive collision, at least up to some practical limit.

An alternative to detecting collisions is called *p-persistent CSMA*. Here all stations with traffic to send monitor the medium and, when it becomes idle, transmit immediately with probability p, $0 < p < 1$. For the remaining proportion of time $1 - p$, the station waits a predetermined time and then tries again. The decision at each attempt, is, of course, made randomly. Thus p is the average rate, and can be varied dynamically with the network loading.

As input, or *offered load*, to a CSMA/CD LAN is increased, the actual *throughput* initially increases proportionally, but eventually the network becomes congested. Further loading actually reduces throughput as more and more bandwidth is consumed by collisions. In the extreme the LAN bandwidth becomes totally consumed by repeated collisions and there is no throughput at all, in other words, a deadlock [2].

Frame, or packet,† size is another key LAN design consideration. Excessively long frames preclude timely access by other stations, while very short frames can be completely transmitted before a collision occurs at the other end of the LAN. This latter case depends on propagation speed, cable length, and data rate.

A typical random access LAN acts like a very fast switch, connecting successive pairs of stations in HDX fashion. Switching order is determined by the access method, and can be considered as a form of asynchronous packet multiplexing. As with other serial systems, it is necessary to recover the bit clock from the received data signal. The most common baseband procedure, known as *split-phase* or *Manchester coding*, adds the clock signal directly to the data coding at the expense of increased bandwidth. Figure 11.2a shows the Manchester encoding of an arbitrary data stream. It can be considered as modulation of the clock by the data, and always results in a *midbit transition* of the transmitted signal. Note that the transition direction is *up* for a data 1 and *down* for a 0; hence if the line polarity is reversed the received data is inverted. For Manchester coding the transmitted baud rate is twice the data bit rate, and hence requires about twice the

† For LANs the terms "packet" and "frame" are often (mis)used interchangeably.

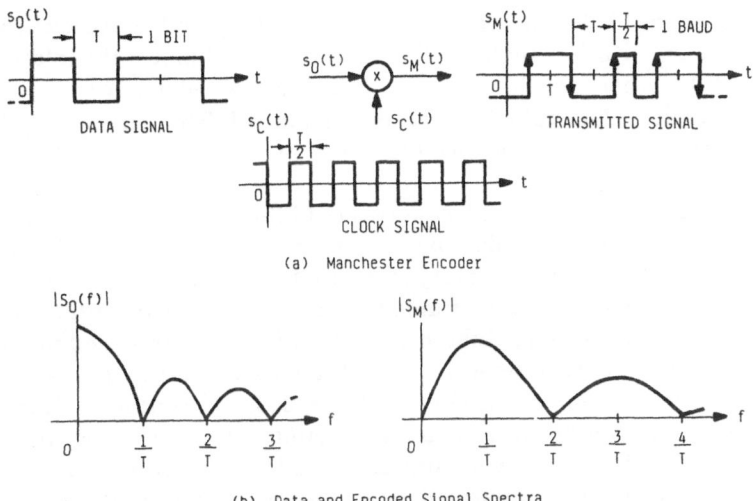

(a) Manchester Encoder

(b) Data and Encoded Signal Spectra

Figure 11.2. Manchester coding.

bandwidth. The corresponding spectra are shown in Fig. 11.2b. Since the transmitted signal has zero average value, transformer coupling is feasible.

The dominant CSMA/CD LAN is *Ethernet* [3], of which the IEEE 802.3 [4] standard is essentially a superset. The Ethernet and 802.3 LANs operate at 10 Mbps at baseband on a special 50-Ω coaxial cable in a bus topology. Mechanical connection of stations to the cable is with "vampire taps," which punch through the outer shielding to contact the center conductor. Up to five 500-m segments may be connected by repeaters, but only three of these may have stations (up to 100) connected to them. Bridges are used to connect to additional networks of the same type.

The 802 standard is rather formal and goes into considerable detail to divide the data link layer into two sublayers. The *Media Access Control* (MAC) sublayer depends on the specific access method and defines such network-specific things as frame format, access and collision procedures, error control, and station addressing. This sublayer interfaces with the *physical layer* which handles the Manchester coding, cable driving, collision detection, and general signal quality. On the other hand the upper data link sublayer, called *Logical Link Control* (LLC), is common to all 802 LANs. It may or may not be used with a particular LAN implementation. Thus we shall relegate LLC to the later section on LAN protocol.

Ethernet and 802.3 frames are configured as datagrams. Access is of course CSMA/CD, with *truncated binary exponential backoff*. This means that the collision backoff interval is selected randomly from an interval proportional to $[0, 2^{n-1}]$ for the nth successive access attempt. After 10 collisions ($n = 10$) the interval remains at $[0, 1023]$ until n exceeds 15, at which time an error is signaled. After each successful access the interval is reset to $[0,1]$. When a station detects a collision it immediately sends a few bytes of random data called a *jam*, then shuts off. The jam ensures that all other stations clearly identify the collision and do not mistake it for noise, etc.

Manchester coding is used, so recovery of the imbedded bit clock must be done from the received data by each station. Each frame is effectively broadcast to all stations, and every station must obtain synchronization in order to receive enough of the frame to determine if it is the addressee.

Figure 11.3a shows an Ethernet frame, with the corresponding 802.3 frames in (b). Initially there is a 64-bit *Preamble*, the last byte of which is called the *Start of Frame* (SOF) delimiter in 802.3. Except for a final pair of 1 bits, the preamble contains an alternating 1/0 pattern used by each receiver to acquire clock synchronization and define the beginning of a frame. The frame end is indicated by simply the cessation of the carrier. To prevent signal transients from interfering with the next frame there is a mandatory 9.6-μsec (96-bit) spacing between frames.†

The 48-bit *Destination and Source Address* fields provide 2^{48}, or some 281 trillion, different addresses! This is enough for each station worldwide

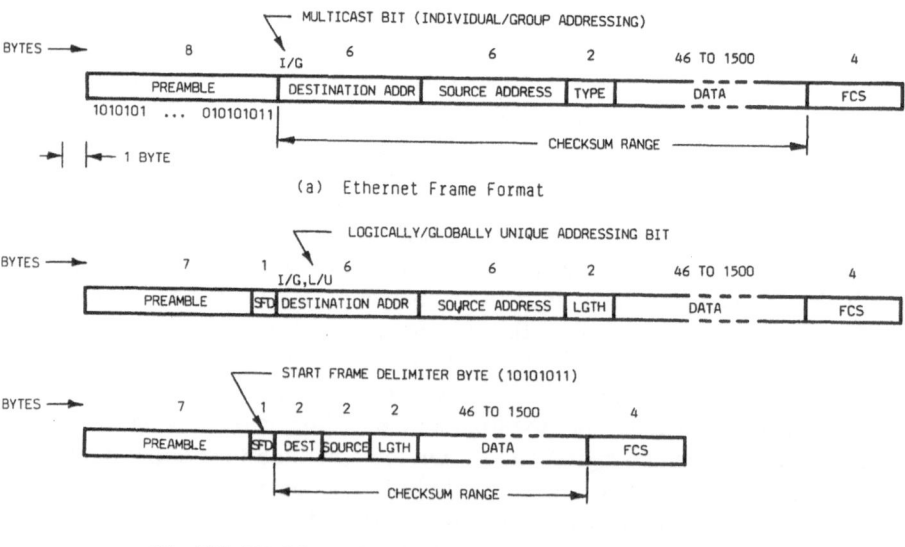

(a) Ethernet Frame Format

(b) IEEE 802.3 Frame Formats with 48 and 16 Bit Addressing

Figure 11.3. Ethernet and IEEE 802.3 frame formats.

† In addition, there may also be a brief (about 1.0 μs) test burst of the collision signal between frames, known informally as the "heartbeat."

to have a *globally unique* address. In fact, the original Xerox concept was to provide satellite links for any Ethernet to communicate with any other, and each Xerox license includes a globally unique block of addresses. The destination address field begins with a *multicast* bit (I/G) indicating an *individual* or *group* address. The broadcast group address is defined to be all ones. In the 802.3 frame the second bit (U/L) indicates whether *globally* or *locally* unique addressing is used. There is also provision in 802.3 for shorter 16-bit addressing as shown.

The Ethernet *Type* field indicates the particular high-level protocol used in the data field and must be included in each frame since there are no virtual circuits. The corresponding 802.3 *Length* field gives the number of LLC bytes at the beginning of the data field. Ethernet frames range in size from 72 to 1526 bytes, allowing a data field of from 46 to 1500 bytes. To send less than this minimum amount, padding must be added to the data. Finally, there is a 32-bit CRC *Frame Check Sequence* (FCS)† that covers everything except the preamble.

Ethernet is currently the most popular LAN for office applications. Actually DEC and Intel were also sponsors of Ethernet, and the latter, along with other vendors, has developed IC chips and cards for controller and transceiver functions. DEC has, of course, incorporated Ethernet into DNA. For time-critical applications such as process control there are some potential drawbacks to CSMA/CD because of the uncertain access times. For example, if a process starts to become unstable then a CSMA/CD LAN connecting sensors and controllers will experience a large increase in loading and hence collisions. This might result in some critical stations failing to gain access in time to correct the problem.

In an office environment, on the other hand, such a congested situation might cause a user to go get a cup of coffee and return later with no adverse effects. Finally, Ethernet has been adopted almost intact as the IEEE 802.3 standard and is widely supported by other LAN vendors and terminal equipment manufacturers.

Besides Ethernet there are numerous other LAN products based on CSMA-type access. We shall consider some of these variations later in the chapter, but we now turn to the completely different token passing access method in the next section.

11.3. Token Access

Random access can be extremely unfair to a station at times, although this should not be the case on average. Furthermore, it is susceptible to

---

† The generating polynomial for this FCS is given in the Appendix to Chapter 7.

throughput degradation with heavy network loads. In many situations a more consistently fair and predictable access method is desirable. The most common deterministic access for LANs is by *token passing.*

Basically, each station gets an ordered turn to transmit one frame. It then must wait until all other stations have had their turns before sending its next frame. The right of a station to transmit, i.e., access, is determined by possession of the "token." It may be a special frame, or perhaps just a flag in a normal data frame. In some cases, particularly WANs, it is possible to have multiple tokens, but we shall restrict our study of LANs to the single token case.

Closely related to token passing is the issue of *physical and logical topology.* Since the token must be passed among stations in a specific order, a *ring*† topology as in Fig. 11.4 is a natural form. The token is passed around the ring from station to station until it reaches one with traffic to send. This sending station removes the token (or resets the token flag), replacing it with a data frame which proceeds from station to station completely around the ring. Each station buffers the frame momentarily to check the address, then transmits it on. However, the addressee also makes a copy of the entire frame for processing and delivery to the end user. The original frame eventually returns to the originating station, which cannot send another frame immediately. Rather, it must replace its old frame with the token. Authorization to transmit in the form of the token then passes to the next station. Before it can recover the token and send a second frame, the original station must wait for this same process to recur at each station having traffic to send.

Clearly token access is fair, with no station getting blocked out and none allowed to hog the network. Furthermore, in the heavily loaded case

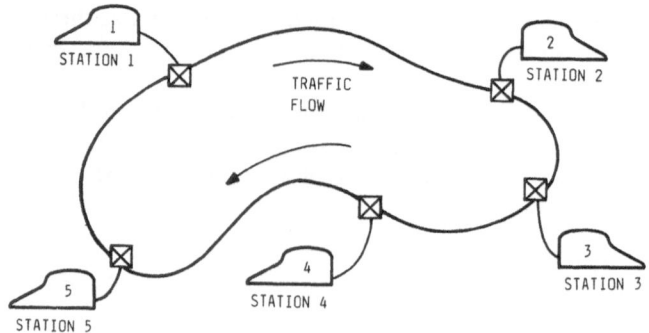

Figure 11.4. Physical ring LAN topology.

---

† *Loop* is sometimes used synonymously with *ring*, but we shall restrict the former to connote centralized control such as polling.

where every station has waiting traffic the maximum time any station must wait to gain access is known and bounded. However, this waiting time is proportional to the number of stations on the ring and thus may limit the practical size of the LAN. As network loading becomes heavy the throughput approaches an asymptotic limit but never decreases; thus congestion should not occur. On the other hand, with very light and uneven loading a station with a long file to send may waste a lot of time waiting for the token to pass idly around the ring after each frame of the file is sent. This is in sharp contrast to random access for light and heavy loading.

Physical ring LANs are inherently HDX, with each station always receiving from, and transmitting to, the same two adjacent stations. This has a number of advantages. For example, the received signal level at all stations can be made consistent, while with bus topologies it may vary enormously from end to end (8.5 dB for Ethernet). A broadcast medium is not required, so fiber optics cables are well suited for rings. Since the data signal is regenerated at each station, long rings are feasible, with the primary limitations being the maximum acceptable access time and the build up of jitter from the station repeaters.

On the other hand, physical rings are completely disabled if the ring is broken anywhere. They also have inherent security problems since every frame must pass through every station. Finally, there is the issue of ring management. Adding and deleting stations or detecting and recovering from failures involves considerable complexity, particularly if the LAN is not to be reinitialized each time.

Fortunately these disadvantages are not insurmountable. For example, data can be encrypted for security, and sufficient intelligence can be designed into each station controller to handle the ring management problems. One solution to the reliability problem is the *dual ring configuration*, proposed by IBM [5] for LANs and illustrated in Fig. 11.5.

In normal operation all traffic is carried by the outer ring, with groups of stations connected at addressable *wiring concentrators*. This is shown in Fig. 11.5a. Should a station fail, the ring management protocol simply tells the wiring concentrator to bypass the station, as shown for station 5 in part (b) of the figure. In some cases, such as loss of station power, relays may effect the isolation automatically. If a section of the outer cable fails, then the wiring concentrators isolate the entire link as shown, substituting the remaining inner loop for it. Although this may increase the propagation delay around the loop, the order of stations remains unchanged.

Token passing access is not limited to physical ring topologies. It can also be used to provide nonrandom access on bus topologies by defining a *logical ring*. As shown in Fig. 11.6 the stations are ordered in some manner such as address values, and then restricted to transmit only to the next station in the ordering. Although all stations hear each transmission, only

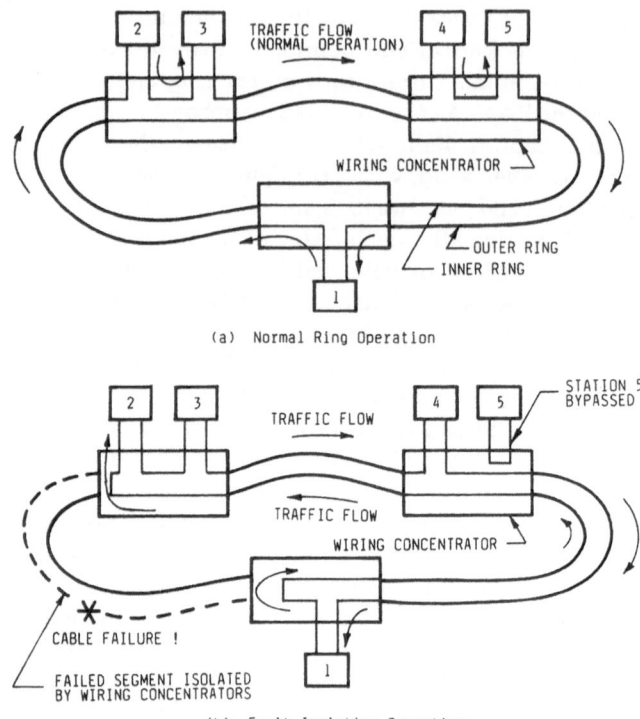

(a) Normal Ring Operation

(b) Fault Isolation Operation

Figure 11.5. Physical ring operation.

the station that follows the source address of a frame may receive it, as indicated by the dotted arrows.

With these preliminary notions about token access on physical and logical rings, we will now consider the two corresponding IEEE 802 standards, namely, 802.5 and 802.4, respectively. Although there is considerable similarity between the token formats and procedures at the data link layer,

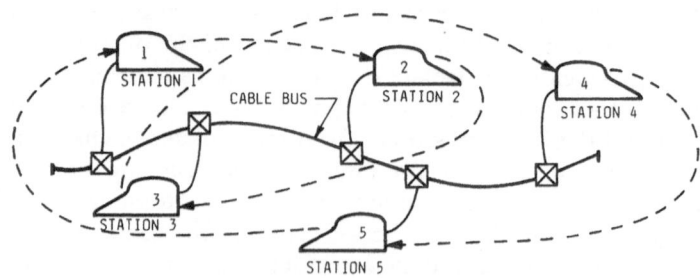

Figure 11.6. Logical ring operation.

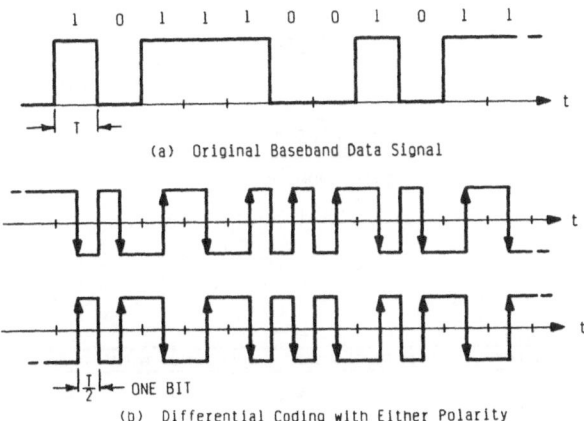

(a) Original Baseband Data Signal

(b) Differential Coding with Either Polarity

Figure 11.7. Differential Manchester coding.

the physical layers are entirely different. Not only do they differ in topology, medium, and speed, but one is baseband and the other broadband.

The IEEE 802.5 standard [6] defines token access on physical rings. It is based heavily on an experimental LAN developed by IBM around 1980 which led to the IBM token-ring product announced in 1985. Starting with the physical layer, *differential Manchester coding* is used for the baseband signals. This method, shown in Fig. 11.7, always has a transition in the middle of each bit for clock recovery. For a data 1 there is no transition at the beginning of the encoded symbol, while for a 0 there is a transition. The main advantage of this variation relative to the conventional Manchester coding used in Ethernet is that it is independent of the wire-pair polarity. Hence the leads at a station can be reversed without affecting the decoded data. Note also, of course, that there is no dc spectral component.

Each 802.5 frame begins with a *Starting Delimiter* (SD) and contains an *Ending Delimiter* (ED) in the trailer. To make these one-byte fields unique, the rules for differential Manchester coding are intentionally violated. As Fig. 11.8 shows, two pairs of symbols do not have the usual midbit

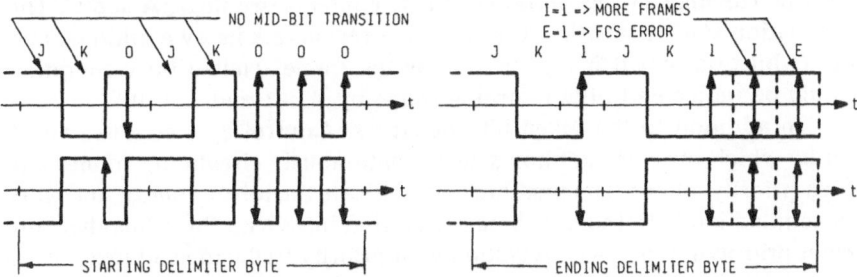

Figure 11.8. IEEE 802.5 starting and ending delimiters with coding violations.

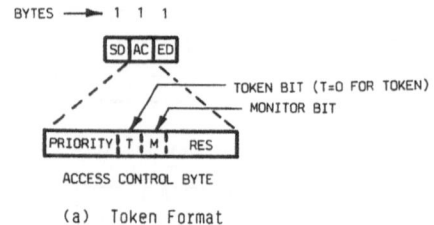

(a) Token Format

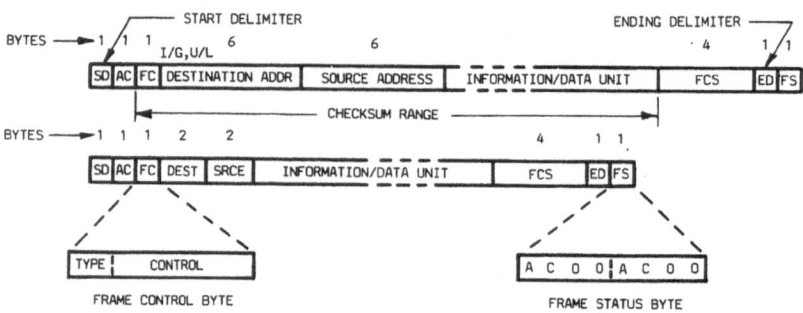

(b) Frame Format with 48 and 16 Bit Addressing

Figure 11.9. IEEE 802.5 token ring formats.

transition. Code violations always occur in JK pairs for clarity and to keep the dc value zero. By definition J has the same polarity as the end of the preceding coded symbol and K has the opposite polarity. The pair of bytes SD, ED together are used to abort a transmission.

Formats for the token and information frames are shown in Fig. 11.9 with both the 16- and 48-bit addressing options. The token is actually indicated by setting the T bit to zero in the *Access Control* (AC) field. As with the 802.3 frame, there is an I/G bit set to indicate group addressing and, in the 48-bit address case, a U/L bit indicating universal or local address significance. Furthermore, the same 32-bit error check procedure is used. The final *Frame Status* (FS) byte contains two flags, A and C. The destination station sets the A bit to 1 if it recognizes its own address, and the C bit also to 1 if it is able to copy the frame. Thus a busy condition would be indicated in the returning frame by $A = 1$ and $C = 0$.

In addition to the token bit, the Access Control byte contains an M bit by which the ring monitor detects continuously circulating frames and high-priority tokens. The first three bits indicate the token *priority*, and work in conjunction with the last three reservation bits. With three bits there are eight priority levels (0–7). A station with priority traffic to send first checks the reservation field of the current frame for existing higher-priority requests.

If the station's desired priority level is greater, it increases the reservation value to that priority; otherwise, it must wait until all higher-priority transmissions are completed. Other stations on the ring with even higher-priority traffic, of course, can further increase the reservation value. The next station to originate a token must now set its priority, and hence that of the ring, equal to the reservation value. This effectively locks all lower-priority traffic out of the ring until either the high-priority traffic is all transmitted, or a timer limits the time one station can continue to send. Other stations on the ring at the same priority can also send, and any arriving higher-priority messages can cause a further increase in the ring priority level.

In the latter case, the station increasing the ring priority saves each previous priority level in a memory stack in order to return the ring to the preceding level when it reinserts the token. Thus the station that raises the token priority must also be the one that lowers it.

The functions of the *Frame Control* (FC) byte are rather complex. The first two bits indicate the *frame type* as either MAC or LLC, and hence the type of data unit contained in the information field. Since LLC is covered in the next section, we will describe only the MAC sublayer functions here. In this case the last six bits of the FC byte indicate six different functions relating to ring management. The information field of each MAC frame contains a *vector*, with one or more subvectors, which contains the MAC function and associated parameters.

Although the ring has no permanent controller, some node is always designated the *active network monitor*. Its function is to ensure proper ring operation and, should it fail, another node will take over as monitor. In normal operation the monitor sets the M bit of any passing token or frame with priority greater than zero. If the bit is still set in the next pass, the monitor replaces the frame or token with a zero-priority token, since it was not properly removed by its originator.

To verify the monitor's presence it periodically broadcasts an *Active Monitor Present* (AMP) MAC frame, in response to which all other stations send *Standby Monitor Present* (SMP). This verifies the monitor and identifies the immediate predecessor of each node, for example, when a station has been removed. A new station entering the ring in normal operation just repeats traffic until an AMP frame comes by. It then captures the token and sends a *Duplicate Address Test* (DAT) MAC frame to ensure no existing station has the same address. If this is found to be the case, the new station next sends an SMP frame to update the adjacent node addresses, after which it begins normal operation.

Should the monitor fail to send an AMP frame on time, the first station to recognize the condition begins sending *Claim Token* frames. If other stations also do this, then the one with the largest address becomes the new monitor. It then reinitializes the ring using *Purge* (PRG) MAC frames, after which it introduces a new token and normal operation ensues. More serious ring failures are identified and isolated with *Beacon* (BCN) MAC frames.

At the physical layer, the 802.5 ring operates at either 1 or 4 Mbps over twisted wire pairs. Each ring can accommodate up to 250 stations and repeaters, and the ring must be able to hold at least one token, i.e., 24 bits. Stations are connected with a relay arrangement that allows the station to be disconnected without breaking the ring continuity. The Manchester coding allows transformer coupling for station isolation. The clock reference is always taken from the active monitor station, with precise limits on accumulated jitter around the ring. An acceptable average error rate of no more than one error per 10^8 bits is also specified.

Despite the conceptual simplicity of token passing, efficient ring operation and management can be quite complex. There were also some initial patent problems which impeded early implementation. There are several additional details of the 802.5 standard that we have not covered, but presenting them would add little to our fundamental understanding. Instead, we shall now introduce the second IEEE LAN standard for token-passing bus access.

The IEEE 802.4 committee has, as the old joke goes, truly created a *camel*. This broadband token-passing bus standard [7] includes something for practically everyone, with a smorgasbord of alternatives. There are three different modulation choices, three data rates, and multiple selections for topology, medium, line symbol coding, and electrical hardware. Fortunately, the same MAC and LLC procedures apply to all variations, so to the actual user the appearance is amazingly consistent.

Most fundamental are the three specified modulation techniques. The simplest choice is called *Phase Continuous FSK*, which operates only at 1 Mbps. Line coding is a cross between conventional FSK and Manchester coding. At 1 Mbps a bit lasts 1 μsec. One-half a bit time, or 0.5 μsec, of the higher 6.25-MHz FSK frequency is denoted by H, and L denotes a similar interval of the lower frequency, 3.75 MHz. In terms of H and L a data 0 is transmitted as HL and a 1 as LH. This results in a *midbit frequency transition* in each bit interval for clock recovery, although with continuous phase there are no amplitude discontinuities. Coding violations for frame delimiters are coded as LLHH over two bit times. The idling pattern is alternating, i.e., HLLHHLLH .... Although cable branching is possible, the recommended topology is a bus using conventional CATV cabling.

The second modulation alternative is called *Phase Coherent FSK*. This is a more bandwidth-efficient technique which allows either 5 or 10 Mbps over the same CATV cable as the previous method. However, in this case a tree topology is recommended. Again there are two FSK frequencies, but now the lower equals the data rate and the higher is exactly twice the data rate. A data 0 is sent as two cycles of the higher frequency, or HH, while a 1 is one cycle of the lower frequency, or LL. Delimiter code violations are encoded HLLH, and the idle pattern is LLLLLL . . . , i.e., all ones. Since this modulation is phase coherent, there is always a midbit zero crossing of the FSK data signal which provides the clock.

The preceding FSK schemes send traffic between logical ring stations in a single frequency band on one bidirectional cable. In contrast, the third broadband alternative sends traffic either over *two separate unidirectional cables*, or else over *two separate frequency bands* on a single cable. In either case, traffic must first go to a *headend station*, where it is either *switched* to the other (receive) cable, or *remodulated* into the other (receive) frequency band, respectively. All stations transmit on the same frequency band (or cable, in the dual-cable case) and receive on the other. The corresponding modulation technique, called *Multilevel Duobinary AM/PSK*, is a combination of baseband precoding and two-phase QASK modulation.

The general duobinary AM/PSK technique is rather complex [8], but basically precodes the binary baseband data into a multilevel signal which is the modulated using a two-phase QASK technique. The 802.4 standard uses a relatively simple version† with only two precoded values (denoted by 0 and 4), but includes more complex and bandwidth-efficient versions as future contingencies. Delimiter code violations are obtained by using an intermediate amplitude level (denoted by 2). The headend station also uses code violations to send a pseudosilence pattern when there is no other traffic, which keeps the broadband modems at each station synchronized. To maintain synchronization with normal traffic, a scrambler identical to that of the Chapter 5 Appendix is used.

Regardless of which 802.4 modulation technique and medium are used at the physical layer, the MAC layer operation is the same. The frame format of Fig. 11.10 is also quite similar to that of 802.5. Note, however, that a conventional MAC frame with no data field is used for the token, and there are no Access Control or Frame Status bytes. Each frame begins with a preamble to provide interframe spacing and to synchronize the receivers for the particular modulation scheme used. The preamble must be at least $2\,\mu$sec long, which requires one byte at 1 Mbps and three at 10 Mbps. Very long preambles which might be confused with modem streaming (jabbering) must be avoided, since otherwise expiration of a *jabber timer* will shut off the LAN modem.

The content of each frame is delineated by *Start* (SD) and *Stop* (ED) *delimiters*, each defined uniquely by the pair of code violation symbols (N) as defined for the particular modulation used. As the last figure shows, the ED ends with the same I and E flags as 802.5 to indicate intermediate frames and FCS errors, respectively. The same frame abort sequence of SD,

---

† This is known in practice as Class 1 Partial Response signaling.

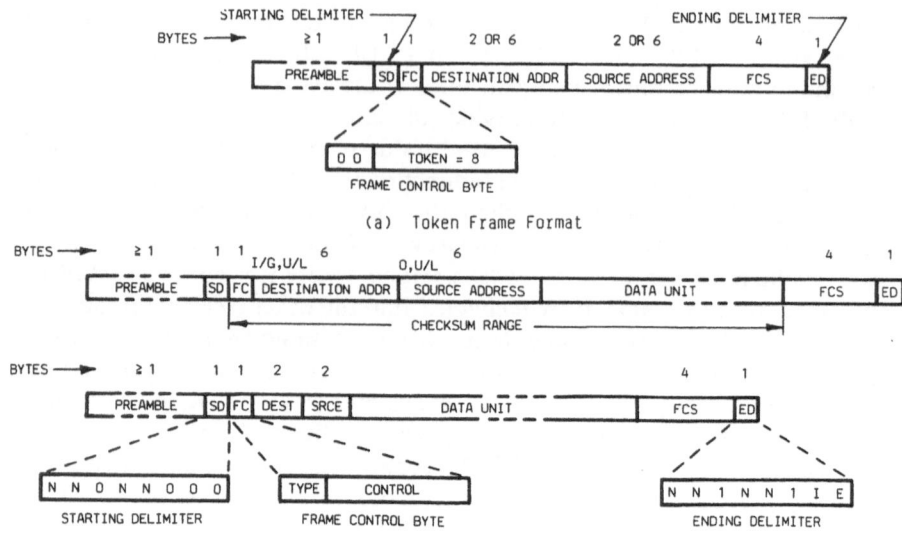

Figure 11.10. IEEE 802.4 token bus formats.

ED is defined. Destination and source addressing, as well as the FCS, are also the same as 802.5, including the I/G and U/L flags and the optional 16-bit local address format.

The remaining *Frame Control* (FC) byte identifies the frame type with the first two bits and supplements this type with the following bits. Again we defer the LLC frames to the next section. In normal operation the token MAC frame is passed among stations in the descending order of addresses. A station with the token may continue to transmit frames until its *token hold* timer expires, at which time it must pass the token. A priority mechanism is defined that allows higher-priority traffic to preempt that with lower priority. It is also possible for a station with the token to, in effect, poll any other station and receive a one-frame response immediately without waiting for the token.

As with the token ring, there are elaborate procedures that employ MAC frames to initialize the ring, manage the token, add and delete stations, detect errors, and recover from various failures. Unlike the token ring, however, there is no specific ring monitor station. This is circumvented because each station can monitor its own, and any other station's, transmissions and thus immediately detect any operational abnormalities.

We conclude this section with some general comments and comparisons between the 802.4 and 802.5 LANs. First the token ring is clearly much simpler from a technical standpoint. At baseband, expensive and tempermental high-frequency modems and headends are not required. Bidirectional, point-to-point operation at baseband also avoids problems with large received signal level variations and signal reflections at connection points which can be significant on bus networks. The physical ring is also ideal for fiber optic cable, which is inherently bidirectional, so much higher data rates and noise immunities are feasible as the network grows. And in the meantime, the 802.5 ring can be implemented on existing, or easily installed, twisted pair wiring.

On the other hand, the token bus as defined by 802.4 is simpler to manage despite its complexity. It also offers a lot of alternatives for implementation, and uses existing CATV hardware to a large extent. Tree topologies are feasible, which can make installation and particularly expansion of the LAN much easier than with ring and bus topologies. Perhaps the biggest advantage, however, is the capacity of broadband cables to carry multiple frequency bands simultaneously. For complex applications requiring several types of traffic, broadband is an attractive alternative to installing multiple networks to cover the same local area. Thus, an 802.4 LAN could conceivably be augmented to carry some voice channels for use in coordination of a process control system carried over the main token bus. Finally, the physical length of token bus LANs is potentially greater than either token rings or CSMA/CD busses, since it is not limited by the need to quickly detect collisions.

So far we have focused on specific LAN media and access methods. It is important to understand the ramifications of these specifics, not only for the designer, but for anyone involved in the acquisition, operation, and maintenance of LANs. Ideally, however, the user would rather not be concerned about the network technicalities, but instead only with its use. Just as a driver interfaces with an automobile through the driver's controls, the LAN user interfaces through high-level protocol which should appear independent of the particular network. In the next section we extend our physical and MAC layer background to include first the LLC, and hence all of the data link layer, and then look at some higher-level protocols applicable to LANs.

11.4. LAN Protocol

For LANs the OSI Reference Model provides an orderly framework for understanding protocol issues. We have already seen that the basic LAN itself provides the physical and some data link protocol, while the remaining layers are generally provided as needed by the station and user devices. In this section we shall first look at the OSI layers for LANs, including gateways. Within this framework we describe several important public and proprietary standards for higher-level LAN protocol. Finally, we introduce the mundane but crucial subject of network management.

At the OSI *physical layer,* LAN protocol is concerned with the technical details of signal characteristics, line coding, cables, connectors, clock recovery, collision detection, signal quality, and component failures. These are highly dependent on each particular network type, and are difficult to generalize. However, the actual terminal interface is often RS-232C, so little or no modification of station equipment or procedures is needed.

The *data link layer protocol* is the real workhorse with LANs. It establishes and disconnects links, and manages the flow of blocks of data over the links. This includes framing, error control, addressing, and access control, e.g., collision recovery. As we have seen, the 802 committee distinguishes these network-dependent *Media Access Control* (MAC) functions from the remaining *Logical Link Control* (LLC) functions common to all 802 LANs.

The LLC protocol is defined by the 802.2 standard [8] and is closely related to HDLC and X.25 LAPB. Since within a single LAN, data links are also *end-to-end* associations, a number of network layer functions are either unnecessary or redundant, so the LLC actually has some network layer characteristics. For example, there are two types of operation: *connectionless* ("datagram") and connection-oriented ("virtual circuit").

Figure 11.11 shows the LLC *Protocol Data Unit* (PDU) imbedded in a frame data field. While the MAC sublayer addresses a station, LLC addresses a *Service Access Point* (SAP) which is typically the logical interface between processes in adjacent protocol layers. The first bit (I/G) of the destination SAP field indicates *individual or group addressing,* while the corresponding bit (C/R) in the source SAP field identifies a *command* or *response* PDU in the HDLC sense. Finally the data field contains whatever information is provided by higher layers.

The connectionless or Type 1 operation is quite simple, with no link establishment or termination, no flow or error control, and no acknowledg-

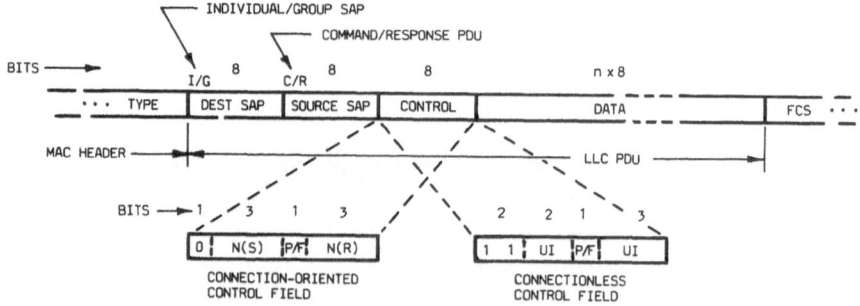

Figure 11.11. IEEE 802.2 LLC protocol data unit format.

ments. PDU control fields are like those of the HDLC UA frames, with no sequence numbers. The only other PDUs for Type 1 operations are XID to confirm and identify a station and TEST for initiating link testing.

In the connection-oriented or Type 2 case, a data link connection is established and terminated with the SABM and DISC commands, respectively, much like in HDLC. This allows error and flow control with modulo 8 sequencing and the conventional RR, RNR, REJ, and FRMR procedures. As Fig. 11.11 shows, the control field format is that of an HDLC I frame, with the P/F bit used just as in the ABM. Frames are acknowledged, of course, by either piggybacking on an I frame or using an RR frame. There are also the usual HDLC timeout and retry parameters.

The 802.2 standard also includes a good deal of formal specification and state diagram detail, which will not be considered here. As a final note on LLC we point out that the 802.2 LLC is not mandatory, and in some cases may be omitted or replaced by other data link protocols.

The *network layer* in most LAN protocol is rather weak compared to WAN protocols. The traditional functions of routing, sequencing, and flow control are frequently done in adjacent layers as the preceding LLC discussion suggested. Clearly the LLC PDUs have some packet characteristics, and are even located in the usual packet position within the frame.

However, one place where a network protocol is essential is with LAN *internets*. Here routing may involve links in many networks, each with different characteristics. There may be alternative internet routes with flow control and end-to-end acknowledgment required. A good example of such a protocol is the Xerox *Internet Datagram Protocol* (IDP) [9]. It is part of the Xerox Network System architecture, which also includes Ethernet and a number of higher layers oriented towards office applications. Originally known as PUP, it is similar to the ARPANET IP discussed in the last chapter.

An IDP *internet datagram* goes into the data field of an Ethernet or other LAN frame. It begins with an optional checksum, followed by the packet length. The transport control field contains the *hop count*, i.e., the number of networks traversed, and datagrams are discarded when this count reaches 16. They are also discarded if the *maximum packet lifetime* is exceeded. Source and destination addresses include subfields for the *network* (32 bits), the *host* or processor (48 bits), and a *socket* or process (16 bits) within the host. There is also a *multicast bit* for distinguishing group and broadcast addressing.

The *transport layer* is as important in the protocol of LANs as in other networks. Recall that its basic function is to add whatever protocol is necessary to provide the required uniform quality of service regardless of the underlying network. This requirement for transparency and reliability can be substantial where there is a weak network layer or an internet of heterogeneous LANs. The transport layer also allows existing high-level protocols to operate without regard for the nature of the underlying network, LAN or otherwise. The preceding Xerox IDP network layer protocol works integrally with the corresponding transport layer protocol. The latter handles end-to-end packet transfers, data integrity, and sequencing, and also does some testing and maintains the routing data base.

Another transport protocol example is found in the General Motors *Manufacturing Automation Protocol* (MAP). This is the result of a major effort led by GM to standardize LAN hardware and protocol for use in automated manufacturing. For the LAN itself, the IEEE 802.4 broadband token passing bus has been selected. For the crucial transport layer protocol only Class 4 of the OSI and NBS standards is used. This class, described in the previous chapter, assumes a worst-case underlying datagram network. It provides full error detection and recovery, including checksums, window flow control, sequencing, lost and duplicate packet resolution, and a variety of error-handling mechanisms. The MAP version also provides for optimal connectionless datagram operation (called *unit data* by NBS), which is oriented primarily towards gateway operations.

The remaining upper protocol layers are not particularly unique for LANs, since the transport layer provides a uniform interface to the network. The particular session, presentation, and application layer functions required depend on the purpose and capabilities of the network, and this can vary greatly between, for example, an office and a manufacturing environment. All of the upper-level protocol discussions in the previous chapter are directly applicable to LANs.

The use of LANs is growing steadily. As this technology becomes more established and pervasive, users will increasingly depend on LANs for critical applications in office, service, military, and manufacturing environments. This dependence, combined with the geographical span of LAN and WAN internets, makes LAN *network management* an issue of growing importance.

A *network manager* has centralized responsibility for monitoring and controlling the network, typically through some type of hierarchial structure with access at all protocol layers. Management of single or small clusters of LANs is facilitated by the short distances spanned, inherent reliability and intelligence of the technology, and single-owner nature of LANs. For more diverse internets, involving also perhaps VANs or leased telephone lines and complex gateways, the management problem is much more involved.

Monitoring functions include the collection of performance statistics for making operational improvements and the early detection of errors. Traffic information may also be collected for routing and flow control algorithm use, as well as for administrative functions like billing and directory maintenance. Another increasingly important role of monitoring is for security analyses to detect unauthorized or improper users. Various kinds of monitoring information are typically accumulated at selected points throughout the network, then dumped on request to the network monitor, where it is processed into a data base for later use.

The network management *control* function includes a number of both routine and exceptional conditions. Error detection and correction is an obvious control function. Even if the normal protocol recovers from an error, e.g., with a reset, the network manager may be used to effect related changes. Errors that cannot be quickly corrected should, if possible, be reported to the users concerned. Control also includes configuration tasks such as network initialization, adding and deleting stations, and topological changes. Routine and nonroutine maintenance of hardware and software is another important function.

Finally there are lots of potential security problems with LANs which must be managed. Since LANs are inherently broadcast systems, any user could conceivably eavesdrop on any traffic on the network. This might also be done remotely via a telephone access port or interconnected network. Network management security functions might include limiting user access through authorization codes, documenting usage with logs and audit trails, and the use of encryption.

Network management procedures for LANs are not yet well established, but several organizations are working on appropriate standards, including the 802.1 committee. Currently for most operational LANs network management is either nonexistent or proprietary, but this will certainly change in the future.

In addition to the general protocol issues discussed in this section there are others worth mentioning. First, since LANs essentially provide the complete physical and data link layers, any network architecture could conceivably be used with them by just substituting for these two layers. As already noted, this is exactly the case with the Phase IV version of DNA. There are also many more or less complete proprietary network architectures for LANs, as exemplified by Intel's iLNA [10], but these are far too diverse and specialized to categorize here. However, in the next section we shall survey some of these proprietary LANs and touch on a few of their more interesting protocol characteristics.

11.5. Other Access Methods and Examples

Although random and token access are used in the great majority of actual LANs and are now well standardized, there are several other techniques of practical interest. In addition, LAN functions can be provided by digital PBXs which allow both voice and data to be distributed over a local area. In this section we shall look at a few of these other LAN access methods, including the PBX approach. Then we include a concise survey of a number of commercial LANs that are both widely used in practice and have interesting and instructive characteristics.

We have previously considered physical and logical ring LANs with only token access, but there is no reason random access methods could not also be used. An interesting variation is the *contention ring*, in which a station with traffic to send first listens for a carrier on the ring. If there is none, then it simply begins transmitting as in pure CSMA. At the end of transmission the station removes its data frame and *inserts the token*. If the token goes completely around the ring and returns, the originating station removes it, the network returns to the contention mode, and any station may then begin transmission randomly.

On the other hand, if a station with traffic senses a carrier on the medium it must wait for the inevitable token before transmitting and the LAN now operates like a token ring. Each data frame is always followed directly by a token. Thus for low loading the LAN operates with random access, and shifts smoothly to token operation under increasing loading. This is the best of both worlds, with no station locked out for indeterminate periods. Note, however, that provision must still be made for such eventualities as collisions and lost tokens.

Another well-known access method is the *slotted ring*, of which the British Cambridge Ring is the most publicized. This LAN contains a series of *message slots*, each containing a *minipacket* which continually circulates around the ring. Thus there are multiple minipackets, and each contains a *full/empty* flag. Stations with traffic just insert it, along with sufficient addressing, into the next empty minipacket and set the full flag. It goes around the ring, is copied by the addressee, and returns to the originator, who then sets the empty flag. There is a parity bit for the error control and the addressee may also set flags in the minipacket trailer to indicate an inability to accept the packet, i.e., busy. The minipacket format is shown in Fig. 11.12. The *monitor bit* is a flag set by the ring controller to identify empty minipackets with their full flags erroneously set.

The question of slot length is worth some comment here. Long slots and/or short rings necessitate the introduction of artificial delays to prevent

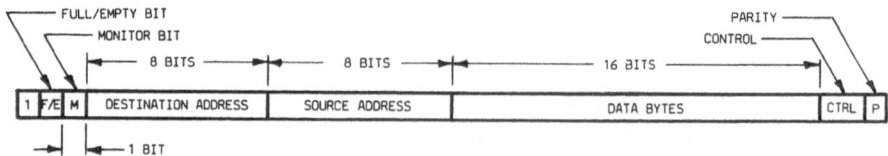

Figure 11.12. Cambridge ring minipacket format.

the slots from "overrunning their tails." But such delays increase buffer cost and access time. Conversely, very short slots hold little user data and have a large overhead. The experimental Cambridge ring [11] operates at 10 Mbps on a twisted-pair cable with each minipacket containing only two bytes of data! However, the network was never optimized for throughput, and can contain several minipackets.† Conventional data link protocol frames are obtained by combining up to 1024 data minipackets with others for the header, checksum, and logical channel information.

Register insertion is another attractive technique, which operates conceptually much like a railroad siding. Each station has a shift register that holds one packet, and which can be shifted in and out of the ring path. A packet to be transmitted is first placed in the isolated register, then connected in the ring either between two consecutive packets or at a time when the ring is idle. This instantly lengthens the ring by the length of the shift register and transmits the message. Any following traffic also goes through the shift register. Once the packet goes all the way around, it eventually returns to exactly fill the register again, at which time the register can be switched out of the ring and its contents replaced by the next packet to be sent. Provision must also be made, of course, for normal buffering and address inspection of all incoming traffic. From an access standpoint register insertion differs from other ring access techniques for two main reasons. First, the length of the ring is continuously varying, and the delay along with it. Second, new packets are inserted in front of existing ones, so that an old packet may be shifted further back at each passing station.

One of the major advantages of *broadband* LANs is their potential capability to carry many different channels simultaneously with FDM. This allows different groups of stations to use access methods most appropriate to their functions. For example interactive office users might share a frequency band over which CSMA/CD is used, while file transfers could be assigned to fixed bands and process control run on a third band with token access. Many attractive combinations are possible, but always at the price of more complexity and financial cost.

Over the years many other access methods have been proposed, some of which are variations on basic reservation and priority approaches. Most have not found widespread use in practice, so we shall not pursue them here. Rather, we shall look briefly at the local networking capabilities of the digital PBX.

The modern third-generation PBX is attractive for local data communications for several reasons. Switching is done digitally under computer control and voice is digitized either at the PBX or the user's handset, usually at 32 or 64 kbps. Thus it is straightforward to extend the hardware to also

---

† One might say it takes small bites, but eats exceedingly fast.

handle data; so simultaneous voice and data can be handled easily by a PBX. Furthermore, the wiring is generally already in place and the equipment is familiar to most users. There are also a number of possible digital PBX interfaces such as X.25 to WANs and to computers.

On the other hand, the PBX is the hub of a centralized star network and, although convenient for management, a failure can disable the entire network. The bandwidth is also presently rather limited because of the dominance of voice over data and the speed of the internal PCM switching hardware.

Despite some portrayal of the PBX vs. LAN issue as a winner-take-all struggle, there is a place for each. A feasible scenario is that the PBX would handle voice and perhaps such data applications as files and continuous monitoring devices. Numerous cable LANs would then carry strictly data traffic for specialized applications such as among departmental office terminals or manufacturing equipment. The cable LANs could then access remote networks and the telephone system via gateways to the PBX. Other scenarios are clearly possible, but the one here illustrates the issues.

We now consider some representative examples of vendor LAN products. Although the LAN industry is still relatively young and volatile, these example LANs all have well established customer bases. They also illustrate the wide range of alternative approaches possible with a sophisticated technology prior to the development of widely accepted standards.

One of the oldest LANs is the Datapoint *Arcnet*, introduced in 1977. It is a baseband token-passing bus operating as a logical ring at 2.5 Mbps over coaxial cable. Line coding is a variation of the established NRZ technique, and the bandwidth is designed to fit easily into a standard 6-MHz TV channel. There may be up to 255 stations connected in different topologies via distribution nodes called hubs. Repeaters are used every 2000 ft on long installations.

The proprietary Arcnet protocol encompasses the physical and data link layers, with elaborate procedures for initialization, station addition and deletion, and management of the logical ring. Frames can carry up to 253 bytes of data. The LAN originally supported only Datapoint office equipment, which performed all necessary upper-level protocol. More recently it has been adapted for use with Radio Shack personal computers, and there are several VLSI controller chips that support it.

The *Hyperchannel* LAN [12] was developed by Network Systems Corporation, also in 1977, to provide communication between large mainframes and their peripherals. It is a baseband coaxial cable network that operates at 50 Mbps. Access is via CSMA with collisions resolved by a proprietary time-slot priority reservation scheme known as collision avoidance, hence CSMA/CA. The bus topology has repeaters every 6000 ft, and there are bridges to other Hyperchannels. Up to 64 stations can be connected to each LAN.

Hyperchannel protocol supports the physical and data link layers using a modified form of SDLC. This LAN is clearly oriented towards sophisticated, high-bandwidth stations. Specialized adapters have been developed for various computers, mass storage devices, and terminals. There are also interfaces for satellite, optical, and radio links, and for other network types including the digital telephone network.

Wangnet is another proprietary LAN designed to support the office product line of its vendor, Wang Laboratories. This LAN operates over a dual broadband CATV cable, and has three distinct and different main bands within its 10–350-MHz bandwidth. A tree topology is used, with each half of the dual cable carrying traffic in opposite directions as shown in Fig. 11.13. All transmitted traffic goes to the network headend where the inbound cable simply loops around to form the outbound cable of the pair. This eliminates the need at the headend for the frequency band translation required on single-cable broadband LANs.

The *Wang Band* is a conventional 12-Mbps, CSMA/CD LAN using a data link protocol similar to HDLC. Thousands of devices can theoretically be connected to this band. Station connections are via a Cable Interface Unit processor which executes all protocol layers up to the session layer. The Wang Band is intended to provide communications among Wang office equipment and computers.

The *Interconnect Band* provides three types of point-to-point data links on a permanent or switched basis. The first type consists of 16 FDX permanent circuits at 64 kbps, and the second, of 32 similar circuits at 9.6 kbps.

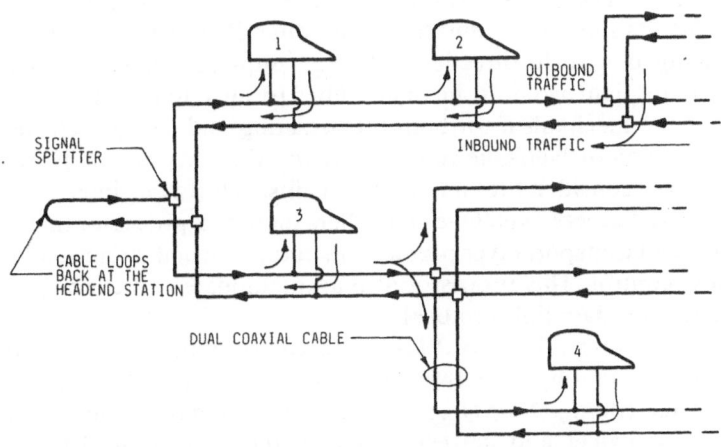

Figure 11.13. Wangnet dual broadband cable LAN.

Then there are 256 FDX 9.6 kbps circuits switched via a separate vendor-supplied digital switch. Access to these channels is with radio frequency modems that are frequency-agile, i.e., they can be programmed by the switch to operate on different bandwidths. This is, of course, necessary since switched circuit bands are separated by FDM.

Finally, there is a *Utility Band* with 42 MHz of free bandwidth. This is sufficient for several TV channels, and can be used for analog video functions such as videoconferencing or security surveillance. Other user-defined applications are also possible. The Wangnet LAN illustrates the versatility possible with broadband. It also circumvents the inherent complexity and reliability problems of single-cable active headends that must transpose inbound and outbound frequency bands.

Net/One is actually a family of LANs that has evolved since the original introduction by Ungermann-Bass, Inc. in 1980. It has since grown to accommodate a wide spectrum of terminal devices and to take advantage of new technologies. The basic network is a standard *baseband* Ethernet with CSMA/CD at 10 Mbps. There is also an "economy" version that spans less distance with smaller cable and standard cable connectors.

The *broadband* version of Net/One uses a single CATV cable with an FDM band for each direction. This midsplit arrangement requires that each transmission go through a central headend station where it is modulated into the other band and broadcast to the receiver. Each network interface is connected via a radio frequency modem. Five channel pairs are available, each carrying 5 Mbps over a 6-MHz bandwidth for distances up to 15 km.

Net/One also comes in a 10 Mbps *fiber optic* version. To implement CSMA/CD on unidirectional fiber cables instead of bidirectional coaxial cables and to perform carrier sensing and collision detection requires some modification. The key innovation is the use of a central *star coupler*, to which all stations are connected with dual-fiber cables with the fibers conducting in opposite directions. The star coupler has the unique capability to distribute light from an incoming fiber to all outgoing fibers. Hence it provides the mechanism for both broadcasting and carrier detection. This optical version of Net/One runs at only the Ethernet rate of 10 Mbps, but it provides considerable immunity against electrical interference and unauthorized access. Net/One controllers include a processor that provides network and transport layer protocol, including virtual circuits, addressing, and flow control. This processor also accommodates a variety of common interfaces and data link protocols.

The Complexx Systems *XLAN* is a good example of a low-end, low-cost LAN. Operating over twisted pairs or coaxial cable at 1 Mbps with Manchester coding, XLAN uses pure CSMA and relies on the data link protocol to detect and correct errors by retransmission. This approach simplifies the transceiver considerably, and the controller includes a four-port statistical multiplexer with one modem port for remote access via telephone. Although the access method leads to congestion as network loading becomes heavy, performance for installations involving dozens of active stations is comparable to much more sophisticated LANs.

Our final LAN example is the relatively new AT&T *Information Systems Network* (ISN) [13]. This is basically a digital PBX system, but is designed for packet-switched LAN operation rather than conventional circuit switching. The basic configuration is a central Packet Controller built around the AT&T Datakit™ packet switch, which also does the network management. Terminals, computers, and concentrators connect in a star topology over either fiber optic cables for high-speed devices or, for slower ones, twisted pairs such as existing voice telephone wiring. There is, naturally, a modem pool to provide access to the telephone network.

Terminals, concentrators, and other similar LANs interface to the packet switch through several types of addressable *interface modules*. When a call is placed by a terminal device, a virtual circuit is set up between source and destination via the interface modules. The switch stores the address of the two concerned interface modules internally in a high-speed *address translator*, thus defining the virtual circuit path.

When terminal traffic is sent to the switch, it is first packetized by the interface module in a proprietary internal format. Once the interface module contends successfully for access, it sends one packet containing up to 16 data bytes to the address translator via the *transmit* (incoming) *bus* at 8.64 Mbps. The packet address is quickly translated from source to destination, and this packet is sent to the destination interface module over the *broadcast* (outgoing) *bus*, also at 8.64 Mbps. It is then converted to the appropriate form for transmission on to the destination terminal.

In the above example the interface modules actually perform a PAD function for asynchronous terminals. Clusters of such devices may also access the switch via a *concentrator*, which packetizes and statistically multiplexes the terminal traffic over a high-speed fiber cable to the switch. Concentrators and interface modules may also provide XON/XOFF flow control and clock timing as required. It is interesting to note that RS-232C connections are used for up to 500 ft at 9600 bps with ISN, which is ten times the specified distance. Longer connections are made with asynchronous line driver units.

The unique bus contention scheme operates over a separate slotted *contention bus* as shown in Fig. 11.14. At the beginning of every time slot each interface module with traffic to switch begins sending, bit-by-bit, a *contention code* made up of a priority code and its own address. So long as the bits agree with the bus the module continues to send more bits, but drops out as soon as disagreement is detected. The bus is so short that propagation delay is negligible. Only the numerically highest contention

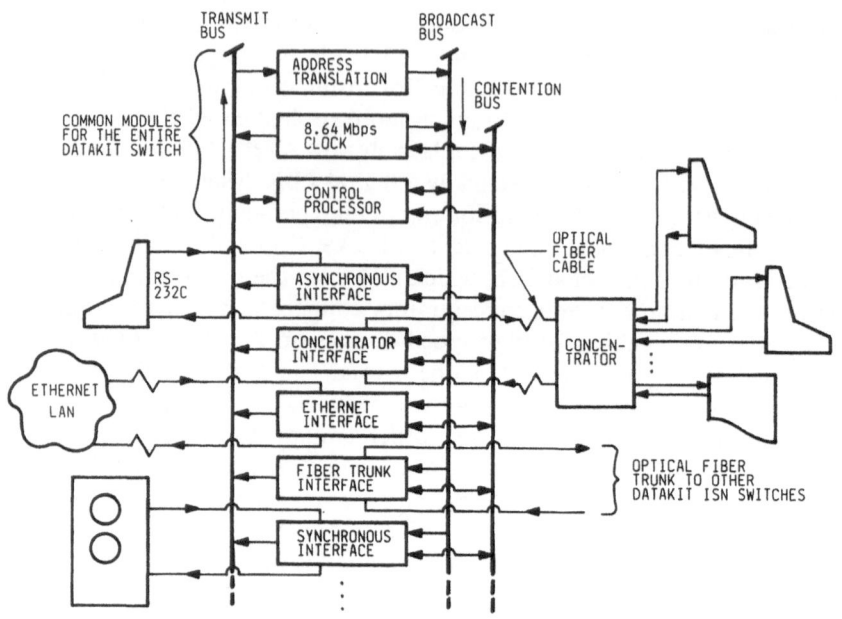

Figure 11.14. AT&T ISN packet-switched LAN.

code in the slot is successful, and the corresponding interface module then sends one packet on the transmit bus in the following slot. To make this collision-free contention process fair, the priority code can be increased for stations that have contended unsuccessfully.

One ISN Packet Controller can switch up to 1200 virtual circuits. For larger applications these LANs can be interconnected via fiber optic cable to produce internets that operate either as a collection of individual LANs or as one combined LAN. Since the internal protocol is used on internodal trunks, no specific internet protocol is needed other than the end-to-end virtual circuit and routing information. There are also interfaces and gateways to other different networks, including X.25, SNA, and the T-carrier system. ISN is clearly an innovative and instructive form of the LAN.

The preceding examples and discussions illustrate the rich diversity of local area networks. Not only is this due to technological ingenuity, but also to the many different applications in which LANs can be used. The predominant interest today is in the office environment, where there is great hope that they will finally improve the long lagging productivity of white collar workers. As appropriate standards solidify, LANs should also become the backbone of automated manufacturing and process control systems. Other important areas include education, home and apartment intercommunications, and highly automated weapons and space systems.

Each LAN approach has its own merits and demerits for a given application. Such issues as broadband vs. baseband, cable vs. PBX, wire pair vs. optical fiber, and random vs. token access are often academic, and we do not intend to get bogged down in them here. Each situation must be judged individually by the ultimate user and operator. One common figure of merit is the *cost per connection*, which assumes the actual cable cost is minor compared to the transceiver and controller. Another important consideration is the ease of expansion, modification, and interconnection of an LAN once it is installed. Regardless of the LAN or WAN selected, it is always subject to unauthorized use in order to gain information, use resources, disrupt normal network operations, or damage physical equipment. In the next chapter we look at the question of how to guard against such eventualities by use of good network security practices.

References

1. N. Abramson, Packet switching with satellites, *Proc. NCC* **42**, 695–702 (1973).
2. L. Kleinrock and F. A. Tobagi, Packet switching in radio channels: Part 1 — Carrier sense multiple-access models and their throughput-delay characteristics, *IEEE Trans. Commun.* **23**(12), 1400–1416 (1975).
3. R. M. Metcalfe and D. R. Boggs, Ethernet: Distributed packet switching for local computer networks, *Commun. ACM* **19**, 395–404 (1976).
4. *CSMA/CD Access Method* (IEEE, Std. 802.3), IEEE, New York (1985).
5. R. C. Dixon, N. C. Strole, and J. D. Markov, A token-ring network for local data communications, *IBM Syst. J.* **22**(1/2), 47–62 (1983).
6. *Token Ring Access Method* (IEEE Std. 802.5), IEEE, New York (1985).
7. *Token-Passing Bus Access Method* (IEEE Std. 802.4), IEEE, New York (1985).
8. *Logical Link Control* (IEEE Std. 802.2), IEEE, New York (1985).
9. *Internet Transport Protocols*, Xerox Corporation (1981).
10. R. Ryan, G. D. Marshall, R. Beach, and S. R. Kerman, Intel local network architecture, *IEEE Micro* **1**, 26–41 (1981).
11. R. M. Needhan and A. J. Herbert, *The Cambridge Distributed Computing System*, Addison-Wesley, London (1982).
12. J. J. L. Pang, *Network Systems Corporation Hyperchannel Local Area Network*, SIT 661 Term Report, Southeastern Institute of Technology, Huntsville, Alabama, February (1983).
13. A. S. Acampora and M. G. Huchyj, A new local area network architecture using a centralized bus, *IEEE Commun. Mag.* **22**(8), 12–21 (1984).

Suggested Readings

W. Stallings (ed.), *Local Network Technology* (1st and 2nd eds.), IEEE Computer Society Press, Washington, D.C. (1983 and 1985).

These two IEEE tutorials contain a large collection of more recent key papers on LANs.

K. J. Thurber and H. A. Freeman (eds.), *Local Computer Networks* (2nd ed.), IEEE Computer Society Press, Washington, D.C. (1981).

A nice tutorial collection of many of the important earlier papers as the LAN technology was developing.

Introduction to Local Area Networks, Digital Equipment Corporation (1982).

A market-oriented survey of LANs that is very easy to read and surprisingly complete.

IEEE Journal on Selected Areas in Communications (Special Issue on Local Area Networks) 1(5), (1983).

A great collection of papers on different vendor LAN products at a solid but easily readable theoretical level.

C. Tropper, *Local Computer Network Topologies*, Academic, New York (1981).

This text presents lots of theoretical analysis for all kinds of access schemes that have been proposed for LANs.

Check Your Understanding of Chapter 11—True or False?

1. The original Ethernet addressing concept was a globally flat address space.
2. Token access is generally more suitable than CSMA/CD for time-critical applications.
3. LANs operate essentially as distributed packet switches.
4. It is not possible to use true token access with a bus topology.
5. The IEEE 802.3 preamble serves to set the receiver clock recovery circuitry.
6. Internetting is only of minor importance with LANs.
7. With ALOHA access the throughput between a pair of stations can never be above 50%.
8. Much of the complexity of the 802.5 protocol deals with ring management.
9. Most modern LANs operate at data rates well below 1 Mbps on the medium.
10. Carrier sense does not work with baseband LANs since there is no carrier to sense.
11. The 802.4 token consists of an entire frame with no data field.
12. Manchester coding inserts clock information into the actual data stream.
13. Cable failures in 802.5 LANs are resolved by switching to an alternative cable.
14. Broadband LANs may use dual cables to avoid frequency translation at the headend.
15. Clock signals are recovered from the data by the LAN controller.
16. The IEEE 802 standards use the same FCS algorithm as used in SDLC.
17. The bandwidth of a broadband LAN may actually be less than that of a baseband LAN.
18. The 802.2 LLC protocol is unique to each 802 LAN MAC standard.
19. For p-persistence with n stations, p should always be greater than $1/n$.
20. In IEEE 802.5 the token is indicated by a single bit in the frame control field.
21. The 802.2 LLC protocol is actually a subset of the HDLC data link protocol.
22. The extensive 802.4 protocol includes a CSMA/CD alternative among its many options.
23. It is possible to use HDLC as the protocol for an LAN.
24. The Xerox XNS and GM MAP are both high-performance sportscars made in Japan.
25. Ethernet collisions are followed by a cram, where all stations dump all their currently buffered data onto the transmission medium forthwith.

12

Network Security

12.1. Security Considerations

Computers and computer data networks are becoming increasingly crucial in the day-to-day operation of our modern society. We are fast becoming dependent on networks in such fundamental areas as banking, retailing, law enforcement, education, government administration, and many more. Most people are barely aware of, much less concerned about, the huge amount of their personal information that is accumulating in various government and private data bases, with much of it conceivably accessible via one or more data networks. Although already widespread, this trend toward storing more information in computers, providing more network access to it, and relying more heavily upon it is accelerating and promises to be pervasive in the near future.

Properly controlled, the potential benefits from the communication of computerized information are great. Most of us enjoy a much higher standard of living today than in past decades, and at least some of it is the result of better communications. On the other hand, the potential adverse

419

effects are also great if such information is misused. Besides the obvious civil rights issues involved, there is also definite potential for crime in the form of information theft, falsification, and destruction.

All of these considerations point out the increasing need for adequate *security* of computer data communications networks. This chapter addresses the topic from a rather general standpoint that is applicable to all types of data networks. We begin by looking at so-called "white-collar" computer crime and some representative ways to attack a network, because understanding the enemy is the place to begin establishing a defense. Next we look at the particular vulnerability of the domestic telephone system since it carries a major share of all data traffic, either directly or through lines leased to most other public carriers. Many private networks also use telephone lines, but may have better control since network use can be more carefully supervised. Once the need for network security is established, we consider some practical techniques for providing it. Many of these techniques are based on plain common sense combined with an awareness of the problem. Finally we briefly examine the fascinating subject of encryption and, in particular, the federal encryption standard that is now widely used.

We shall define *network security* as the protection of the network from damage and of the transmitted information from unauthorized disclosure, modification, or destruction. Our main interest is in protecting data located in the network itself, but an equally important consideration is the security of data files stored at the terminal sites. We shall touch on this latter subject at times since there is considerable overlap between database and network security—if the data is compromised prior to transmission then network security becomes an exercise in futility.

The "risk-to-reward" ratio for the computer criminal today is quite favorable relative to more traditional forms of crime. Computer crime is usually considered an intellectual and "gentlemanly" (or "ladylike") pursuit that is strictly nonviolent and perpetuated against huge impersonal institutions (the "Robin Hood Syndrome") rather than individuals. Motives may range from raw avarice to simply the thrill of outsmarting the system, and infractions can vary from stealing a few minutes of processor time all the way to maliciously erasing a vital data base or embezzling substantial sums of money. Furthermore, the risk of a careful and intelligent criminal getting caught is low and, even if caught red-handed, prosecution is anything but certain. Federal statutes are rather embryonic and some states have no pertinent laws at all for certain computer violations.

There are many ways to attack a data network. Most publicity relates to the more esoteric approaches such as sophisticated wiretaps and skilled code-breakers, but outside of international espionage most computer crime is the result of insider personnel helping themselves to improperly protected data. In many cases a bored or disgruntled employee can print out a financial or customer file or reconnect a personal terminal just out of curiosity, and then be unable to resist temptation when the value of the resulting information is realized. With the proliferation of personal computers having internal modems, it is also extremely simple to dial up the company computer and copy files at home if such access is not restricted.

More formally, we can distinguish two major categories of network attack: passive and active. A *passive attack* involves only monitoring the network to get information, perhaps with a simple wiretap. The result might be actual message content, encrypted data for later analysis, or perhaps just traffic patterns regarding message direction, size, and frequency. Passive attacks are generally hard to detect but easy to neutralize with encryption and dummy messages. An *active attack*, by contrast, attempts to interfere with the network traffic in some manner. This might involve changing the numerical value of a bank transaction, removing traffic, reinserting previously recorded messages, or even impersonating one terminal to the rest of the network by intercepting traffic and substituting false and deceptive messages. This last approach is known as *spoofing*. Active attacks are generally difficult to foil but relatively easy to detect, since the attacker must actually insert a terminal directly in the original circuit. Other attacks are also feasible, such as *scavenging*, i.e., searching randomly for information in wastebaskets or by talking informally to knowledgeable people.

An important and instructive network for security consideration is the telephone system. Not only does it carry a huge amount of data directly on leased and dial-up lines, it also provides lines to a large proportion of private and value-added data networks and to practically every home and office. Since it is a public network, data security is not a paramount network concern; rather, the system is designed for maximum physical survivability in the face of natural disaster or civil disturbance. It must serve the general public more or less impartially, so there is very limited control over users or their terminal devices. Furthermore, much of the network is exposed as overhead or underground wiring, or as radio and satellite beams.

Since local feeder and distribution cables are periodically exposed at wiring access boxes, either above ground or in manholes, there is ample opportunity to connect a tap. In high-rise buildings telephone wiring goes through cableways with wiring closets on each floor where all wiring is exposed. A specific circuit can often be located by simply connecting a modem and printer across different lines until an interesting printout is obtained. Alternatively, a target phone can be called with a short tracer signal which can then be quickly identified by the tap.

Calls on trunk routes can also be intercepted easily in many cases. Since trunks only connect telephone offices, it may be difficult to locate a specific dial-up call. However, for voice channel signaling the signaling tones immediately identify the *called* party. Similar results hold for groups of calls with common channel signaling, but the actual user traffic goes over a separate channel and perhaps even another trunk so identification is more difficult. However, this signaling channel provides considerable traffic analysis. Trunks of wire pairs can be tapped at various access points, even when buried. Radio and satellite channels are easily intercepted with hidden or innocuous antennas, such as modified direct-broadcast TV dishes. Often there are transmitter antenna sidelobes that allow covert reception from behind the main transmission direction. Coaxial cables are more difficult, since they often contain pressure alarms and reflectometers to detect faults due to environmental effects. Lightguide cables are, at present, difficult but not impossible to tap optically along the cable. However, repeaters must be located periodically on long-haul fiber optic cables, and are subject to conventional electronic taps. Figure 12.1 illustrates a number of points at which the domestic telephone system is vulnerable to unauthorized access.

Other public carriers are also vulnerable to security violations. Since many use leased telephone lines, they are subject to the same penetrations as the telephone systems. This is particularly true if fixed routing is used, since a pair of users will always be assigned the same internal lines. Besides taps, unauthorized access is possible through dial-in connections, provided a viable identification code can be stolen or guessed. This is not always difficult, since user-selected passwords often are chosen to be familiar names or dates, while network-assigned codes usually are written down and occasionally misplaced or thrown away.

Terminal equipment can also be an attractive place to attack. Processor-based equipment contains high-frequency clock signals, traces of which can radiate along cables and power lines. The radio frequency signals are easily received and may provide information about the equipment operation. This is particularly true of local digital PBXs, since all leased and dial-up off-premises communications goes through them.

There are many other considerations and techniques for attacking data networks, but our discussion has illustrated some of the more common and straightforward. The point of the material is not to encourage wiretapping, but rather to develop an awareness of the varieties of approach that are readily available to a determined perpetrator. One of the most fundamental problems in network security is simply a *lack of awareness* that it is needed. Passive taps can run for years without detection if nothing is done to check for them. Having now introduced the security problem, in the next sections we describe some techniques for solving it in part or in full.

12.2. Security Techniques

Once the need for network security has been acknowledged and the general nature of potential threats understood, the effecting of security

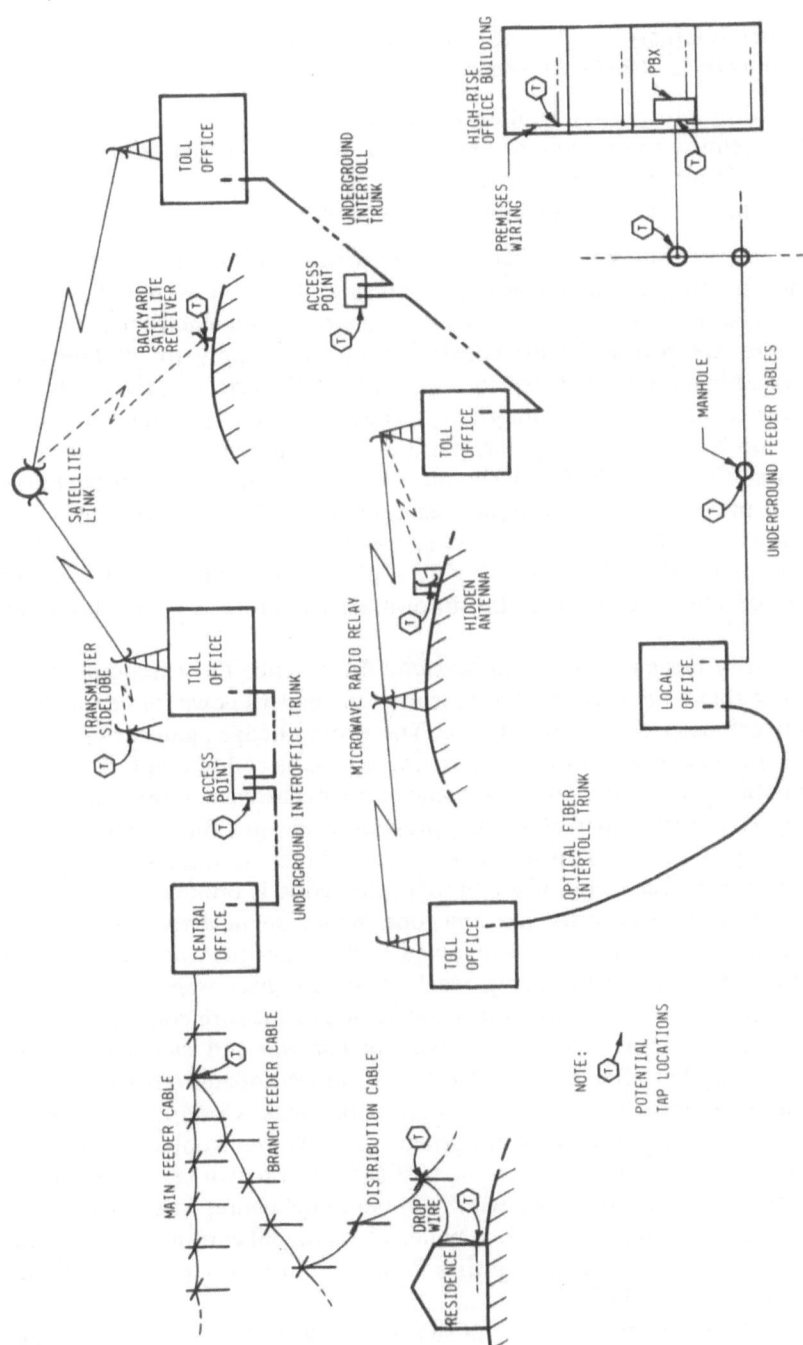

Figure 12.1. Telephone system vulnerability.

techniques is largely a matter of common sense and diligence. The conventional objectives of network security are to

o minimize the chance of successful attack;
o maximize the chance of detecting a successful penetration;
o minimize the damage from an attack;
o minimize the time and effort for recovery.

Good security thus goes beyond trying to prevent successful attacks; it includes the proper reaction and recovery as well. It is naive to believe that perfect security is possible and that any network is impenetrable.

There are several major considerations that apply to all kinds of networks. First, the *level of security* should match the *perceived threat*. Overly elaborate techniques are simply not needed to protect normal household records on a personal computer, particularly if the cost is far more than the value of the records. Similarly, there is little point in an industrial firm trying to prevent the FBI from accessing its network. The government has the knowledge, resources, and usually also the legal basis, to tap any domestic network and to break highly sophisticated codes. If the security costs more than the value of the information it protects—you shouldn't do it!

Security measures *must not be static*. In practice there is no absolute guarantee that a network is totally secure. One never knows for certain that the network has *not* been penetrated. The way to hedge against this eventuality is to change such things as passwords, access codes, crypto keys, and safe combinations both periodically and at the occurrence of random events like key employee departures. This procedure ensures that only intended personnel gain access to the network and also limits the amount of information that can be recovered if a cryptographic code is broken.

Good network security, like any good management, always has *definite goals* that are known to those involved and always provides feedback to determine if the goals are being reached. It also *fixes responsibility*, both general and specific, on individuals and indicates the sure consequences of irresponsible actions. It includes adequate *back-up* and *emergency repair procedures* so that after a successful attack normal operations can resume with minimal inconvenience to the network users. Contingency cryptographic codes are particularly important in recovering from a compromise of current codes. However, even the best security procedures are worthless unless they are faithfully executed. Such mundane chores as changing keys and passwords, inventorying documents, checking all entrants into and out of secure areas, and backing up files daily should be done religiously, no matter how boring they seem.

In many cases some degree of *indirect security* is obtained from other operating practices. For example, the use of time-division multiplexing and data concentration techniques can make it difficult to locate a specific channel within the composite bit stream. Data compression coding with run-length or transform codes can reduce redundancy that is vital to breaking ciphers. The use of little-known proprietary character codes and protocols can also make it difficult to locate and identify information within intercepted data. Even such common practices as scrambling data in modems and bit or character stuffing can be helpful against simple attacks. Finally, many large network protocols such as X.25 have provisions for closed user groups and one-way calling which are effective against other network users.

Indirect security can also result from good *personnel management*. People involved in computer operations tend to be highly intelligent and well educated, and usually take great pride in their personal and professional skills. They may be extremely capable, but may also be egotistical, independent, and just plain weird. A heavy-handed, autocratic, "Theory X" management style is not only unnecessary for such people, but can quickly alienate them, with disastrous results to the organization. The computer trade is full of horror stories about "logic bombs," "superzappers," and "trapdoors" that have caused great losses of information long after the responsible party has departed the firm. On the other hand, fair management policies that respect the individual as a person can create values such as loyalty, pride, and job satisfaction that are essential for good network security.

At a more concrete level, there are many *direct security* techniques for networks. An obvious measure includes limiting access to network centers and nodes to only those with specific need for access and then verifying that they are currently authorized for such access. This can be done with verification devices, or with locks and guards. Personnel verification techniques are commonly placed in three categories. They may depend on some *special knowledge* such as a password or lock combination; upon some *physical item* like an identification card or a key; or upon some unique *personal characteristic* like fingerprints, voiceprints, hand shape, or handwriting motion. Some of these techniques, such as voiceprints, are not always reliable but can be used very effectively in combination with others.

Passwords are the most common technique for limiting network access. As already indicated they are often compromised inadvertently and require special care to be an effective security measure. A password should be random in nature, changed at appropriate times, never written down, and never displayed when entered on a terminal. At the computer, the password file should be carefully protected and perhaps encrypted to prevent theft. Password security can be strengthened by requiring multiple levels of password access and also by using *call-back*. In this latter procedure a caller requesting access to a central computer is automatically disconnected; the computer then immediately calls the terminal back, thus at least verifying the location if not the actual user. This is awkward for large public networks because of the reversal of long-distance toll charges. If multiple attempts are made to gain network access with incorrect passwords the corresponding terminal should be disconnected until a satisfactory explanation is obtained, since such activity generally indicates illegal attempts to guess passwords by so-called "hackers."

Another important network security consideration is *environmental threat*. Fires, flooding in cableways, and rough handling of equipment can be as detrimental to network functioning as intentional attacks. Power surges from heavy equipment like welding and x-ray machines can also cause serious sporadic network failures. Rugged equipment, protected conduits, and suitable alarms are valuable assets in managing environmental threats. In the extreme, such disasters as fires may be intentionally created by perpetrators, who then take advantage of the ensuing confusion to steal information. Obviously, emergency actions must be planned and rehearsed to both handle these emergencies and maintain security. Combining fire drills with security penetration drills can be very effective in combating this combined threat.

Good network security is closely related to *network control*. Network monitoring can reveal unusual traffic patterns and access attempts. Accurate network logging and audit procedures can reveal unsuccessful penetration attempts, and also allow complete analysis of successful ones. When compromises do occur, good network control should effectively manage the recovery procedure.

For networks carrying sensitive organizational information of significant value to competitors or enemies, some type of *security classification* scheme is indicated. This may range from a simple *company confidential* classification up to the elaborate myriads of secruity levels and compartments used by national governments. It is convenient to think of security classifications in general as a matrix, or *lattice*, in which information is restricted in two ways. The classification *level* depends on the *consequences to the organization* or unauthorized disclosure; this is the familiar confidential, secret, top secret hierarchy. In the other dimension restriction is made according to a *need to know* criterion. Here a user may be authorized access to information only in certain areas in which he must work, for example engineering but not accounting, and then only up to his allowed classification level. With a lattice approach only a very few people should ever get a detailed overall picture of what is going on in the organization, and isolated security breaches will hopefully be confined to relatively minor blocks of information.

So far we have discussed network security in terms of physical, personnel, and procedural measures that can be taken. These are extremely important and fundamental techniques. However, particularly with public networks and leased telephone lines, the network user simply cannot ensure the physical security of his traffic or the integrity of network equipment and personnel. In such cases the only recourse is encryption of the data during transmission and perhaps also for storage. In the next section we look at the subject of data encryption.

12.3. Encryption

Cryptography has a long and fascinating history [1], much of it shrouded in mystery and international intrigue. Until fairly recently, encryption was little used in the private sector, and its use by the government for diplomatic, military, and intelligence purposes was classified and seldom discussed openly. Over the past decade or so this situation has changed dramatically.

First, with the growing reliance on electronic data processing and storage, private as well as government organizations are finding that valuable information can easily be copied electronically through data networks. This can often be done in very short time intervals and without leaving any trace. In addition, much of the historical mystique has been removed from practical data encryption through active publication and debate in the open literature. Most of this activity has been motivated by the publication of a federal data encryption standard giving complete implementation details. As is so often the case in data communications, a market need and an electronic technology have converged to produce effective and affordable encryption capabilities.

Cryptography (crypto) is the process of transforming information in such a way that it can be easily understood by the knowledgeable intended receiver but will be useless to anyone else. This is accomplished by providing the intended receiver with special information, usually in the form of a *crypto key code*, with which the message can easily be decrypted to recover the transmitted information. Keys must generally be kept secret from all unauthorized parties. For practical encryption techniques it is always theoretically possible to decrypt messages without a key but this should be so difficult that it will not be worth the effort. Thus for national security, very complex and expensive encryption may be justified, while for commerce and industry less elaborate methods will be entirely adequate.

Modern cryptography is based on families of algorithms, or *ciphers*, that operate by converting messages from *plaintext* into *ciphertext*, and conversely. The particular algorithm used is determined by the particular secret key selected. Modern crypto keys are typically just a long string of bits. Good security dictates that the encryption procedure be changed frequently, e.g., weekly, daily, hourly, or even per session. This makes changing the entire family of algorithms impractical, and so common practice is to simply change the secret key and not rely on the secrecy of the family of algorithms. If there are enough potential keys, i.e., if the *key length* is sufficiently long, then it will be impractical to break the cipher by trial and error even if the algorithm family is known. With this approach security depends on the secrecy of the key, and recovery from compromises can be accomplished by simply changing to a new key.

Besides solving the *privacy* problem of transmitted information being stolen, encryption can also provide *authentication* that the information received is valid, i.e., has not been modified en route or initiated by an imposter. So long as the spoofer does not have the correct key, any message falsification is immediately detected at the destination. Obviously secret keys must never be distributed over the transmission channel in the clear (as plaintext) since they could be intercepted. Key distribution is a complex topic, but common procedures are to use couriers or registered mail to distribute *master keys* which are then used to encode the *operational keys* used for actual session encryption. In *public key* systems the encryption keys are made publicly available since *different* (and secret) keys must be used for decryption.

Encryption may be done in theory at any protocol layer, although conceptually it is a presentation layer function. There is no concensus on which layer is best, but an obvious distinction is between encrypting on a link-to-link or an end-to-end basis. A related issue is where to actually locate the encryption hardware or software. The prevalent choices are either as a "black box" between the modem and terminal (DCE and DTE) or within smart terminal equipment such as intelligent terminals, nodal switches, computers, FEPs, and cluster controllers.

With *link-to-link* encryption there is a separate key for each link. Hence at each intermediate node a message or packet is decrypted, then reencrypted with a different key for the ongoing data link. The originator only needs one key for each node to which he is connected, but the message is converted to plaintext at intermediate nodes over which he has no control. Clearly the network itself must be secure, and this may be difficult to ensure with a public carrier. However, on the links the entire message, including header and checksum, is encrypted. This can make traffic analysis very difficult, particularly if additional dummy messages are intentionally sent.

Security of the plaintext at nodes is of major concern with link-to-link encryption. One feasible solution to this problem is the *security module* concept. This is a physically protected area where all decryption and reencryption is done, and the plaintext never leaves this "red" area. All other areas of the node are "black," i.e., they only handle encrypted traffic. Security modules may be constructed to be completely automated and sealed so that no operator need ever handle plaintext passing though the node.

Operational keys can even be downloaded to the module so that only the master key is loaded manually.

With *end-to-end* encryption only the information field is encrypted, so much useful traffic analysis can be done by an eavesdropper. However, again dummy messages can be intentionally sent to mask tell-tale changes in operational traffic. With this kind of encryption it is desirable that sequence numbers and error check fields be included in the encrypted data to detect spoofing. A major problem with end-to-end encryption is *key distribution*. Since different keys must be used between *every* potential pair of users, an enormous number of keys can be required in a large network—and the problem is even worse with internets.

With the further requirement for frequent key changes, direct end-to-end key handling and distribution can easily get out of hand. However, for small networks in which only a limited number of users require encryption, it is entirely feasible. In larger networks one solution is to establish a *key control center* (KCC) to distribute keys. Here an end user wishing to send encrypted traffic to another user first requests a key from the KCC. The center then verifies the sender and destination, generates a session key, and distributes this key to each user with each user's particular masterkey. Consequently each user only needs one permanent key to communicate with the KCC, but the security of the KCC is now paramount. Common locations in the protocol for end-to-end encryption are at the presentation, session, and transport layers, with the lower of these allowing both data and higher-layer protocol information to be encrypted.

Given the above considerations, the best approach when very high security is required may be to always encrypt on a link-to-link basis and also encrypt end-to-end whenever the security requirements justify it. This adds more overhead, and still does not completely solve the key distribution problem, but it may be a reasonable price to pay under the circumstances.

In 1977 the U.S. NBS announced a federal *Data Encryption Standard* (DES) [2], which was later adopted by ANSI. The DES began with considerable controversy over the National Security Agency reducing the original 128-bit key length proposed by the developer (IBM) to 56 bits.† In retrospect, however, the only publicly known consequence has been a consensus that by around 1990 technology should be advanced to the point where brute force searches for the key will be feasible, although expensive. The DES is specified for government operations other than defense and diplomatic traffic. It is also being adopted by an increasing number of commercial firms engaged in banking, manufacturing, and information collection and retrieval.

---

† Crytographic devices with key lengths greater than 56 bits cannot be legally exported from the U.S.

The theory behind the DES goes back to a 1949 Bell Laboratories paper [3] suggesting that good encryption could be obtained by many repeated basic operations to thoroughly scramble the data. These basic operations are *permutation*, the reordering the message bits, and *substitution*, the replacement of message bits with entirely new bits. In the DES there is one initial and one final permutation stage of each 64-bit block of data, with 16 stages, or *rounds*, of substitutions in between. The permutations are always the same, so the security of the DES depends on the substitution rounds, which, in turn, depend on the 56-bit key code. The original key is processed by a DES function called the *key schedule* to produce a separate 48-bit *subkey* for each of the 16 rounds, as shown in Fig. 12.2.

The figure also shows the detail of round two which, except for the subkey, is identical to the others. The 64 bits between stages are divided into 32 bit L and R groups within stages, with R of round n becoming L of round $n + 1$. At round 2 the 32 R bits are first expanded to 48 bits and added modulo 2 (XOR) to the 48-bit subkey K_2. The result is then divided into six-bit blocks and each block transformed according to one of eight highly nonlinear substitutions (S boxes) defined by the standard. The resulting four-bit outputs are recombined into a 32-bit word, permuted, and added modulo 2 with the 32 L bits to produce R for the next round. Since all the table operations are fixed by the standard, it is clear that secrecy depends entirely on the key.

There are several important operational considerations when using the DES. First, it can be implemented in either software or hardware. It is also

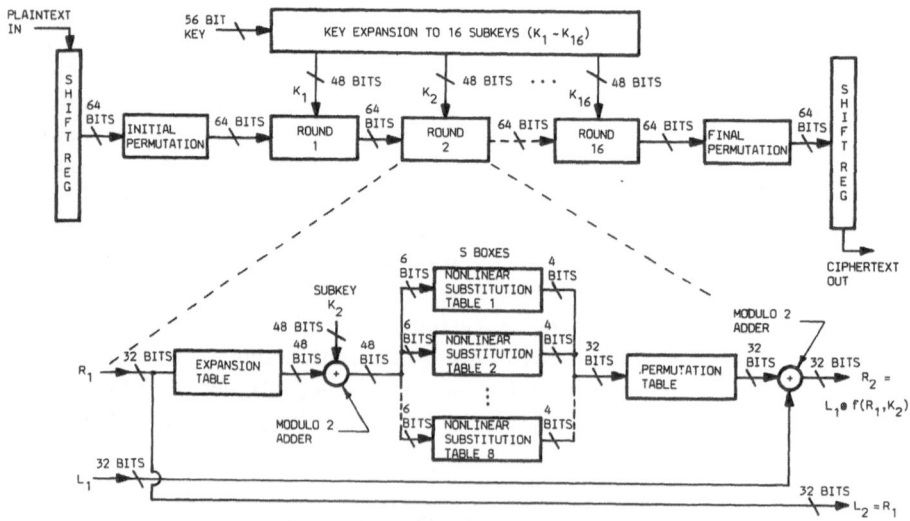

Figure 12.2. Structure of the DES.

reversible so decryption is performed with the same key and hardware. The basic mode of DES operation is on 64-bit blocks. This *block mode*, also called the *Electronic Code Book* (ECB) or key-autokey (KAK) mode, has some disadvantages for data communications traffic. Since each block is encrypted independently, it is possible that redundant messages and protocol formats could be exploited in an analysis to determine the key. However, for short one-time messages like key distributions the block mode is appropriate. In fact, 56-bit keys are usually formatted as 64-bit words by including eight parity bits for error protection.

Good encryption techniques should have the property of *error propagation*, which means that any bit change in the cipher text results in a large number of errors, ideally 50%, in the decrypted block. This provides protection against attacks by using small perturbations. Although the DES has this property within each block, with data transmission it is also desirable to have some dependence among encrypted blocks. Otherwise, for example, a spoofer might be able to remove or rearrange a message by blocks and not be detected. Besides ECB, the NBS recommends three other modes of operation [4], of which the first two provide this interblock dependence.

In the *Cipher Block Chaining* (CBC) mode 64-bit blocks are encrypted and transmitted, but are also fed back and added modulo 2 to the next such block. As Fig. 12.3 shows, any error in a transmitted block will cause error propagation in that block, and this will affect the next block; but following blocks are not affected. The *initializing variable* is a secret random bit pattern, often transmitted with the key, which is used only for the first block. Its purpose is to prevent attackers from using known headers, such as with SDLC, to compare plaintext with the corresponding ciphertext. In addition there is a 64-bit delay in both the encryption and decryption devices. Note, however, that the data and encryption rates are the same; e.g., for 9600 bps data the DES must execute at 9600 bps, too.

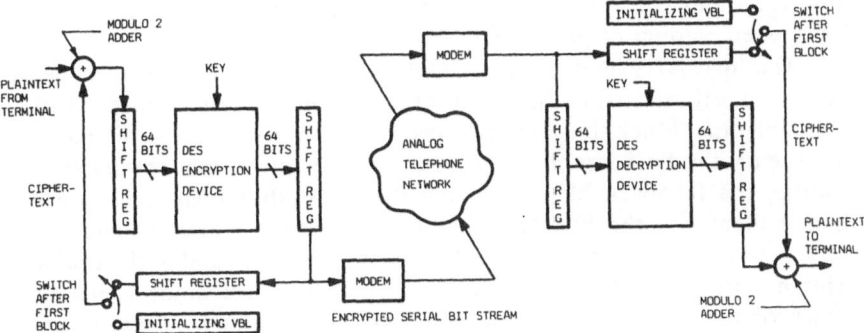

Figure 12.3. DES cipher block chaining mode.

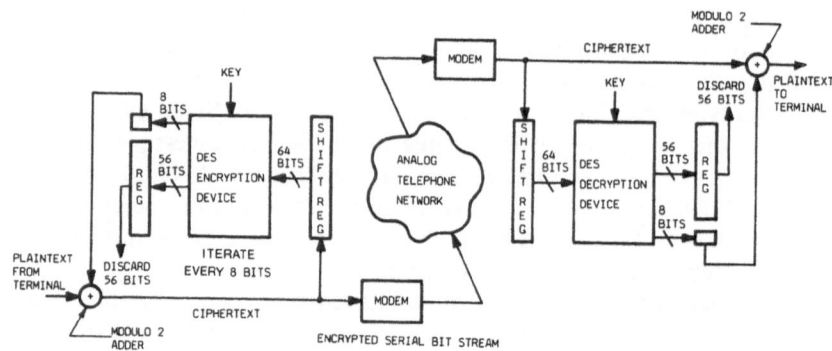

Figure 12.4. DES cipher feedback mode.

The third DES mode is known as *Cipher Feedback* (CFB) or sometimes ciphertext autokey (CTAK). As Fig. 12.4 shows, the data stream is encrypted and decrypted with no delay. Thus CFB is sometimes called a stream cipher. The DES is used in a feedback loop to essentially generate a random bit stream which is added modulo 2 to the data stream. The figure shows the configuration for processing eight-bit characters, but any other sizes such as one or sixteen are also feasible. An initialization variable can be used with CFB just as with CBC, and error checksums should be included in the plaintext.

Note that with eight-bit CFB, the DES must process 64 bits for every eight data bits, so the data rate is only one-eighth of the DES rate. For the eight-bit CFB configuration, a bit error will be propagated through the current and following seven characters, but then will have no further effect. Thus it is self-synchronizing, much like a modem scrambler. The CFB mode of DES operation is appropriate for character-mode transmission or for bit-mode where no buffering is possible. Data rates are also typically lower than ECB and CBC since the DES device must iterate many more times for a given amount of plaintext.

The final DES mode is called *Output Feedback* (OFB). This mode of operation is illustrated in Fig. 12.5 for the case of eight-bit feedback blocks, although other block sizes are again possible. Once started with an initializing vector, the DES simply keeps generating its own inputs by iteratively shifting the feedback blocks through the input shift register completely independent of the plaintext and the ciphertext streams. In effect, the DES is now used as a random pattern generator at each end of the channel. This has two important consequences. First, the OFB mode is *not* self-synchronous and must be reinitialized if synchronization is ever lost during transmission. Second, there is no error extension since transmission bit

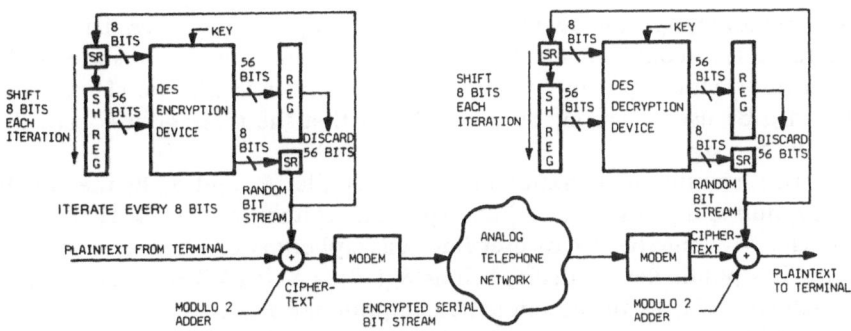

Figure 12.5. DES output feedback mode.

errors are never fed back through the algorithm; a one-bit transmission error in the ciphertext results in a one-bit plaintext error. This property clearly precludes the use of OFB for authentication but, on the other hand, may allow transmission to continue in many cases where an occasional bit error can be tolerated at the receiver. As with CFB, the DES algorithm must run much faster than the data rate when operating in the OFB mode.

There are numerous commercial hardware devices and software packages for implementing the DES, and it is a standard option on some newer equipment. Wide speed ranges are available. There are also other possible configurations, but we have described the major ones. With good physical security and key control practices, the DES should provide adequate encryption for applications below the national security level for many years. However, it is only as secure as the key, and the importance of careful key management and frequent key changes cannot be overemphasized.

Another entirely different approach known as *public key cryptography* was introduced in 1976 [5]. It is based on an elegant mathematical theory that gets around the key distribution issue altogether by generating two entirely different keys for encryption and decryption. Furthermore, the encryption key will not decrypt its own message, nor is it practical to derive the decryption key from it. Thus it can be made public, e.g., published in a *directory of public keys*. Of course, the decryption key is kept secret by its creator. To send a secure message the sender simply looks up the recipient's public key and uses it to encrypt the message. Only the corresponding secret key will decrypt and it is known only to the recipient who generated both keys. Thus the key distribution issue is moot.

However, in making the key public it now becomes possible for anyone, including imposters, to send an encrypted message. Thus the *authentication of the originator* becomes of paramount importance. For this problem public key cryptography can be used to produce *digital signatures*, i.e., short messages that could only have come from a sender with the secret key. Such information not only verifies that the sender is who he claims, but can also be used as proof by the recipient that the message was actually sent by the signer.

To illustrate the operation of public keys, let P_A and S_A be the public encryption and private (secret) decryption algorithms of user A. Then some message M would be sent by user B to A as ciphertext C, where $C = P_A(M)$ using A's public key. A decrypts it as $S_A(C) = S_A P_A(M) = M$. For digital signatures we make the additional assumption the $P_A S_A(M) = M$, so either key can be used for encryption and the other one will decrypt the ciphertext. Now to "sign" M, user B first encrypts with S_B and then again with P_A; hence B sends to A the ciphertext $C = P_A S_B(M)$. Upon receipt, assuming A knows the claimed source is B, A first decrypts with his private key $S_A P_A(S_B(M))$ to get ciphertext $S_B(M)$. A then decrypts again with B's *public* key to get the plaintext $M = P_B S_B(M)$. This process can only occur with knowledge of both private keys and hence authenticates B as the true sender (unless S_B has been compromised).

Several viable public key systems have been proposed, of which the RSA system [6] is currently the most popular. It is based on the product of two large prime numbers, which is very difficult to factor in order to break the cipher. We shall omit the theory since it is rather involved, but a nice tutorial treatment is available [7]. RSA is a block cipher, but the blocks can be much longer than the 64 bits of the DES. Blocks can also be chained for very long messages.

Public key cryptography is conceptually appealing since key distribution is trivial, and intellectually appealing because of its mathematical basis. It is now used in many practical applications. However, there is always the lingering fear that some equally elegant theory will be discovered (or has already been?) that will render the security of it invalid. It has yet to be proven that such a theory does or does not exist. Consequently, the public key approach entails some potential risk, particularly for large networks and highly critical traffic. For smaller applications, however, where the investment is relatively small and recovery from a compromise can be made quickly, public key cryptography is definitely viable.

The subject of network security, and particularly cryptography, is fascinating and there are many details that we have not included in this rather introductory chapter [8]. What we have covered are the essentials, and they will provide a good foundation both for putting security in the proper context and for further study. Network security is fast becoming commonplace on many networks and there is every indication that this trend will continue in the foreseeable future. In the next chapter we look at some additional future trends in data communications networks.

References

1. D. Kahn, *The Code Breakers*, McMillan, New York (1967).
2. *Data Encryption Standard* (FIPS Pub. 46), National Bureau of Standards, Washington D.C. (1977).
3. C. E. Shannon, Communications theory of secret systems, *Bell Syst. Tech. J.* **28**, 656–715 (1949).
4. *DES Modes of Operation* (FIPS Pub. 81), National Bureau of Standards, Washington, D.C. (1980).
5. W. Diffie and M. E. Hellman, New directions in cryptography, *IEEE Trans. Information Theory* **22**(6), 644–654 (1976).
6. R. L. Rivest, A. Shamir, and L. Adleman, On digital signatures and public key cryptosystems, *Commun. ACM* **21**(2), 120–126 (1978).
7. D. W. Davies (ed.), *The Security of Data in Networks*, IEEE Press, New York (1981).
8. W. Diffie and M. E. Hellman, Privacy and authentication: An introduction to cryptography, *Proc. IEEE* **67**(3), 397–427 (1979).

Suggested Readings

L. D. Smith, *Cryptography the Science of Secret Writing*, reprint of 1943 version by Dover, New York (1955).

This is a very entertaining introduction to the historical approaches to cryptography for the layman.

IEEE Communications Magazine (Special Issue on Communications Privacy) **16**(6), (1978).

This is an excellent collection of tutorial papers by key practitioners at the time the DES and public key cryptography were first introduced.

D. E. Davis (ed.), *The Security of Data Networks*, IEEE Press, New York (1981).

This thorough IEEE tutorial not only contains most of the key crypto papers through 1980, but also a copy of the DES standard and one of the best discussions of public key cryptography available anywhere.

IEEE Communications Magazine (Special Issue on Computer Security) **23**(7), (1985).

Several good tutorial papers on the incorporation of security features in data communications protocols.

M. D. J. Buss and L. M. Salerno, Common sense and computer security, *Harvard Business Rev.* **62**(2), 112–121 (1984).

An interesting discussion of some management issues and strategies regarding corporate computer crime.

Glossary for Computer Systems Security (FIPS Pub. 39), National Bureau of Standards, Washington, D.C. (1976).

A convenient terminology collection applicable to data network security.

Check Your Understanding of Chapter 12—True or False?

1. The key to secure encryption is to secure the encryption key.
2. Spoofing refers to inserting illegitimate traffic into a network.
3. Most public network security is so good that recovery from violations is of no concern.
4. Key distribution can be a serious problem in large networks with end-to-end encryption.
5. A telephone company employee working in the toll plant would make an excellent spy.
6. The DES CFB mode produces a stream cipher with no error propagation.
7. Passive wiretaps are generally easier to detect than active ones.
8. Call-back modems provide complete protection against hackers and crackers.
9. For a given key, the substitutions in each round of the DES are identical.
10. In most cases network security is mainly a matter of common sense and diligence.
11. With the CBC mode, each DES-encrypted 64-bit block is independent of all others.
12. Arrogant, unfair, inconsiderate management practices can cause many security problems.
13. Digital encryption is far superior to analog scrambling for modern data networks.
14. DES encryption is seldom worth the cost for nongovernmental organizations today.
15. The U.S. telephone network was completely hardened against penetration since divestiture.
16. Encryption can theoretically be done at any OSI protocol layer.
17. The best security practices, once in place, should never be changed unless penetrated.
18. The DES OFB mode directly combines the cleartext with a random bit stream to get ciphertext.
19. Telephone taps can always eventually be detected from the local central office.
20. Transmission errors never propagate with the DES OFB mode of operation.
21. Most reported "white-collar" computer crime is just management paranoia.
22. Lack of sender authentication makes public key crypto just an academic curiosity.
23. A PBX is an attractive location to tap a corporate telephone system.
24. Public hanging is the only final solution for computer crime.
25. The federal penalty for disclosure of the DES algorithm is lifetime imprisonment without parole.

Future Network Trends

13.1. Future Networks

In each of the dozen preceding chapters we focused on one particular aspect of *Practical Computer Data Communications*. Looking back, it should now be apparent that these topics are by no means isolated but rather fit together to give a coherent picture of data networks, much like the tiles of a mosaic. We have studied essentially all of the key technical aspects of modern data networks as they exist today: how they evolved, how they are designed, how they operate, and how they can be used.

In this final chapter we shall shift our focus from the present to the future, relax our technical formality, and briefly consider how data networks might continue to evolve over the next few decades. Predicting the future is a risky business, even for the "professionals"† and we have no intention

---

† The *First Law of Futures Forecasting* is well known to all *experienced* forecasters: "If you're ever right—never let'em forget it."

of attempting it here. Rather, we shall look in a general way at some technical and nontechnical issues that could have major impact on data networks in the future. The technological issues center on the *integration* of digital voice and video with data for transmission over a common digital network. Much of the foundation for such integrated networks has already been laid and there is great interest by the telephone companies, particularly in the ISDN.

The nontechnical issues are much broader and more ambiguous. The potential benefits from intelligent integrated computer communications networks are truly enormous. As automation and mechanization take over the bulk of our manual work, knowledge and information will increasingly become the tools with which people will earn their livings. The very infrastructure of this advancing civilization should be the system that provides, stores, transports, and processes this information. From our perspective today, that system would appear to be a worldwide collection of interconnected computer data networks.

In *business* and *commerce* computer networks could offer mail, conferencing, information storage and retrieval, marketing, design, and production services far in excess of what firms have today. Specialized information could be made instantly accessible to professional practitioners, and also on a limited basis to the layman. *Educational services* by the best teachers could be provided to people of all ages via data networks, and formal learning could take place at any convenient time and place. There are also clear benefits to *government organizations*, from providing military command and control to collecting taxes and distributing government benefits. In the *home*, networks could provide entertainment, security, shopping, and a host of advisory services. The list of potential benefits seems endless.

However, for such benefits to accrue and be shared equitably by all of our people, many pieces must fall into place. First, the necessary *technology* must be developed. Given the recent rate of innovation in communications technology worldwide, this does not appear to be a major limitation. Huge *capital investments* will be required to develop sophisticated networks, so it is imperative that the networks offer the services that users want. Creating new services based on sophisticated technology with little regard for the ultimate *market* has never worked well in the past. This lesson will be even more applicable in the future. Future networks must provide services that are really needed at prices that are really affordable by practically everyone.

Finally, there is the issue of *regulation*. For data communications to fulfill its promise, it must be made available wherever people need to use it. This universal service can occur in two basic ways: either many small networks can be connected into one ubiquitous internet, or there can be one (or a few) major network. In the former case standards for interfaces, service quality, and cost must be established and adhered to, either voluntarily, because of government authority, or through competition. With a single network, regulation is obviously needed to control monopolistic practices. In general, the regulatory policy must create, or at least allow, an environment in which sufficient innovation and growth can occur in the direction of market needs.

Data network evolution clearly is a complex question that will require great effort and wisdom; the enormity of the task and immense potential benefits to our society make careful planning and management imperative. Currently there is a fair consensus that integration of voice and data will be the next major step in future digital network evolution, and so in the following section we consider integrated networks. We shall first look at several practical digital speech techniques which require less bandwidth than the historical PCM approach now widely used, then review some of the more promising initial approaches towards realization of the ISDN.

13.2. Integrated Networks

For centuries people have sought to develop means for synthesizing, storing, and transmitting human speech. The analog telephone was, of course, the major historical innovation for speech transmission and it has served admirably for well over a century. During the last quarter-century digital speech processing has become viable. There are now many commercial devices for synthesizing and recognizing speech, numerous services are available that store and recall spoken messages, and a large proportion of the voice telephone traffic is now transmitted digitally within the telephone system.

In the past, voice and data transmission have used separate networks. Practically everyone has used the domestic telephone system for voice traffic, while data has been carried by all kinds of private and public networks, including point-to-point links, VANs, satellite carriers, and the telephone companies themselves. This situation is beginning to change for several reasons. First, the technology now exists to process, transmit, and switch digital speech efficiently. For example, fiber optics now is providing the large transmission bandwidths necessary for digital speech, and the corresponding switching capability is also being developed.

The proportion of data traffic relative to voice is steadily increasing, and will eventually dominate. For large data users separate networks for voice, data, and a variety of other traffic types are inefficient and expensive and integration appears to be an attractive solution. For large network providers, particularly the telephone companies, integration will allow them to utilize their newer digital plants more fully and also to participate in the fast-growing data market. Finally, even small consumers are attracted by services like stored voice messages, data base access, and home monitoring and alarm services that could be provided along with voice service by an integrated digital carrier.

There are other arguments in favor of digital speech in addition to integrated transmission. Digital signals are less susceptible to noise than their analog counterparts, and digital speech encryption is much more secure than analog speech scrambling. Digital speech can be stored, broadcast, and processed in various ways such as for error correction. Finally, digital speech can potentially be used with data communications protocols such as for packet switching [1].

Techniques for speech digitization in the telephone network go back to the introduction of PCM speech transmission in 1962, and the use of 64-kbps companded PCM for digital speech has been dominant in the telephone T-carrier system for many years. Only recently has a more efficient 32-kbps adaptive coding technique been standardized and implemented. The other speech coding techniques that are widely used in practice are *delta modulation* (DM) and *linear predictive coding* (LPC).

Figure 13.1a illustrates speech coding with DM. The basic idea is to track the analog waveform by taking a series of small discrete steps of size Δ at a rate much faster than the Nyquist rate. Whenever an upward step is made a data 1 is transmitted, and for each downward step a data 0. At the receiver the staircase waveform approximation to the speech is reconstructed by simply accumulating, i.e., increasing the cumulative value by Δ for each received 1 and decrementing by Δ for each 0, then smoothing with an LPF. The step size should be large enough to track fast changes in the speech waveform but small enough to prevent noticeable distortion (granular noise) when there is little speech signal change. One solution to this step size problem is to make Δ variable. This variation can be made continuously dependent on the rate at which the speech signal is changing, in which case it is called *continuously variable slope DM* (CVSD). CVSD is used in many commercial and military applications to digitize voice-grade speech at 16 or 32 kbps.

Speech waveform samples, even at the Nyquist rate, tend to have a high correlation; for example, if one sample is large then the following sample will usually also be large, etc. As a result, PCM samples tend to be somewhat redundant. One way to reduce this correlation, and hence also the data rate, is to only transmit the difference between adjacent samples. Further improvement can be obtained by using several past samples to *predict* the next sample value, and then transmitting only the *prediction error*. This is the basic idea behind *differential PCM* (DPCM), as shown in Fig. 13.1b. The predictor is typically a digital transversal filter. Note that now sampling is done according to the Nyquist criterion and the prediction error is quantized and binary encoded before transmission. At the receiver the predicted value is reconstructed and corrected by the decoded error

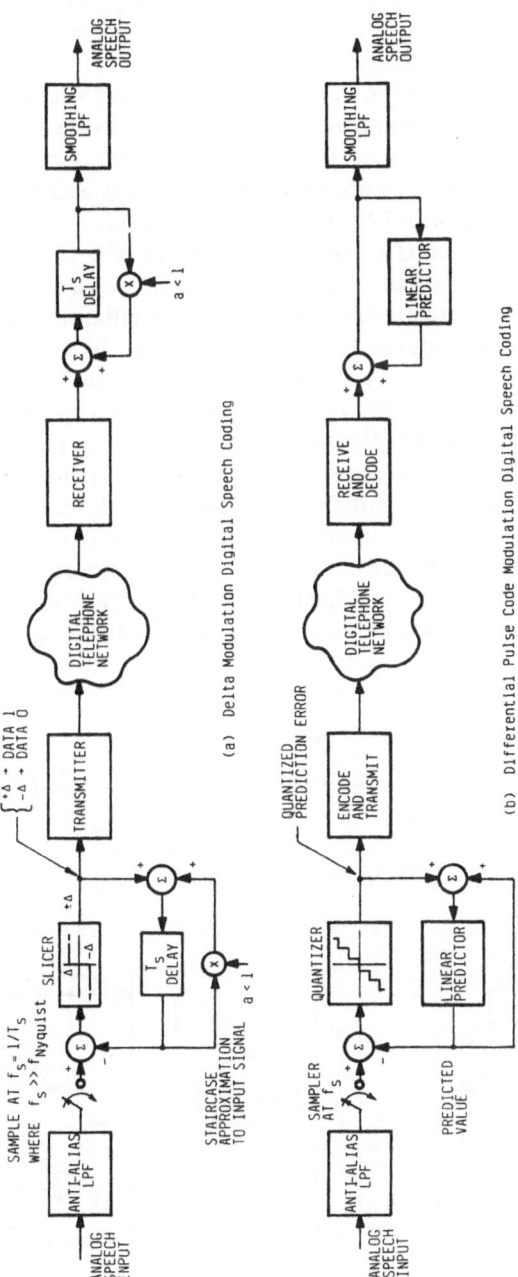

Figure 13.1. Waveform coding techniques for digital speech.

value, then smoothed by filtering to get the original speech signal. DPCM can also be made adaptive by making the predictor coefficients (filter tap gains) and/or the quantizer step size dependent continuously on the speech signal. This *adaptive* DPCM (ADPCM) approach is used in the 32-kbps telephone system standard mentioned in the previous section [2].

The other widely used speech coding technique we shall mention is LPC [3]. This is also a predictive method but, instead of coding the speech waveform directly, with LPC the predictor *parameters* are transmitted. Such parametric speech coders are called *vocoders*. The LPC approach is based on a linear model of the human vocal tract shown in Fig. 13.2a. Small intervals of sound are produced by either an oscillator for *voiced*, e.g., vowel, sounds or a noise source for *unvoiced* sounds like "hisses." The human mouth and nasal cavity that then filter these sounds are modeled by a predictive filter as shown. In digital speech transmission the transmitter adaptively adjusts its predictive filter coefficients for minimal error between the actual and predicted samples, then transmits these coefficients along with corresponding amplitude, pitch, and sound source information. The actual speech waveform is never transmitted, only the model parameters. Such frames typically contain around 60 bits and are sent every 25 msec, hence a data rate of only 2400 bps. This is much lower than the previous waveform coders, but LPC produces relatively lower quality speech and is

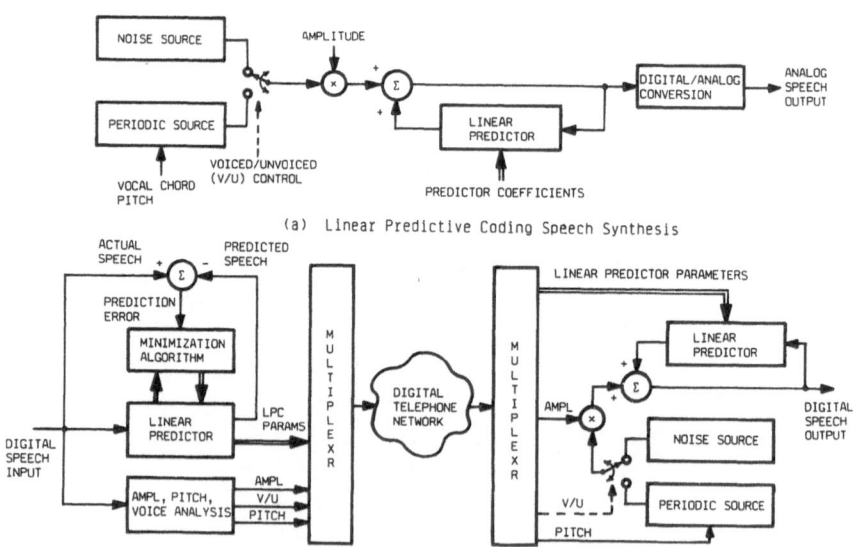

Figure 13.2. Linear predictive speech coding.

very susceptible to bit errors. However, it is used satisfactorily in many applications requiring narrow bandwidths and encryption for secure voice.

Digital voice and data can be integrated for transmission over a data network in several basic ways. The simplest is to use conventional TDM and assign separate slots to each voice and data channel. This is exactly the approach taken by AT&T with the T-carrier system. A T1 frame is divided among data, signaling, and either 64-kbps or 32-kbps digital voice, and the traffic mix can now be reconfigured by the user.

For packet switching, several voice and data integration schemes have been proposed. Although different in detail, the basic idea is to divide each packet into a voice and a data section, with a variable boundary separating them. Since voice traffic is real-time, it is given priority and arriving voice traffic is immediately placed in the packet first so long as the boundary allows. Data is then added up to the maximum allowed packet size and the packet is transmitted with the usual protocol. When traffic is heavy, data can be buffered or flow controlled but voice traffic is usually just discarded. This loss has little effect on intelligibility so long as it is only sporadic.

A third integration approach for asynchronous point-to-point traffic, known as burst switching, operates by inserting short bursts of data in between intervals of speech activity, or talkspurts. This concept has been used for many years on analog undersea cables to increase the number of voice channels carried, where it is known as Time Assignments Speech Interpolation (TASI).

Having introduced the rudiments of digital speech and its integration with data, we now look at some of the more interesting developments that are intended to lead eventually to the *Integrated Services Digital Network* (ISDN). Practically all telephone companies worldwide are involved to some degree with the ISDN. As we have already seen, the ISDN is of enormous importance to telephone companies, particularly those operating in a competitive environment. This impetus, combined with an ever more favorable technology and a steadily increasing customer demand for more sophisticated services, is providing a strong driving force for the ISDN development. There are potentially large economies of scale in integrating not only voice and data but also facsimile, mail, telemetry, videoconferencing, monitoring, etc. The alternative to the ISDN may be a large number of specialized networks with significant constraints on their interoperation.

In the 1984 Plenary Session CCITT adopted a set of new I.-series recommendations for the ISDN. As is typically the case, these were not intended to be final and complete operating standards, but rather define the framework of the ISDN, leaving the details to be worked out during the 1984–1988 study period. Initial ISDN development should roughly parallel the standards refinement effort, with the complete standards and first significant ISDNs operational around 1988.

Many of the key components of ISDN are already in place, or soon will be. For example, in the United States *digital switching* and *digital short-haul trunks* are now widely used, while *fiber optic long-haul trunks* and *digital common channel signaling* (SSN7) are becoming commonplace. The final area to be integrated will probably be the local loop, but even here there is much current work. CCITT is considering two main approaches for sending integrated traffic over two-wire loops. In Europe the *echo canceling* approach is favored, with simultaneous transmissions from both ends being separated by adaptive digital canceling filters. The North American alternative is called *time-compression multiplexing* (TCM) [4]. Here short bursts of data are alternatively sent in each direction at speeds slightly over twice the original rates. For example, for 56-kbps data the "ping pong" loop rate is 144 kbps.

The ongoing ISDN effort at CCITT has produced several recommendations [5] of particular interest to potential users, particularly with regard to user interfaces. Although there are slightly different versions for Europe and North America, the basic concepts are the same, so we shall only describe the latter here. There are two available rates. The *basic rate* (2B + D) consists of two 64-kbps *B channels* for the user's integrated traffic and one 16-kbps *D channel* which can be used for signaling, as an X.25 data link, or for various other data functions. In operation the user signals over the D channel to set up each B channel connection, then any type of traffic desired can be sent through the resulting "bit pipes." The signaling protocol, LAPD, is similar to LAPB of X.25.

For larger users there is the *primary rate* (23B + D) of 1.544 Mbps (in the North American version). In addition to 23 B channels of 64 kbps each, there is a 64-kbps D signaling channel. The remaining 8 kbps corresponds to the current framing bit in the AT&T DS1 frame, as discussed in Chapter 4. There are also possible subrates H_0 (384 kbps) and H_1 (1.344 Mbps). When the ISDN interface is fully defined, it could become the dominant digital interface not only for ISDN but for many other networks as well, and particularly those now specifying analog and X.21 interfaces.

There are several scenarios for the ISDN transition, but however it occurs it should segue gradually from the existing plant because of the huge capital investments involved. Initial ISDN users will probably be predominantly large customers, since many small business and residential users will be unable to pay significant costs for bundles services, many of which they can do without. Eventually, however, the classical supply and demand relationship should determine the mix of services and prices that most consumers will accept.

The CCITT ISDN effort is not the only such program but it is certainly the most important one. Other organizations, primarily government and military, are also developing and using integrated capabilities. However, we have at least touched upon most of the key ISDN concepts and hopefully have put ISDN in the proper perspective relative to data networks in general. In addition to the United States, many other countries are well along with pre-ISDN networks such as the Japanese Information Network System and the British System X. In fact, there are actually three major simultaneous ISDN efforts worldwide in Europe, North America, and Japan. Although there is considerable similarity between the three, significant differences also exist among both the organizations providing the services and the customer bases that will eventually use them. Again the need for strong standardization is apparent for harmonious international networking, with the CCITT taking the lead. It should be exciting to watch the global ISDN evolution unfold and to participate fully in the many benefits that it promises.

13.3. Conclusions

We now come to the end of our study of *Practical Computer Data Communications*. We have progressed from an initial background that presupposed no knowledge of data communications, through the telephone system, data links, wide area networks, local area networks, and some specific considerations of network security and integration. The original goal of this book was to provide a complete, practical introduction to the fascinating field of computer data communications at a suitable level for a broad range of professional practitioners. To accomplish this objective we have introduced many techniques from the assorted disciplines upon which the field is based. Now all the technical pieces are in place and there should be a feeling of continuity and closure about the subject matter. The material we have covered should be of immediate practical value both for solving problems and pursuing further studies. For those (few?) who may not agree we can only quote Mark Twain's classic remark concerning his father† and trust that time will bear out our assertion.

Aside from the technical merits of this particular book, computer data communications will continue to be an exciting as fast-growing field offering opportunities of all types—entrepreneurial, industrial, government, and private practice. It finds application in practically every line of human endeavor—manufacturing, banking, government, public service, military, education, and many more. It is a field that offers deep professional benefits—challenging work, opportunity for growth, ample financial rewards, good working conditions, and just plain fun.

But with every advantage goes a commensurate responsibility, and in our case part of that responsibility as professionals in the field is to see that

---

† "When I was fourteen, I thought my father was an idiot. But when I was twenty-one I was amazed at how much the old man had learned in just seven years."—Mark Twain

this great new technology is used to provide maximum benefit, not just for a few elite consumers, but for all people worldwide. The great U.S. domestic telephone system became the finest in the world based on the 1934 Communications Act mandate of universal, affordable telephone service. Perhaps we will soon need a similar mandate for computer data communications. Each of us must draw his or her own conclusions; the important thing is to do so, and then have the courage to express them in the appropriate forums.

References

1. I. Gitman and H. Frank, Integrated analysis of voice and data networks: A case study, *Proc. IEEE* **66**(11), 1549–1570 (1978).
2. *32 kbit/s Adaptive Differential Pulse Code Modulation (ADPCM)* (Recommendation G.721), CCITT, Geneva (1984).
3. L. R. Rabiner and R. W. Schafer, *Digital Processing of Speech Signals*, Chap. 8, Prentice-Hall, Englewood Cliffs, New Jersey (1978).
4. *Bit Compression Multiplexing* (AT&T Preliminary Technical Reference Pub. 54070), American Telephone and Telegraph Company, New York, January (1984).
5. N. Q. Duc and E. K. Chen, ISDN protocol and architecture, *IEEE Commun. Mag.* **23**(3), 15–22 (1985).

Suggested Readings

IEEE Communications Magazine (Special Issue on ISDN) **22**(1), (1984).

> Several very good articles on the technical considerations in moving to the future ISDN.

N. S. Jayant and P. Noll, *Digital Coding of Waveforms*, Prentice-Hall, Englewood Cliffs, New Jersey (1984).

> This is an outstanding treatment of the various forms and extensions of PCM, and particularly of the theory behind ADPCM coding for 32-kbps digital telephone speech.

R. W. Schafer and J. D. Markel (eds.), *Speech Analysis*, IEEE Press, New York (1979).

N. R. Dixon and T. B. Martin (eds.), *Automatic Speech and Speaker Recognition*, IEEE Press, New York (1979).

> These two IEEE collections reprint a large portion of the key papers in digital speech synthesis and analysis at the time much of the fundamental theory and technology was developed.

IEEE Journal on Selected Areas in Communications (Special Issue on Packet Switched Voice and Data Communications) **1**(6), (1983).

> A collection of technical papers on integrated packet switching for a variety of network types, including LANs and satellite networks.

Check Your Understanding of Chapter 13—True or False?

1. The ISDN 23B + D primary rate equals the DS1 rate in the U.S.
2. Synthesizing human speech can now be done digitally.
3. Divestiture has caused some delay in the evolution of the ISDN in the U.S.
4. CVSD and ADPCM are both adaptive speech coding techniques.
5. Digital voice accounts for well over half of the LAN traffic today.
6. Vocoders achieve very low data rates at the expense of speech quality.
7. Most integrated packet networks employ some form of the moving boundary packet idea.
8. A fundamental component of the information age will be computer data networks.
9. Digitally encoded speech cannot be transmitted with asynchronous modems.
10. Many diverse components must all come together for the ISDN to become universal.
11. Telephone quality speech can now be encoded at 32 kbps rather than 64 kbps.
12. Historically, voice and data have been carried over separate networks.
13. Data communications is now a very mature, slow growth industry.
14. The ISDN could eventually provide most homes with monitoring and alarm service.
15. LPC speech encoding actually sends only vocal tract parameters and not the actual speech.
16. Once the actual ISDN physical interface is well defined it could be adopted by X.25.
17. Today technology is the major obstacle to implementation of the worldwide ISDN.
18. The sound of the letter "z" has both voiced and unvoiced characteristics.
19. Digital common channel signaling is needed for full ISDN evolution.
20. Speech must be sampled below the Nyquist rate for proper DM encoding.
21. ISDN would allow AT&T to get back into the highly profitable local telephone market.
22. Integrated networks were declared unconstitutional by the U.S. by the Supreme Court in 1954.
23. The ultimate objective of the FCC today is universal, affordable deregulation.
24. The ISDN B channels are intended to carry Bulk (junk) electronic mail to every home.
25. The second law of futures forecasting is: "If you know it, flaunt it."

Answers to True/False Questions

Chapter 2.   1. F, 2. T, 3. F, 4. F, 5. T, 6. T, 7. T, 8. F, 9. T, 10. F, 11. F, 12. T,
13. T, 14. F, 15. F, 16. T, 17. F, 18. T, 19. T, 20. F, 21. F, 22. T,
23. F, 24. T, 25. F

Chapter 3.   1. F, 2. F, 3. T, 4. F, 5. F, 6. T, 7. F, 8. F, 9. F, 10. T, 11. T, 12. T,
13. F, 14. F, 15. T, 16. T, 17. F, 18. T, 19. T, 20. T, 21. F, 22. T,
23. F, 24. T, 25. T

Chapter 4.   1. T, 2. F, 3. F, 4. T, 5. F, 6. T, 7. T, 8. F, 9. T, 10. F, 11. T, 12. T,
13. F, 14. F, 15. T, 16. T, 17. F, 18. F, 19. T, 20. F, 21. F, 22. F,
23. T, 24. T, 25. F

Chapter 5.   1. T, 2. F, 3. T, 4. T, 5. T, 6. F, 7. T, 8. T, 9. F, 10. T, 11. T, 12. F,
13. F, 14. F, 15. F, 16. F, 17. T, 18. F, 19. T, 20. T, 21. T, 22. F,
23. F, 24. F, 25. T

Chapter 6.   1. T, 2. F, 3. F, 4. T, 5. T, 6. T, 7. F, 8. T, 9. T, 10. F, 11. F, 12. F,
13. T, 14. F, 15. F, 16. F, 17. T, 18. T, 19. T, 20. F, 21. F, 22. F,
23. T, 24. F, 25. F

Chapter 7.   1. T, 2. T, 3. F, 4. T, 5. T, 6. T, 7. T, 8. T, 9. T, 10. F, 11. T, 12. F,
13. T, 14. F, 15. F, 16. F, 17. T, 18. T, 19. F, 20. T, 21. T, 22. T,
23. F, 24. F, 25. F

Chapter 8.   1. T, 2. F, 3. T, 4. F, 5. T, 6. T, 7. T, 8. T, 9. T, 10. T, 11. T, 12. T,
13. T, 14. F, 15. T, 16. T, 17. T, 18. T, 19. F, 20. T, 21. T, 22. T,
23. T, 24. F, 25. F

Chapter 9.   1. T, 2. F, 3. F, 4. T, 5. T, 6. T, 7. F, 8. T, 9. T, 10. T, 11. T, 12. F, 13. F, 14. T, 15. F, 16. F, 17. T, 18. F, 19. T, 20. T, 21. F, 22. T, 23. T, 24. T, 25. T

Chapter 10.   1. F, 2. T, 3. F, 4. F, 5. T, 6. T, 7. T, 8. T, 9. F, 10. T, 11. F, 12. T, 13. T, 14. F, 15. T, 16. T, 17. T, 18. T, 19. T, 20. F, 21. T, 22. F, 23. F, 24. F, 25. F

Chapter 11.   1. T, 2. T, 3. T, 4. F, 5. T, 6. F, 7. F, 8. T, 9. F, 10. F, 11. T, 12. T, 13. T, 14. T, 15. F, 16. F, 17. T, 18. F, 19. F, 20. T, 21. F, 22. F, 23. T, 24. F, 25. F

Chapter 12.   1. T, 2. T, 3. F, 4. T, 5. T, 6. F, 7. F, 8. F, 9. F, 10. T, 11. F, 12. T, 13. T, 14. F, 15. F, 16. T, 17. F, 18. T, 19. F, 20. T, 21. F, 22. F, 23. T, 24. F, 25. F

Chapter 13.   1. T, 2. T, 3. T, 4. T, 5. F, 6. T, 7. T, 8. T, 9. F, 10. T, 11. T, 12. T, 13. F, 14. T, 15. T, 16. T, 17. F, 18. T, 19. T, 20. F, 21. F, 22. F, 23. T, 24. F, 25. F

Index